AF520863

ODONATA

BIOLOGY OF DRAGONFLIES

Related Books

Titles	Authors
— A Manual of Practical Entomology 3rd Ed	*M.M. Trigunayat*
— Beekeeping : A Compressive Guide to Bees and Beekeeping	*Dharam P. Abrol*
— Changing Faunal Ecology in the Thar Desert	*B.K. Tyagi*
— COMMERCIAL BEEKEEPING Dharm Singh	
— Dr Bastiaan Kiauta: Odonatologist and Polymath	*B.K. Tyagi*
— Dr Ronald Ross Mosquito, Malaria, India and the Nobel Prize	*B.K. Tyagi*
— Dynamics Of Insect Behaviour	*T.N. Ananthakrishnan*
— Ecological Entomology: Insect Life In Odd Environments	*T.N. Ananthakrishnan*
— Entomology Ecology & Biodiversity	*B.K. Tyagi*
— Entomology in the Doon Valley	*B.K. Tyagi*
— Evolution of the Vertebrates 3rd Ed	*E.H. Colbert*
— Emerging Crop Pest Problems: Redefining Management Strategies	*P. Parvatha Reddy*
— Handbook of Insects: Photographic Guide for Identification	*V.B. Awasthi*
— Gall-inducing Arthropods on Forest Trees	*John Prasanth Jacob*
— Insect Biodiversity: Functional Dynamics and Ecological Perspectives E/B	*T.N. Ananthakrishnan*
— Introduction to General and Applied Entomology, *3rd Edition*	*V.B. Awasthi*
— Insect Physiology	*V.B. Wigglesworth*
— Insect Pathology Text Book And Practical Manual	*Nitesh K. Maru*
— Insect Physiology (21st Century Biology and Agriculture: Textbook Series)	*K.P. Sanjayan*
— Insect, Mite And Vertebrate Pests and Their Management In Horticultural Crops	*P. Parvatha Reddy*
— Medical Entomology *2nd Ed*	*B.K.Tyagi*
— Mosquito Hunters (A History Of Hostilities Against Man's Deadiest Foe - The Mosquito - Since 1881)	*B.K.Tyagi*
— Organic Evolution	*B.L. Chaudhary*
— Principles of Insect Behaviour *2nd Ed*	*V.B. Awasthi*
— Pests of Forest Importance and their Management	*B.K. Tyagi*
— Scientist: A Unique Species and the Backbone of A Nations Economic and Intellectual Prosperity	*B.K. Tyagi*
— The Invincible Deadly Mosquitoes	*B.K. Tyagi*
— Technological Innovations In Integrated Pest Management: Biorational and Ecological Perspective	*Dharam P. Abrol*
— Vector-Borne Diseases: Epidemiology and Control	*B.K. Tyagi*

ODONATA

BIOLOGY OF DRAGONFLIES

Editor

Dr. B.K. Tyagi

MSc, DPhil, FRES, FISCD, FISP, FAZRA

Officer in-Charge

Centre for Research in Medical Entomology (ICMR)

Madurai 625002, TN, India

President

Indian Dragonfly Society

&

President, Jodhpur Natural History Society

(Formerly the Officer in-Charge, Desert Medicine Research Centre, Jodhpur; Member in-Charge in South Asia, Odonata Specialist Group of the Species Survival Commission, IUCN)

Published by:

SCIENTIFIC PUBLISHERS (INDIA)
5A, New Pali Road, P.O. Box 91
Jodhpur 342 001, India
E-mail: info@scientificpub.com
Website: www.scientificpubonline.com

Print: 2023

ISBN: 978-81-72334-82-6
eISBN: 978-93-87913-35-6

Printed in India

This festschrift
is
fondly and respectfully dedicated to

Professor Dr. Bastiaan Kiauta

Ph.D., D.Sc.

(Photographed soon after he was knighted by the Queen on 26th April, 2002)

The doyen of odonatology
&
The Patron of International Odonatological Foundation (SIO)

BEFORE AND AFTER 1971: A PERSONAL TRIBUTE TO BASTIAAN KIAUTA

I am old enough to have studied dragonflies for over thirty years before Bastiaan and his colleagues launched the Societas Internationalis Odonatologica (SIO) in 1971 and published the first volume of Odonatologica in 1972. I was interested in dragonflies as a boy but there was no popular book to help me to identify them. Cynthia Longfield's book "The Dragonflies of the British Isles" changed all that, and later she introduced me to the literature on dragonflies on a world scale. However I did not know personally any of those working on dragonflies outside the British Isles. Numerous other odonatologists in other countries were in a similar situation. The initiatives of Bastiaan and his colleagues opened up a new world. Not only was there now a new journal devoted entirely to dragonflies, but it provided a most useful literature abstraction service.

Yet perhaps the most important service that Bastiaan and his colleagues provided was the initiation of biennial International Symposia all over the world, with visits to wonderful dragonfly habitats where all of us lucky enough to attend could in addition enjoy dragonflies together in the field, and form friendships which stood the test of time. I for one will always be grateful to Bastiaan's leadership, so ably supported by Marianne and tirelessly assisted by Janny van Brink, for this enrichment of my life. Thank you, Bastiaan, for your pioneering work.

Best wishes for the future.

20th November, 2006

Prof. Dr. Norman Moore
Ex-Chairman
Odonata Specialist Group
Species Survival Commission (IUCN)

To
My Teacher and Guide
Professor Dr. B. Kiauta

FOREWORD

I am much honoured to write the Foreword to the Festschrift for Professor Dr. Bastiaan Kiauta on occasion of his seventieth birthday. He was born in Slovenia, and moved to the Netherlands, where he studied odonatology, especially as an expert in invertebrate cytology and cytogenetics. He engaged in research and education as a professor at the University of Utrecht.

The First Colloquium of Dutch Dragonfly Workers was held in Utrecht, the Netherlands in 1970 organized by Dr. Kiauta. This led to the venue of the First European Symposium of Odonatology in Ghent, Belgium, in 1971 on which occasion the first explicitly international association named Societas Internationalis Odonatologica (S.I.O.) started. He played the central role in the foundation of this long standing international society. He started the publication of the quarterly journal ODONATOLOGICA, Journal of the Societas Internationalis Odonatologica (S.I.O.) and its semiannual journal NOTULAE ODONATOLOGICAE. The punctuality of publication and the high quality of the published papers led the former journal to be indexed in Current Contents, Science Citation Index and Research Alert, and covered by most of the major abstracting services.

Dragonflies occur worldwide, and the odonatologists need to know their own odonate fauna as well as those of other areas. On this account the international communication and mutual help are indispensable. What Dr. Kiauta has done and continues to do is indispensable for the current and forthcoming research in odonatology alike.

He was knighted by the Queen of the Netherlands on 26th April 2002 in recognition of his remarkable management activities and his highly valued contributions to science. I think his knighthood award is an honour to all of us engaged in odonatological research.

On the occasion of the 1997 Symposium of Odonatology held in Maribor, Slovenia, S.I.O. was reorganized into the "International Odonatological Foundation S.I.O." The reorganization brought some difficulties into the management, but his enthusiasm and generosity opened the way for further development.

When we speak of his activity, we cannot forget to mention the devoted help of his lovely wife Mrs. Marianne Kiauta-Brink. She is always helping him or co-working with him in editorial and administrative matters. On Bastiaan's 70th birthday our heartfelt congratulations also go to her; and the odonatologists, his colleagues and friends, thank her for all she did for us through her help to him! She is also the author of many excellent poems, especially in the field of "haiku", a poetry form, in a pattern of approximately 5, 7 and 5 syllables with reference to the natural world.

Our friend Dr. Kiauta always calls his colleagues in odonatology a "family". We help any researchers, when asked, in the same way as if requested by our brothers and sisters. He is so to speak the "Father of the large family of odonatologists worldwide". We, as the members of this family, look forward to a continuation of our Father's valued contribution to odonatology in the years that lie ahead.

Last but not least, I should like to tender my cordial thanks to Dr. B.K. Tyagi for his initiative to publish this dedication book, and for his kindness to give me the opportunity to write the foreword. I admire his effort and talent which enabled him to publish such a great scientific book written by many authors worldwide.

Date: 10th December, 2006

Dr. Kiyoshi Inoue
President
International Odonatological Foundation S.I.O.
5-9, Fuminosato 4-chome, Abeno-ku,
Osaka, 545-0004, Japan
(e-mail: ks-inoue@mx2.nisiq.net)

PREFACE

Dragonflies (Odonata) are unique insects. They are complex and intriguing creatures. Their life cycle is enigmatic, not least because of the remarkable development from water-dwelling larvae to the aerial life of adult – which necessitates a complete adjustment of the internal respiratory mechanism. With each generation, the dragonfly seems to retrace the evolutionary path that led some animals to leave aquatic life for good and adapt to land living. Quoting Professor Dr. B. Elwood Montgomery, the world famous American odonatologist, Professor Dr. B. Kiauta, Executive Editor of *ODONATOLOGICA*, and the former Director of the Netherlands Centre for Alpine Biological Research (N.C.A.B.R.), Utrecht, has justly given the brief and pointed characterization of the dragonflies, which, in my opinion, could not be styled more clearly and beautifully.

"In more than one feature they differ, at the very first glance, from all other insect superorders. The mayflies (Ephemeropteroidea) and dragonflies (Odonatopteroidea) are the only members of the cohort (or pterygote subclass ?) of Paleoptera and were the first to split off from the pterygotan stock – an event that occurred probably as early as in the Lower Devonian. The two sperorders are but very remotely allied, and do not at all show any structural affinities with the other pterygotan groups. Both superorders characterized by a number of extremely archaic features, have evolved but very little in the course of the past geological epochs, and are, therefore, justly considered as a kind of living fossils."

Evolutionarily speaking dragonflies present a world of information on the pathways insects have gone along to metamorphose into their present form and shape. The living dragonflies, represented by over 6000 known species organized into 28 families under the three living suborders, represent but one order (Odonata) of a once flourishing group, the other order of which, Protodonata, became extinct already in the Upper Permian. In insect phylogeny dragonflies also occupy a unique position. Since the Upper Paleozoic they have not changed essentially from the structural point of view and are considered as " living fossils" par excellence. In no other insect group is the paleontological evidence as complete as in dragonflies; therefore it is not surprising that the classification of the superorder is nearly entirely based on those morphological features that are liable to preservation in fossil material, most particularly on wing venation. Nothing is known on the direct ancestors and the earliest dragonfly history. In the Lower Carboniferous the two orders, Protodonata and Odonata, were already flourishing. While the former became extinct towards the end of the Paleozoic, three out of seven odonata suborders survived to the present time, viz., Zygoptera, Anisozygoptera and Anisoptera. Anisozygoptera were the dominant suborder of the Mesozoic era and are represented in the present day fauna only by two relic species, one of which is endemic in the Himalaya (both in Darjeeling, India and in Shivapuri Hills, Nepal). The Zygoptera and Anisoptera, on the other hand, are the dominant groups of our Kenozoic times.

The dragonflies comprise one of the most important groups of aquatic animals. The ecological significance of dragonflies and their importance in environmental monitoring have often been emphasized. Being voracious predators in both immature (aquatic) and adult (aerial) stages they are important elements of all, except the drier (or high alpine) environments in temperate and tropical regions, occupying a position at the apex of the food chain of invertebrate life. In addition to their biological importance in such environments, these relationships render them the ultimate accumulators of persistent compounds in polluted waters. They are thus ideal organisms to be used as indicators of water pollution and contamination. Many species serve as intermediate hosts of fluke parasites of birds, and thus are important in the transmission of parasitic diseases, especially of domestic poultry and wild ducks. Because of their unique morphology and physiology, dragonflies are used extensively in the study of many biological phenomena.

Dragonflies present manifold applications in biomedical sciences, too. They are a good biological control agent for several disease transmitting mosquitoes, especially *Aedes aegypti*, the vector for deadly dengue, yellow fever and chikungunya. Although distantly conjectured, yet these insects have lately been associated with the forensic science. In more recent times it has been realized to develop nano-technologically a highly sophisticated, effective and durable eye lens for human use on the basis of knowledge of the dragonfly's magnificent compound eyes. Also, dragonflies veritably find a place in human social and cultural development. There are regions in some continents of the world where dragonflies are offered as a delicacy on the menu, while in others they are source of music composition, folk dance and folk tales. They are beauty *nonpaeil* and, therefore, dragonflies appear in print on textile and ornamental products of varying nature.

Dragonflies have attracted attention of some of the world's most celebrated entomologists, to mention a few, like Drs. R.J. Tillyard, P.P. Calvert, G.D.H. Carpenter, A.E. Gardner, C.H. Kennedy, E.B. Williamson, J.G. Needham, S. Asahina, R.M. Gambles, M.J. Westfall, C. Longfield, B.E. Montgomery, F.C. Fraser, M.A. Lieftinck, P.S. Corbet and, of course, B. Kiauta. Their attraction to these taxonomically and evolutionarily highly significant, yet absolutely harmless, creatures is a testimony of dragonflies' enormous value in our knowledge-based ecosystem. Prof. Bastiaan Kiauta, to whom this festschrift is being most fondly and respectfully dedicated, for his lifelong outstanding services to the science of odonatology, including dragonfly conservation and the S.I.O., unparalleled in the annals of dragonfly science, is one of those greatest ever scientists who by their energetic character not only practically educate the budding researchers in habits of industry, but by the example of diligence and perseverance which they set before them, largely influence the scientific activity in all directions and contribute in a great degree to form the national character. I hereby choose to record what many of his odonatological colleagues, friends and students across the world, with some of whom I had an opportunity to communicate recently, think of him: **Dr. Angelo B.M. Machado**, "*I am a very good friend of Bastiaan ..... nobody has contributed so much for the development of the world odonatology than him.*"; **Dr. Milen Marinov**, "*I'm happy to have your Invitation and it will be an honour for me to write anything to celebrate Prof. Kiauta's birthday. In fact he is the one who supported me in the beginning of my researches with dragonflies and gave me every important advices and magazines as well.*"; **Dr. Theischinger Gunther**, "*Bastiaan has always been a very dear friend*"; **Dr. Jan von Tol**," *Our good and esteemed colleague Professor Bastiaan Kiauta*"; **Dr. Robert Baker**, "*I am very appreciative of all that Professor Kiauta has done for Odonatology*"; **Dr. Gerhard Jurzitza**, "*His (Dr. Kiauta) 70th birthday will be in January 2007 ....If I shall be able I shall send him a personal letter .... I would be glad to get a copy of the congratulation book.*"; **Dr. René Hoess**, "*Our honorable colleague Bastian Kiauta.*"; **Dr. Do Manh Cuong, "***I would be very happy if I have published something in the festschrift in proud honor of*

Prof. Dr Bastian Kiauta.";. **Dr. León Andrés Pérez Gutiérrez**, "*Festschrift ...in proud honour of Prof. Dr Bastian Kiauta .... my paper.*" These are but a few of the scores of similar highly appreciative communications I have received, and all these show enormous respect and honour in the hearts of Dr. Kiauta's thousands of young and old associates all over the world!

I hope that this book, written originally by as many as 23 specialist odonatologists from different countries in all continents of the world, covering a vast range of subjects as diversified as taxonomy, ecology, zoogeography, biology, cytogenetics and cytotaxonomy, ethnoentomology and many more, offer plenty of novelties in the ever growing science of odonatology and will serve a good purpose to expand knowledge globally by both the professional and amateur scientists and scholars alike.

Finally, writing this rare but most significant festschrift in the profound honour of 70th glorious birthday of Professor Dr. B. Kiauta has been a great and highly rewarding experience for me. The kind of help, support and encouragement showered on me in accomplishing this task on time by the numerous extremely busy and preoccupied odonatologist authors of this festschrift, at a very short notice, is a huge learning experience for me about the values of commitment, perseverance and punctuality. Without their, and very many others', kind cooperation and attention an important responsibility like the preparation of this festschrift would have been almost impossible to be realized on time! My heartfelt thanks are always due to all of these dragonfly friends for everything they have done to the science of odonatology, to the festschrift of Dr. Kiauta and to keep the faith in the institution of "good odonatological family", a thesis always so close to the heart of Prof. Kiauta. Indeed, we feel extremely obliged and blessed in dedicating this festschrift to Professor Dr. B. Kiauta – the doyen of odonatology.

Madurai **_Dr. B.K. Tyagi_**

20th January, 2007

ACKNOWLEDGEMENT

Foremost I express my deepest sense of gratitude to the various authors who appreciated wholeheartedly this noble project of writing the festschrift in proud honour of the glorious 70th birthday of Prof. Dr. B. Kiauta, and instantaneously accepted to contribute highly valuable manuscripts even at a very short notice. This dream of offering a festschrift to Dr. Kiauta could become a reality largely due to their, and of scores of other odonatologists from all continents of the world, unparalleled sense of commitment, great camaraderie, unflinching cooperation and incessant encouragement. I wish to particularly thank Mrs. Marianne Kiauta for helping with the compilation of Prof. Kiauta's odonatological bibliography, one the most important compositions of the festschrift, and for kindly providing help with certain other details. My heartfelt thanks are also due to Dr. K. Inoue for his generous gift of Prof. Kiauta's knighthood photograph.

As ever in the past, my wife Ajita and daughters Anupama and Akansha were a great source of energy and determination in my undertaking the Festschrift project. They have been both my greatest critic and supporter.

Mr. Pawan Kumar, Director, Scientific Publishers (India), Jodhpur deserves my very special thanks for his faith and confidence in me, and for bringing out the book on scheduled date.

LIST OF CONTRIBUTORS
(in alphabetical order)

Dr. B.K. TYAGI[1, 2]
[1]Centre for Research in Medical Entomology (ICMR),
4-Sarojini Street, Chinna Chokkikulam,
Madurai 625005 (Tamil Nadu), India
[2]President, Indian Dragonfly Society,
D-4, Saraswati Nagar,
"Shivendralaya", P.O.: KUMB Ki Kothi,
Jodhpur 342005, India
bk_tyagi@sify.com

Dr. M.A.J.E. KIAUTA
Editoral Office of ODONATOLOGICA
Post Box 256, AG 3720
Bilthoven, The Netherlands
mb.kiauta@12move.nl

Dr. HENRI J. DUMONT
Animal Ecology, Ghent University,
Ledeganckstraat,
9000 Ghent, Belgium

Dr. MILEN MARINOV
P.O.Box 134, BG-Sofia, Bulgaria
mg_marinov@yahoo.com

Dr. BURKHARD GREBE
Oberdorfallee 7a, D-53909 Zülpich, Germany
Burkhard.Grebe@t-online.de

Dr. YORDAN KUTSAROV
18 "P. Kiselov" Str., BG-7600 Tutrakan, Bulgaria
kalimok@gmail.com

Dr. JAN J. BEUKEMA
Royal Netherlands Institute for Sea Research,
P.O. Box 59, 1790AB
Den Burg, Texel, The Netherlands

Dr. L. PÉREZ[1],
Zoology and Ecology Aquatic Laboratory LAZOEA
Biologic Sciences Department,
Universidad de los Andes Bogotá, Colombia
le-perez@uniandes.edu.co

Dr. D. MONROY
Zoology and Ecology Aquatic Laboratory LAZOEA
Biologic Sciences Department,
Universidad de los Andes Bogotá, Colombia
le-perez@uniandes.edu.co

Dr. E. REALPE
Zoology and Ecology Aquatic Laboratory LAZOEA
Biologic Sciences Department,
Universidad de los Andes Bogotá, Colombia
le-perez@uniandes.edu.co

Dr. J. OTT
L.U.P.O.GmbH,
Friedhoftstrasse 28,
D – 6705 Trippstadt, Germany,
L.U.P.O.GmbH@t-online.de

Dr. T. BROCKHAUS
An der Morgensonne 5,
D-09387 Jahnsdorf/Erzgebirge

Dr. S. G. BUTLER
'Red Willow' AllStretton, Shropshire.

Dr. R. G. KEMP
5 Maltings Close,
Ackleton, Wolverhampton, WV6 7WB.

Dr. G.S. VICK
'Crossfields' Little London Basingstoke, Hampshire.

Dr. J.B. IRUSTA
Programa de Pós-Graduação em Psicobiologia, Departamento de Fisiologia - Centro de Biociências, Universidade Federal do Rio Grande do Norte. Caixa Postal 1511 – Campus Universitário, 59078-970 Natal, RN, Brasil. <*banuelos@ufrnet.br*>

Dr. A. ARAÚJO
Programa de Pós-Graduação em Psicobiologia, Departamento de Fisiologia - Centro de Biociências, Universidade Federal do Rio Grande do Norte. Caixa Postal 1511 – Campus Universitário, 59078-970 Natal, RN, Brasil. <*arrilton@cb.ufrn.br*>

Dr. R. STOKS
Laboratory of Aquatic Ecology,
KULeuven, Deberiotstraat 32,
B-3000 Leuven, Belgium
robby.stoks@bio.kuleuven.be

Dr. M. DE BLOCK
Laboratory of Aquatic Ecology,
KULeuven, Deberiotstraat 32,
B-3000 Leuven, Belgium

Dr. LILIANA M. MOLA
Depto Ecología, Genética y Evolución; Facultad de Ciencias Exactas y Naturales; Universidad de Buenos Aires. Int. Güiraldes 2620,
Ciudad Universitaria, Pabellón II,
4° Piso. (C1428EHA) Ciudad Autónoma de Buenos Aires.
limola@ege.fcen.uba.ar

Dr. E. GONZÁLEZ SORIANO
Instituto de Biología,
Departamento de Zoología,
UNAM. Apartado Postal 70-153,
C.P. 04510 México, D.F.
esoriano@ibiologia.unam.mx

Dr. R. NOVELO GUTIÉRREZ
Instituto de Ecología, A.C., Departamento de Entomología. Km 2.5 carr. Antigua a Coatepec # 351, Congregación El Haya, 91070 Xalapa, Veracruz.
rodolfo.novelo@inecol.edu.mx

Dr. O.N. POPOVA
Laboratory of insect ecology,
Institute of Animal Systematics & Ecology, Siberian Branch, Russian Academy of Sciences, 630091,
Frunze, 11, Novosibirsk, Russia

Dr. A.Yu. HARITONOV
Institute of Animal Systematics and Ecology, Siberian Division of Russian Academy of Sciences, Frunze str. 11, Novosibirsk 630091, Russia
pc@eco.nsc.ru

Dr. AMIT MITRA
Senior Lecturer, Sherubtse College,
Kanglung, Bhutan
amitodonata@yahoo.com

Dr. ANGELO B.M. MACHADO
Departamento de Zoologia,
Instituto de Ciências Biológicas,
Universidade Federal de Minas Gerais,
Belo Horizonte, Minas Gerais,
Brasil, Caixa Postal 486, 31.270-901.
angelo@icb.ufmg.br

Dr. DO MANH CUONG
Hom thu so 16, Buu Dien 10210 – 35
Thai Thinh, Ha Noi, Vietnam
cuongdm@hotmail.com;
docuong@gmail.com

Dr. KIRAN C.G
Mayooram, Pulari Nagar, Kodunganoor, P.O, Thiruvananthapuram, Kerala, India 695013
cgkiran@gmail.com

Dr. F.K. KAKKASSERY
Reader, Research and Postgraduate Department of Zoology, St. Thomas' College, Trichur, - 680001, Kerala
kakkassery@yahoo.com

Dr. ROBERT A. CANNINGS
Royal British Columbia Museum, 675 Belleville Street, Victoria, British Columbia , Canada V8W 9W2

Dr. LEAH R. RAMSAY
British Columbia Conservation Data Centre, British Columbia Ministry of Environment, 395 Waterfront Crescent, Victoria, British Columbia, Canada V8W 9M2

Dr. SYDNEY G. CANNINGS
Nature Serve Yukon, Yukon Territorial Government, Box 2703, Whitehorse, Yukon, Canada Y1A 2C6

Dr. G. THEISCHINGER
NSW Department of Environment and Conservation, PO Box 29, Lidcombe NSW 1825 Australia
Gunther.Theischinger@environment.nsw.gov.au

Dr. S. J. RICHARDS
Vertebrates Department, South Australian Museum, North Terrace, Adelaide, S. A. 5000 Australia
richards.steve@saugov.sa.gov.au

Dr. MILOŠ JOVIĆ
Natural History Museum Belgrade, 11000 Belgrade, Serbia
milos.jovic@nhmbeo.org.yu

Dr. K.A. SUBRAMANIAN
Centre for Ecological Sciences, Indian Institute of Science, Bangalore- 560 012, India
subbu@ces.iisc.ernet.in

Dr. P.L. MILLER
Department of Zoology
South Parks Road
OXFORD OX1 3PS, United Kingdom

Dr. DAN BARTA
Havanska 8,
Prague – 7,
170 00 Czech Republic
dnlbrt@seznam.cz

CONTENTS

Odonata : Biology of Dragonflies
Ed.: B.K. Tyagi
Scientific Publishers (India), 2007, pp. 1-24

1

PROFESSOR BASTIAAN KIAUTA: AN EXTRAORDINARY AND OUTSTANDING ODONATOLOGIST

Dr. B.K. TYAGI[1,2]

[1]Centre for Research in Medical Entomology (ICMR), 4-Sarojini Street, Chinna Chokkikulam, Madurai 625005 (Tamil Nadu), India
[2]President, Indian Dragonfly Society, D-4, Saraswati Nagar, "Shivendralaya", P.O.: KUMB Ki Kothi, Jodhpur 342005, (India)
e.mail: bk_tyagi@sify.com

PART - I

GLIMPSES INTO PROF. DR BASTIAAN KIAUTA'S LIFE AND ODONATOLOGICAL ACHIEVEMENTS

"Who comprehends his trust, and to the same
Keeps faithful with a singleness of aim;
And therefore does not stoop, nor lie in wait
For wealth, or honour, or for worldly state;
Whom they must follow, on whose head must fall,
Like showers of manna, if they come at all."

The life of Professor Dr. Bastiaan Kiauta, an utterly indefatigable odonatological genius *nonpareil*, truly reflects the spirit radiated from these famous lines by William Wordsworth. Dr. Kiauta is one of those rare indefatigable scientists who by their energetic character not only practically educate the budding researchers in habits of industry, but by the example of diligence and perseverance which they set before them, largely influence the scientific activity in all directions and contribute in a great degree to form the national character. The national progress, it is aptly said, is after all the sum of individual industry, energy and uprightness and Dr. Kiauta is a glorious example in our own lifetime. No wonder, therefore, Her Majesty – the Queen of the Netherlands felt herself honoured in having bestowed upon him in 2002 the highest civilian honour of the country: *Knighthood In The Order Oranje-Nassau*! Dr. Kiauta is the ever shining star of our odonatological constellation.

Born on 20th January, 1937 in Slovenia, he obtained his early education under the affectionate, yet watchful and disciplined parental solicitude. While pursuing his education in biological sciences he soon learned and mastered several of the European languages including French, German, Italian, Russian and, of course, Dutch as well as English. He took to teaching and research as his lifelong profession, and in course joined the Department of Animal Cytogenetics & Cytotaxonomy at the Institute of Genetics in Utrecht University, The Netherlands in the early 1960s. An impressive personality full of charms, he developed very fruitful research associations with several entomologists across the world. Some of the most remarkable and everlasting companionships he had developed, to mention a few, were those with Prof. Dr. J.M. van Brink, a world renowned scientist on insect cytogenetics, with whom he was jointly editing the famous journals, *Genetica* and *Genen en Phaenen*; Dr. M.A. Lieftinck, an internationally acclaimed odonate taxonomist and a specialist *extraordinaire* on the southeast Asian fauna of dragonflies, in particular; Prof. Minter J. Westfall Jr., the world's one of the finest taxonomists and a specialist on American Odonata; Dr. S. Asahina, the unparalleled universally acclaimed odonatologist from Japan; and, Prof. Norman W. Moore, the great ecologist and dragonfly conservationist.

Owing to his prolific linguistic prowess, inexhaustible energy and rare insight into the science of dragonflies (Odonatology), Dr. Kiauta emerged much earlier in his professional career as a champion of odonatological researches across the globe. Consequently, when a group of celebrated scientists conceived an idea of founding a truly international society of odonatology, the *Societas Internationalis Odonatologica* (SIO), for the advancement of both the dragonfly knowledge and the researchers, in 1972, he was naturally voted unanimously for the most challenging job of the Executive Editor of *Odonatologica*, the society's newly commenced quarterly journal, in addition to much enviable responsibilities of the Coordinator for organizing the SIO's biennial international odonatological symposia and Chief Executive of nearly half a dozen SIO National Offices in various different parts of the world – all at a young age of 32 years! Later on, he added up to these, the editorial and publication responsibilities of equally significant *Notulae odonatologicae* (semi-annual), *Advances in Odonatology* (biennial), and the time-specific *SIO Rapid Communications* and *Odonata Specialist Group* series of the Species Survival Commission under the International Union for Conservation of Nature & Natural Resources, IUCN, with which the SIO collaborated on conservation issues. In brief, Dr. Kiauta's indelible imprint of editorial skills can be summarized as follows:

1970-1974 *CONTACTBRIEF NEDERLANDSE LIBELLENONDERZOEKERS / BULLETIN OF THE NETHERLANDS ODONATA WORKERS*, Association of the Netherlands Odonata workers, Utrecht. Irregular intervals. – [Dutch & English] ***Editor***, Pts 1-13 and Supplements

1971 *ABSTRACTS OF PAPERS READ AT THE FIRST EUROPEAN SYMPOSIUM OF ODONATOLOGY*, Gent/Belgium. – [Western congress languages and Russian] ***Editor***

1972-2006 *ODONATOLOGICA*. Journal of the Societas Internationalis Odonatologica (SIO), Utrecht. Quarterly. International Editorial and Referee Boards. Indexed in *Current Contents, Science Citation Index* and *Research Alert*. – [Western congress languages] ***Executive Editor***

1973 *ABSTRACTS OF PAPERS READ AT THE SECOND INTERNATIONAL SYMPOSIUM OF ODONATOLOGY*, Karlsruhe/Germany. – [Western congress languages] ***Joint Editor***

1978-2006 *NOTULAE ODONATOLOGICAE*. Bulletin of the Societas Internationalis Odonatologica (SIO), Utrecht. Semian-nual. International Editorial Board. – [Western congress languages and Spanish] ***Editor***

1980-2006 *SOCIETAS INTERNATIONALIS ODONATO-LOGICA RAPID COMMUNICATIONS* and *SUPPLEMENTS*. Irregular intervals – [Western congress languages] ***Editor***

1994-2006 *EXUVIAE*. Journal of the Slovene Odonatological Society, Ljubljana. Irregular intervals – [Slovenian & English] ***Member of the Editorial Board***

As a researcher, Dr. Kiauta delved deeply into the then evolving new science of insect cytotaxonomy and phylogeny, but his love for Odonata as a group was unmatched. Dragonflies became his lifelong companion – the object of his studies and the objective of his irrepressible love and attention for their worldwide protection and conservation. Soon he began to win for him exotic research projects of great challenges. When he became the Director of the Alpine Biological Expedition to Nepal during early 1970s, he wrote several significant monographs as teaching aids in insect cytotaxonomy which, along with several other publications particularly his annotated dragonfly chromosome catalogue, continue to be the reference works for a serious research student even today! I can never forget my early days of learning under his watchful patronage. Early in 1976 I had just stumbled upon an interesting phenomenon of sex-determining mechanism in two dragonflies, *Onychogomphus saundersi duaricus* and *O. schmidti*, but had no clue as to how the decreased number of chromosomes in both mitotic and meiotic stages could be explained! Dr. Kiauta then came to my help; he studied my chromosome slides, photographed and developed pictures in his laboratory, critically perused my manuscript submitted to *ODONATOLOGICA*, suggested a whole lot of changes in my write-up (by virtually having pains of re-writing the whole manuscript) and, allowed publication of my first ever scientific (odonatological, of course) paper absolutely gratis (because I was unemployed and had no money to pay the costs)! This research paper, published in 1977, became a mile stone in my career and continue to be the source of great inspiration of learning science in the finest tradition.

A true scientist and a great human being, Dr. Kiauta possesses priceless virtues and qualities that earned him international reputation both as a scientist and a teacher. There are many events in the annals of both odonatology and the life of SIO whose brilliant execution clearly owes, to a very great extent, to his irrepressible sense of hard work and sincerity to the advancement of the science of odonatology. His odonatological knowledge and editorial skills lent him good support to elevate *Odonatologica* above the rest of the world's entomological journals dedicated to a single insect Order. Its popularity could be gauged from the simple fact that the SIO soon registered a membership largest ever for a similar activity within the Class Insecta. This phenomenal success was achieved only through the dint of his perseverance, far-sightedness, and focused dedication for more than three decades; all the issues of various journals have been incessantly brought out on exactly the scheduled dates of publication! *Odonatologica*, in particular, demands a greater amount of care and attention as its articles are written in either of the four prominent European languages, that is, French, Italian, English and German – the languages no one else in odonatological realms knew better!

Execution of great responsibilities also demands the investment of an enormous amount of energy and labour. In 1985, when I went to Paris for attending the Eighth International Symposium of Odonatology, on the invitation of Dr. Kiauta, I was amazed to see him speaking to various delegates in different languages – all at the same time! I keenly observed that he, as a top executive of the SIO Council, participated in all meetings that continued till late hours in the night, but still he was there on the dot of time when the next day's proceedings at 9 am sharp would commence. His sense of valuing significance of time in all kinds of deeds is phenomenal. He shows great respect to the punctuality of time - a character the budding scientists have to emulate in their life. With a colossal stature amongst the world's leading odonatologists, Dr. Kiauta is nevertheless a symbol of humility and camaraderie to his thousands of friends and students world over. So, when an Indian scientist colleague of mine, Dr. D.B. Tembhare from the Zoology Department of Nagpur University, went to attend the seventh SIO Symposium held in Chur,

Switzerland, he was courteously invited by Dr. Kiauta to stay at his house. Dr. Tembhare was simply taken aback by the huge energy Dr. Kiauta possessed in executing the various responsibilities with absolute precision and thoroughness. He works uninterruptedly for 14-16 hours every day without any lapse, rising early at 5 am and going to sleep not until past midnight, leaving for tomorrow no work of today! "Still he finds time to look after his garden!", Tembhare told me with a clear shine of amazement in his eyes! I am reminded of the most famous and inspiring lines by Robert Frost –

The woods are lovely, dark and deep
But I have promises to keep,
And, miles to go before I sleep,
And, miles to go before I sleep.

For the young and budding odonatologists in all continents, Dr. Kiauta is an epitome of hard work and application. His deep sense of sincerity for commitment is exemplary. One particular anecdote is worthwhile to be told here as an example - out of myriads of such examples experienced by hundreds of his friends and students - that I often tell everybody as the finest example of the custodian of science!! It was during the Paris symposium of SIO that the rare idea of penning a bibliography of Col. Dr. F.C. Fraser (a great odonatologist, particularly an authority on Indian dragonflies, and best known for his *magnum opus*, "ODONATA" volumes I-III, published under the famous *Fauna of British India* series, dawned upon me. I immediately placed my thought before Dr. Kiauta who, holding his favourite pipe in his left hand close to his mouth, appreciated my proposal but cautioned me saying that Dr. Fraser had published his papers in a large number of journals, but sometimes by himself, and there was no authentic listing prepared on his publications. "But," Dr. Kiauta, judging my disappointment, came to my rescue almost immediately and suggested, "A rather incomplete manuscript of Dr. Fraser's publication existed in Dr. B.E. Montgomery's archives and perhaps you could use that as the launching pad." Then, it is a history of great learning for me, and many other students of science, to acknowledge the fact that though I was doing my own bit to complete a rather less significant part of the whole exercise, Dr. Kiauta took upon himself the major burden of accomplishing the task and regularly searched the titles in the big library of Amsterdam, several kilometres away from his home in Utrecht, guiding me all the way to dig out additional information on Fraser's life and works from his contemporaries like Dr. R.M. Gambles and Dr. E.C.G. Pinhey, both of whom had made extensive correspondence with him. In spite of his very busy teaching and research schedules at the Utrecht University, he consulted libraries in Amsterdam, Utrecht and Bilthoven, and completed the daunting task of check-list of Fraser's publications numbering over four hundred. All this he did for me, and for Dr. Montgomery's memory who of course also solely authored the booklet, courtesy scientific genius of Dr. Kiauta. It was a great learning for me in upholding the scientific tenets in the best tradition. Dr. Kiauta fulfils all his commitments and promises (even when made in the lightest sense) – virtues of greatness in all the geniuses of the world!

One of the finest characters that all geniuses are said to have is to attend the correspondence promptly, on the day as it is received. Dr. Kiauta's legendary prowess in this respect is something the young generation of odonatologists must emulate – for developing the quality of leadership. He is absolutely immaculate and meticulous in attending to a communication. Paucity of time, excess of work and growing age never hindered his endless energies and strictly disciplined life. Even in those days when e-mails did not yet exist, his promptness in replying to a communication, however trivial or even intricate for that matter, can be gauged from the simple fact that the inquirer was calculatingly bound to get his reply on the 7th day across 6000 km from Europe to any place in the world! Today, of course, his reply is there on my computer monitor when I enter my office next morning at 9 am sharp. His correspondence is always like an ocean of knowledge, both scientific and management. I was able to build the SIO National Office in India with this kind of rare training in science and the results were obvious: a series of national symposia at two-year intervals, an international symposium of odonatology of the SIO in India, publication of the annual journal INDIAN ODONATOLOGY, semi-annual newsletter FRASERIA, need-based irregularly published OCCASIONAL PUBLICATIONS and several other isolated publications of enormous value. His teachings have always been the guiding source to his innumerable students all over the world. It is this indefatigable spirit of learning that the Indian dragonfly lovers have once again mustered in enough courage to advance the science of odonatology and formed the Indian Dragonfly Society in 2006.

Dr. Kiauta is immortalized in the annals of odonatology by having had several dragonfly species' names given after him: *Protosticta kiautai* Zhou, 1986; *Dythemis kiautai* De Marmels, 1989 (currently in *Elasmothemis*); *Risiocnemis kiautai* Hämäläinen, 1991; *Paragomphus kiautai* Legrand, 1992; *Macromia kiautai* Zhou, Wang, Shuai & Liu, 1994; *Navicordulia kiautai* Machado & Costa, 1995; *Telebasis bastiaani* Bick & Bick, 1996; *Argia kiautai* Steinmann, 1997 (nomen novum for *Argia apicalis* Matsumura, 1913; actually *Rhipidolestes okinawanus* Asahina, 1951) and *Rhipidolestes bastiaani* Zhu & Yang, 1998. Besides, a keen taxonomist himself, Dr. Kiauta has also described *Croatocordulia platyptera* (Charpentier) gen. nov. in 1969 and *Crocothemis servilia mariannae* ssp. nov. in 1983. This depicts an immeasurable popularity and love and respect Dr. Kiauta is held with by thousands of his friends and associates all over the world. He is the true king of hearts of dragonfly lovers accross the world, which in the words of Samuel Smiles is crowned like this –

'tis not in mortals to command success,

'll do more, deserve it.

With several honours and decorations won (1981- Societas Internationalis Odonatologica (S.I.O.): MEMBERSHIP OF HONOUR; 1987 - Societas Internationalis Odonatologica (S.I.O.): PRESIDENT; 1993 - Slovene entomological Society: GOLDEN MEDAL; 2002- Queen of the Netherlands: KNIGHTHOOD in the order Oranje-Nassau; and 2005 - The Netherlands Association of Odonatologists: GOLDEN DRAGONFLY MEDAL), Dr. Kiauta is always extremely polite, soft-spoken and generous to all his friends, colleagues and students. All these laurels and honours bestowed on the great Dr. Kiauta owe to a very dear and respected person - Marianne Kiuata, his wife and herself a highly dedicated odonatologist, who is the greatest source of inspiration and support for Dr. Kiauta.

Dr. Kiauta is a complete scientist – with dragonflies, and only dragonflies - in his thoughts and actions. He is only 70 years young now and may the Almighty shower on him another 70 years of youthful good health, love and affection.

PART - II

ODONATOLOGICAL BIBLIOGRAPHY OF Dr B. KIAUTA, 1954-2006

M.A.J.E. KIAUTA

Editoral Office of ODONATOLOGICA
Post Box 256, AG 3720, Bilthoven, The Netherlands,
e.mail: mb.kiauta@12move.nl

(*Note*: Titles of papers written in non-congress languages are given in the language of the summary or in English translation. Book reviews, minor notes, forewords etc. in various odonatological books are omitted.)

1954a The Odonata of the surroundings of Ljubljana. *Proteus, Ljubljana* 16(8): 220-222, fig. 1. – [Slovenian]

1954b Biogeographically alien elements in the odonate fauna of the Ljubljana Moor. *Proteus,Ljubljana* 17(2): 44-48, figs 1-4. – [Slovenian]

1954c A biologist's impressions from the Mala Paklenica Gorge. *Proteus, Ljubljana* 17(4/5): 115-118, figs 1-2. – [Slovenian; includes notes on the Odon. of the Velebit Range, northern Dalmatia]

1959 On the Pleistocene dragonfly relicts. *Proteus, Ljubljana* 21(10): 260-263, fig. 1. – [Slovenian]

1960 La faune des libellules (Odonata Fabr.) dans le Parc national du Triglav. *Varstvo Spom.* 7: 376-377.

1961a Beitrag zur Kenntnis der Odonaten-Fauna Sloweniens. *Biol. Vest.* 8: 31-40.

1961b Beitrag zur Kenntnis der Libellen-Fauna (Odonata Fabr.) in der Umgebung von Škofja Loka. *Loški Razgl.* 8: 174-182, figs 1-6.

1962 Dragonfly fauna of the Triglav National Park and its surroundings (Odonata Fbr.). *Varstvo Narave* 1: 99-117, figs 1-7.

1963a Copper engravings in Scopoli's Entomologia carniolica. *Kronika, Ljubljana* 11(1): 57- 60, fig. 1. – [Slovenian; the origin and fate of the plates are traced, the odon. figs are analyzed in depth, and the odon. inventory is critically discussed]

1963b Lindenia tetraphylla v.d. Lind. und Somatochlora metallica meridionalis Nielsen aus Nordwestistrien (Jugoslawien) (Odonata: Gomphidae, Corduliidae). *Beitr. naturk. Forsch. SüdwDtl.* 22(1): 65-66.

1963c Additions to the Cyprus list of Odonata, with some notes on the annotated list given by K.J. Valle. *Commentat. biol.* 26(8): 1-6.

1963d A note on an unusual habitat of Calopteryx virgo (L.) (Calopterygidae). *Tombo, Tokyo* 6(3/4): 25-26, fig. 1.

1963e The dragonfly fauna (Odonata Fbr.) of the Kamniška Bistrica river basin (Slovenia, Yugoslavia). *Kamn. Zb.* 9: 137-153.

1964a A note on the discovery of Somatochlora alpestris in the Slovenian Alps. *Plan. Vest.* 20(8): 359-360. – [Slovenian]

1964b Notes on some field observations on the behaviour of Leucorrhinia pectoralis Charp. (Odonata: Libellulidae). *Ent. Ber., Amst.* 24(5): 82-86, figs 1-2.

1964c On the occurrence of Somatochlora arctica (Zetterstedt, 1840) in the Netherlands (Odonata: Corduliidae). *Ent. Ber., Amst.* 24(11): 235-238, fig. 1.

1964d (Van Brink, J.M. & –). Notes on the chromosome behaviour in the spermatogenesis of the damselfly Enallagma cyathigerum (Charp.) (Odonata: Coenagrionide). *Genetica* 35: 171-174, 2 pls (figs 1-9) excl.

1964e On the dragonfly fauna of the Peel peatbog area, the Netherlands. *Levende Nat.* 67(1): 12-17, figs 1-3. – [Dutch]

1964f On a migration of Libellula quadrimaculata L. (Odonata: Libellulidae). *Levende Nat.* 67(5): 59-63, figs 1-3. – [Dutch]

1964g Beobachtungen betreffs der Biologie, der Ökologie und der Ethologie der Quelljungfern im Bergland von Škofja Loka (Odonata: Cordulegastridae). *Loški Razgl.* 11: 183-192.

1965a Notes on the odonate fauna of some brackish waters of Walchern island. *Ent. Ber., Amst.* 25(3): 54-58, fig. 1.

1965b Notes sur le dépouillement de Cordulia aenea (L.) (Odonata: Corduliidae). *Ent. Ber., Amst.* 25(6): 111-113, fig. 1.

1965c On the odonate fauna of Warmbad Villach in southern Carinthia. *Beaufortia* 13(152): 35-46, figs 1-4.

1965d The chromosome behaviour in spermatogenetic meisosis of Anax imperator Leach (Odonata: Aeshnidae). *Tombo, Tokyo* 7(3/4): 18-21, figs 1-5.

1965e The behaviour of the ultimate instar larva of the dragonfly Anax imperator Leach (Odonata: Aeshnidae). *Levende Nat.* 65(2): 35-40, figs 1-4.

1965f A 1964 Libellula quadrimaculata migration in the Netherlands. *Levende Nat.* 65(2): 35-40, figs 1-4.

1965g Dragonflies – victims of traffic and modern civilisation. *Proteus, Ljubljana* 27(7): 196-197. – [Slovenian]

1966a [Descriptive review] "On a probable centre of origin and on primary directions of the dispersal of holarctic Odonata", by B.F. Belyshev. *Ent. Ber., Amst.* 26(7): 152.

1966b Some nature conservation aspects of the odonatological program of the RIVON, *Ann. Rep. St. Inst. Nat. Conserv. Res., Zeist* 1965: 35-36, 62, 1 col. pl. excl. – [Dutch]

1966c The chromosome behaviour in spermatogenetic meiosis of the dragonfly Sympetrum striolatum (Charp.) (Odonata: Libellulidae) from Luxembourg. *Bull. Soc. Nat.Luxemb.* 69: 54-60, figs 1-12.

1967a Considerations on the evolution of the chromosome complement in Odonata. *Genetica* 38(4): 430-446, textfigs 1-4, 1 pl. (figs 1-3) excl.

1967b Meiotic chromosome behaviour in the male damselfly, Calopteryx virgo (Linnaeus), with a discussion on the value of chromosome numbers and karyotype morphology in odonate systematics. *Dt. ent. Z.* (N.F.) 14(3/4): 339-348, pls 1-2 (11 figs) excl.

1967c A new hypothesis on the karyotypic evolution in Odonata. *Tombo, Tokyo* 10(1/4): 29-33, figs 1-4.

1968a Variation in size of the m-chromosome of the dragonfly Calopteryx virgo (L.) and its significance for the chorogeography of the Calopteryx virgo superspecies. *Genen Phaenen* 12(1): 11-16, fig. 1.

1968b Additions to the list of Odonata of the Dutch Wadden islands, with an account of the water quality data of larval habitats, and a review of the dragonfly fauna of the Dutch and German Northsea islands. *Biol. Jaarb. Dodonaea* 36: 88-114, textfig. 1, pls 1-2 (figs 2-6) incl., 2 fold. tabs excl.

1968c Morphology and kinetic behaviour of the odonate sex chromosomes, with a review of the distribution of sex determining mechanisms in the order. *Genen Phaenen* 12(2): 21-24, figs 1-3.

1968d Annotations on the Dutch Odonata, 1. Crocothemis erythraea (Brullé, 1832) in the Netherlands (Libellulidae, Sympetrinae). *Ent. Ber., Amst.* 28(8): 156-160, fig. 1.

1968e (– & J.M. Van Brink) Sex chromosomes and sex determining mechanism in the order Odonata. *Proc. 12th int. Congr. Genet., Tokyo* 1: 206.

1968f Variation in size of the dragonfly m-chromosome, with considerations on its significance for the chorogeography and taxonomy of the order Odonata, and notes on the validity of the rule of Reinig, *Genetica* 39(1): 64-74, textfig. 1, 1 pl. (figs 2-5) excl.

1968g The chromosome numbers of eight Old World dragonflies (Odonata). *Chrom. Inf. Serv., Tokyo* 9: 3-4, figs 1-8.

1968h The chromosomes of the male dragonfly Cordulegaster boltonii (Donovan, 1807) (Odonata: Cordulegastridae). *Biol. Vest.* 16: 87-94, figs 1-18.

1969a Sex chromosomes and sex determining mechanisms in Odonata, with a review of the cytological conditions in the family Gomphidae, and reference to the karyotypic evolution in the Order. *Genetica* 40(2): 127-157, textfigs 1-6, pls 1-8 (figs 1-51) excl.

1969b Autosomal fragmentations and fusions in Odonata and their evolutionary implications. *Genetica* 40(2): 158-180, pls 1-5 (fisg 1-37) excl.

1969c *Methods for study of dragonfly chromosomes.* Inst. Genet., Univ Utrecht. 4 pp.

1969d The chromosomes of the Hawaiian endemic dragonflies, Megalagrion oahuense (Blackburn) (Coenagrionidae: Pseudoagrioninae) and Nesogobia blackburni (McLachlan) (Libellulidae: Sympetrinae), with a note on the cytotaxonomic affinities between the genera Nesogonia Kirby and Sympetrum Newman (Odonata). *Proc.Hawaii. ent. Soc.* 20(2): 429-433, figs 1-6.

1969e The chromosomes of eight dragonfly species from continental Africa and Madagscar (Odonata). *Arnoldia, Rhod.* 4(15): 1-8, figs 1-7.

1969f *Studies on karyotypic evolution in Odonata.* Bronder, Rotterdam. 112 pp., 105 figs.

1969g Survey of the odonate fauna of the Autonomous Region Friuli-Venezia Giulia (northern Italy). *Atti Mus. civ. Stor. nat. Trieste* 26(6): 177-247, figs 1-18.

1969h The odonate collection from the karst of Inner Carniola and the Slovene Littoral in the Natural Museum of Triest, with a review and zoogeographical considerations on the dragonfly fauna of this and the adjacent territories. *Biol. Vest.* 17: 101-111, fig. 1.

1969i A proposal for the protection of some Slovene dragonfly species that are rare or threatened with extinction (Odonata). *Varstvo Narave* 6: 121-130, figs 1-2.

1969j Croatocordulia platyptera (Charpentier, 1843) gen. nov., from the Lower Miocene of Radoboj in Croatia (Odonata, Anisoptera, Corduliidae, Coduliinae). *Geologija* 12: 85-90, figs 1 (excl.)-2.

1970a The chromosomes of four neotropical dragonflies from Mexico. *Chrom. Inf. Serv., Tokyo* 11: 8-9, figs 1-4.

1970b Some remarks on the evolution of the centromere, based upon the distribution of centromere types in insects. *Ent. Ber., Amst.* 30(4): 71-75. – [A preprint was published in *Genen Phaenen* 13(1969): 1-6]

1970c Colloquium of Dutch dragonfly workers. *Selysia* 5(1): 3. – [Report on the proceedings of the first of the two Colloquia which were forerunners of the International Symposia of Odonatology, and at which the idea of founding the Societas Internationalis Odonatologica (SIO) was born]

1970d The karyotype of the damselfly Epallage fatime (Charpentier, 1840), with a note on the cytotaxonomic affinities in the superfamily Calopterygoidea. *Genetica* 41(4): 525-531, pl. 1 (figs 1-4) excl.

1970e Bibliography on the Swiss odonate fauna. *ContactBr. ned. LibellenOnderz.* 2(Suppl.): 1-6.

1971a Studies on the germ cell chromosome cytology of some cytotaxonomically interesting or hitherto not studied Odonata from the Autonomous Region Friuli-Venezia Giulia (northern Italy). *Atti Mus. civ. Stor. nat. Trieste* 27(2): 65-127, fig. 1, pls 1-12 (figs 1- 84) incl.

1971b Predation by ants, Formica fusca L. and F. rufa polyctena Bondr., on the emerging dragonfly, Aeshna juncea (L.), and its teratological consequences. *Tombo, Tokyo* 14(1/2): 2-5, figs 1-4.

1971c An unusual case of precocious segregation and chromosome fragmentation in the primary spermatocytes of the damselfly Calopteryx virgo meridionalis Selys, 1873 as evidence for a possible hybrid character of some populations of the Calopteryx-virg-complex (Odonata, Zygoptera: Calopterygidae). *Genen Phaenen* 14(2): 32-40, figs 1-10.

1971d Cytotaxonomic peculiarities in the neotropical odonate genera Leptagrion Selys, Orthemis Hagen and Macrothemis Hagen. *Abstr. Pap. 1st Europ. Symp. Odonatol.*, Gent. pp. 27-28.

1972a Scientific results of the Yugoslav 1969 Himalaya Expedition. Odonata. *Biol. Vest.* 20: 109-119, figs 1-9.

1972b The First European Symposium of Odonatology. *Ent. Ber., Amst.* 32(2): 23-24.

1972c (– & G. Jurzitza) Erstes Europäisches Symposium für Odonatologie und Gründung der "Societas Internationalis Odonatologica". *Beitr. naturk. Forsch. SüdwDtl.* 31: 185.

1972d (– & J.M. Van Brink) Societas Internationalis Odonatologica. *Vakbl. Biol.* 52(1): 14. – [Dutch]

1972e (– & J.M. Van Brink) Editorial. *Odonatologica* 1(1): 1.

1972f The karyotype of the damselfly Leptagrion macrurum (Burmeister, 1839) and its possible origin, with a note on the cytotaxonomic affinities of the genus (Zygoptera: Coenagrionidae). *Odonatologica* 1(1): 31-35, figs 1-2.

1972g Synopsis of the main cytotaxonomic data in the order Odonata. *Odonatologica* 1(2): 73-102, figs 1-2.

1972h (Dumont, H.J. & –) [Obituary]. Dr Filimon Cirdei. *Odonatologica* 1(3): 165-166.

1972i Mechanisms of karyotypic evolution in dragonflies. *Abstr. 14th Int. Congr. Ent., Canberra*, p. 107.

1972j (– & J.W. Boyes). Cytology of ten South American Libellulidae, with cytophylogenetic considerations on the genera Orthemis Hagen and Erythrodiplax Brauer (Odonata, Anisoptera). *Genetica* 43(3): 407-421, textfig. 1, pl. 1 (figs 1-19).

1972k Notes on new or little known dragonfly karyotypes, 1. The germ cell chromosomes of three Latin American species: Argia funebris (Hagen), Megapodagrion contortum (Selys) (Zygoptera: Coenagrionidae, Megapodagrionidae) and Castroraeschna castor (Brauer) (Anisoptera: Aeshnidae). *Genen Phaenen* 15(1): 23-26, figs 1-10.

1972l Notes on new or little known dragonfly karyotypes, 2. Male germ cell chromosomes of four East Mediterranean species: Lestes barbarus (Fabricius), Calopteryx splendens amasina Bartenev (Zygoptera: Lestidae, Calopterygidae), Caliaeschna microstigma (Schneider) and Orthetrum taeniolatum (Schneider) (Anisoptera: Aeshnidae, Libellulidae). *Genen Phaenen* 15(2/3): 95-98, figs 1 -11.

1972-2006 Odonatological Abstracts, Nos 1-16147. *Odonatologica* 1[1972]: 53-62, 119-130, 167-179, 263-278; 2[1973]: 49-64, 125-137, 175-189, 335-347; 3[1974]: 79-92, 143-151, 191-201, 273-292; 4[1975]: 35-53, 101-124, 191-206, 277-297; 5[1976]: 79-95, 181-192, 291-307, 391-404; 6[1977]: 35-54, 119-141, 211-226, 297-312; 7[1978]: 91-114, 167-186, 287-307, 391-414;

8[1979]: 71-89, 135-154, 227-244, 333-350; 9[1980]: 107-124, 197-215, 259-278, 329-352; 10[1981]: 65-83, 159-177, 237-257, 333-354; 11[1982]: 63-87, 167-187, 251-270, 347-368; 12[1983]: 93-113, 179-200, 297-313, 401-421; 13[1984]: 169-185, 317-334, 481-500, 603-623; 14[1985]: 65-80, 163-175, 261-278, 369-380; 15[1986]: 137-155, 223-243, 353-371, 469-491; 16[19887]: 103-118, 209-223, 315-334, 407-428; 17[1988]: 65-83, 147-177, 289-311, 435-458; 18[1989]: 107-132, 219-236, 297-324, 397-420; 19[1990]: 91-115, 195-221, 309-332, 403-426; 20[1991]: 127-149, 253-279, 375-399, 483-509; 21[1992]: 119-140, 253-273, 385-408, 515-542; 22[1993]: 101-127, 243-271, 379-397, 511-535; 23[1994]: 77-101, 193-214, 303-327, 431-454; 24[1995]: 129-148, 249-267, 361-382, 485-505; 25[1996]: 87-107, 197-220, 313-333, 405-425; 26[1997]: 95-120, 227-247, 347-373, 483-506; 27[1998]: 125-147, 267-286, 391-414, 495-512; 28[1999]: 93-115, 187-208, 293-315, 439-455; 29[2000]: 75-94, 167-194, 273-294, 369-392; 30[2001]: 123-144, 233-254, 351-373, 457-474; 31[2002]: 95-115, 211-228, 323-343, 415-434; 32[2003]: 99-118, 183-204, 303-320, 393-411; 33[2004]: 105-127, 217-236, 333-352, 443-462; 34[2005]: 83-102, 189-209, 313-334, 415-434; 35[2006]: 89-116, 205-224, 305-327.

1973a First European Symposium of Odonatology. *Selysia* 6(1): 7-8.

1973b Notes on new or little known dragonfly karyotypes, 3. Spermatocyte chromosomes of four nearctic anisopterans: Aeshna californica Calvert (Aeshnidae), Cordulia shurtleffi Scudder (Corduliidae), Sympetrum internum Montgomery, and S. madidum (Hagen) (Libellulidae). *Genen Phaenen* 16(1): 7-12, figs 1-12.

1973c Notes on new or little known dragonfly karyotypes, 4. Spermatocyte chromosomes of Calopteryx s. splendens (Harris) (Zygoptera: Calopterygidae), Gomphus pulchellus Selys and Libellula depressa Linnaeus (Anisoptera: Gomphidae, Libellulidae) from southern France. *Genen Phaenen* 16(2): 55-60, figs 1-19.

1973d Infraspecific karyotypic variability in some oriental dragonflies. *Abstr. Pap. 2nd int. Symp. Odonatol.*, Karlsruhe, p. 16.

1973e A note on the dragonfly folk names in Nepal. *Odonatologica* 2(1): 29-32.

1973f (Chatterjee, K. & –) Male germ cell chromosomes of two Calopterygoidea from the Darjeeling Himalaya (Zygoptera: Chlorocyphidae, Euphaeidae). *Odonatologica* 2(2): 105-108, figs 1-2.

1974 *Introduction to insect cytotaxonomy*. Lectures delivered at the Tribhuvan University, Kathmandu, Vol. 1. Nepal Research Centre, Kathmandu. xii+81 pp., fig. 1 – [Reprint edition published in 1978]

1975a *Cytotaxonomy of dragonflies, with special reference to the Nepalese fauna*. Lectures delivered at the Tribhuvan University, Kathmandu, Vol. 2. Nepal Research Centre, Kathmandu. xiv+78 pp., textfigs 1-5, pls 1-3 (figs 1-42).

1975b On a small dragonfly collection from Nepal. *Kailash, Kathmandu* 3(3): 289-294.

1975c (Haritonov, A.Yu. & –). To Dr Boris Feodorovich Belyshev on his 65th birthday. *Odonatologica* 4(4): 207-218, figs 1-3, portraits. – [Biography, appreciation of odonatological work, and annotated bibliography]

1975d (– & M.A.J.E. Brink) Cytotaxonomic notes on the Sympetrum pedemontanum complex (Anisoptera: Libellulidae). *Odonatologica* 4(4): 249-254, figs 1-3.

1975e (– & M.A.J.E. Kiauta-Brink) The chromosomes of the dragonfly Sympecma annulata braueri (Yakobson & Bianki, 1905) from the Netherlands, with a note on the classification of the family Lestidae (Odonata, Zygoptera). *Genen Phaenen* 18(2/3): 39-48, figs 1-4.

1976a (Belyshev, B.F. & –) To Nikolai Nikolaevich Akramowski on his 70th birthday. *Odonatologica* 5(4): 309-312, fig. 1. – [Biography, appreciation of odonatological work and annotated bibliography]

1976b (– & M.A.J.E. Kiauta) The chromosomes of some dragonflies from the Langtang Valley, central Nepal. *Odonatologica* 5(4): 347-354, figs 1-12.

1977a Notes on new or little known dragonfly karyotypes, 5. The male germ cell chromosomes of Macromia moorei Selys from Nepal (Anisoptera: Corduliidae). *Genen Phaenen* 19(2/3): 49-51, figs 1-8.

1977b (Van Brink, J.M. & –) Annotations on the chromosome complements of some dragonflies from southern Africa. *Abstr. Pap. 4th Int. Symp. Odonatol.*, Gainesville/FL, pp. 18-19.

1977c To Mrs Leonora K. Gloyd on her 75th birthday. *Odonatologica* 6(3): 143-149, figs 1-5, portrait. – [Biography, appreciation of odonatological work and bibliography]

1978a (– & J.M. Van Brink) Male chromosome complements of some Florida dragonflies, United States. *Odonatologica* 7(1): 15-25, figs 1-40.

1978b An outline of the history of odonatology in Switzerland, with an annotated bibliography on the Swiss odonate fauna. *Odonatologica* 7(3): 191-222, figs 1-10, 3 pls incl.

1978c Two cytotaxonomically interesting cases of irreversible autosome fusion in dragonflies: Argia moesta (Hagen) (Zygoptera: Coenagrionidae) and Anaciaeschna isosceles (Müller) (Anisoptera: Aeshnidae). *Notul. odonatol.* 1(1): 7-9, figs 1-4.

1979a (Ferreira, A., – & A. Zaha) Male germ cell chromosomes of thirty-two Brazilian dragonflies. *Odonatologica* 8(1): 5-22, figs 1-54.

1979b (– & G.A. Boon Von Ochssée) Some dragonfly karyotypes from the Voltaic Republic (Haute Volta). *Odontologica* 8(1): 47-54, figs 1-12.

1979c The karyotypes of some Anisoptera from Surinam. *Odonatologica* 8(4): 267-283, figs 1-56.

1979d The karyotype of Ischnura pumilio (Charp.) (Zygoptera: Coenagrionidae). *Notul. odonatol.* 1(3): 47-48, fig. 1.

1979e (– & M.A.J.E. Kiauta) The karyotype of Libellula fulva Müll. from Switzerland (Anisoptera: Libellulidae). *Notul. odonatol.* 1(4): 73-74, figs 1-2.

1979f (– & M.A.J.E. Kiauta) Cytotaxonomy of the genus Argia Rambur (Zygoptera: Coenagrionidae). *Abstr. Pap. 5th Int. Symp. Odonatol.*, Montreal, pp. 17-19.

1979g (Van Leeuwenhoek, A. & –) Letter No. 140 (85), dated Nov. 30, 1694, addressed to Petrus Rabus. *In*: L.C. Palm, [Ed], *The collected letters of Antoni van Leeuwenhoek. A complete edition of the works of Van Leeuwenhoek, annotated and considered in the perspective of the history of science and medicine, with special reference to general and biological microscopy*, Vol. 10 (1694-1695), pp. 152-165. Swets & Zeitlinger, Lisse. – [Annotations and comments]

1980a (– & M.A.J.E. Kiauta) Introduction to the cytotaxonomy of the odonate genus Argia Rambur (Zygoptera: Coenagrionidae). *Odonatologica* 9(1): 35-56, figs 1-38, 1 col. pl. incl.

1980b (Boyes, J.W., J.M. Van Brink & –) Sixteen dragonfly karyotypes from the Republic of South Africa and Swaziland, with evidence on the possible hybrid nature of Orthetrum julia falsum Longfield (Anisoptera: Libellulidae). *Odonatologica* 9(2): 131-145, figs 1-37.

1980c (– & M.A.J.E. Kiauta) On a small collection of dragonfly karyotypes from the Philippines. *Odonatologica* 9(3): 237-245, figs 1-19.

1980d (– & M.A.J.E. Kiauta) Considerations on cytotaxonomy and speciation in the dragonfly genus Rhinocypha (Odonata: Chlorocyphidae). *Abstr. Pap. 16th Int. Congr. Ent.*, Kyoto, p. 19.

1980e (– & M.A.J.E. Kiauta) The karyotypes of Aeshna subarctica elisabethae Djak. and Somatochlora alpestris (Sel.) from Switzerland (Anisoptera: Aeshnidae, Corduliidae). *Notul. odonatol.* 1(6): 104-105, figs 1-3.

1981a *Annotated catalogue and bibliography of taxa introduced in Odonata from 1971 to 1980.* Soc. Int. Odonatol., Utrecht. 64 pp.

1981b A review of the main achievements in odonate taxonomy and systematics from 1971-1980. *Abstr. Pap. 6th Int. Symp. Odonatol.*, Chur, pp. 22-25.

1981c Ten years of the International Odonatological Society: history, development and perspectives. *Abstr. Pap. 6th Int. Symp. Odonatol.*, Chur, p. 72.

1981d (– & M.A.J.E. Kiauta) The karyotype of Risiocnemis incisa Kimmins, 1936 from Luzon, the Philippines (Zygoptera: Platycnemididae). *Odonatologica* 10(2): 151-154, figs 1-5.

1981e (– & M.A.J.E. Kiauta) The Engadine Post-Symposium Tour. *In*: *Field Trips in the framework of the Sixth International Symposium of Odonatology*, Chur, pp. 13-20. Soc. Int. Odonatol., Utrecht. – [Includes a checklist of the Engadine Odonata]

1981f (Kiauta, M. & –) Aspirator for collecting of small zygopterans close above water surface and among dense vegetation. *Notul. odonatol.* 1(8): 133-134, fig. 1.

1981g [Obituary] In memoriam Luitenant Kolonel B.D. Dr Cornelis de Jong. *ContactBr. ned. LibellenOnderz.* 2: 6. – [Dutch]

1982a (– & M. Kiauta) The chromosome numbers of sixteen dragonfly species from the Arun Valley, eastern Nepal. *Notul. odonatol.* 1(9): 143-145, figs 1-5.

1982b (– & M. Kiauta) The chromosome numbers of eleven dragonfly species from Singapore. *Notul. odonatol.* 1(10): 164-165, figs 1-7.

1982c (– & M. Kiauta) *List of species, with chromosome numbers and preliminary notes on the karyotypes of the Odonata, collected in May 1979 and August 1980 by the members of the Kansai Research Group of Odonatology and examined by B. and M. Kiauta.* Report for the Kansai Research Group of Odonatology, Osaka. 8 pp.

1983a [Editorial] In memoriam Professor Basil Elwood Montgomery. *Odonatologica* 12(1): 1-4, 14 figs.

1983b (– & M. Kiauta) Further notes on the Philippine odonate karyotypes. *Notul. odonatol.* 2(1): 14-15, fig. 1.

1983c (– & M. Kiauta) The chromosome numbers of some Odonata from Thailand. *Notul. odonatol.* 2(2): 27-28, figs 1-6.

1983d The status of the Japanese Crocothemis servilia (Drury) as revealed by karyotypic morphology (Anisoptera: Libellulidae). *Odonatologica* 12(4): 381-388, figs 1-15. – [C. s. mariannae ssp. n.]

1983e Über das Vorkommen der Südlichen Heidelibelle, Sympetrum meridionale (Selys), in dem Engadin. *Jber. naturf. Ges. Graubünden* 100: 151-156, fig. 1.

1984a (– & D.C. Geijskes) To Dr h.c. Maurits Anne Lieftinck on his 80th birthday. *Odonatologica* 13(1): 1-4, portrait incl.

1984b (Geijskes, D.C. & –) Annotated catalogue of taxa introduced in Odonata by M.A. Lieftinck, with his complete bibliography (1919-1983). *Odonatologica* 13(1): 21-50.

1984c Aktuelle Probleme der Zytotaxonomie, erläutert bei südasiatischen Prachtlibellen (Odonata: Chlorocyphidae) und schweizerischen Köcherfliegen (Trichotpera: Limnephilidae), mit Bemerkungen über die Bedeutung der Zytotaxonomie für die Umweltforschung. *Opusc. zool. flumin.* 1: 1-20, figs 1-6.

1984d Breeding habitat of Planaeschna risi in Taiwan. *In*: P.S. Corbet, Current topics in dragonfly biology, pt 1, p. 12. *Soc. int. odonatol. rapid Comm.* (Suppl.), Vol. 2, Utrecht.

1984e (Corbet, P.S., C.M. Scrimgeour & –) Author index for Odonatological Abstracts 1-4225 from Odonatologica volumes 1-12. *Soc. int. rapid Comm.* (Suppl.) 4(A), viii + 64 pp.

1984f (Corbet, P.S., C.M. Scrimgeour, J.E. Holmquist & –) A topic index for Odonatological Abstracts 1-4225 from Odonatologica volumes 1-12. *Soc. int. odonatol. rapid Comm.* (Suppl.) 4(B), vii + 64 pp.

1984g (– & M. Kiauta) Crocothemis erythraea (Brullé) recorded from canton ST Gallen, Switzerland (Anisoptera: Libellulidae). *Notul. odonatol.* 2(4): 65-66.

1984h (– & M. Kiauta) The true chromosome number of Aka-tombo, Sympetrum frequens (Sel.) of Japan, with a note on the karyotype of S. depressiusculum (Sel.) (Anisoptera: Libellulidae). *Notul. odonatol.* 2(4): 66-67, figs 1-2.

1985a Foreword. *Proc. 1st Indian Symp. Odonatol.*, Madurai, p. ix.

1985b The identity of F. Ris's correspondent and odonate collector "H. Spoerry jun. in Flums", Switzerland. *Notul. odonatol.* 2(6): 103.

1986a (– & M. Kiauta) The dragonfly fauna of the Flumserberg region, canton St. Gallen, eastern Switzerland (Odonata). *Opusc. zool. flumin.* 3: 1-14.

1986b [Editorial] To Professor Dr Minter Jackson Westfall, Jr on his 70th birthday. *Odonatologica* 15(1): 1-4, portrait.

1986c (– & M. Kiauta) Beitrag zur Kenntnis der Libellenfauna des St. Galler Rheintales zwischen Altenrhein (Bodensee) und Bad Ragaz, Schweiz (Odonata). *Ber. bot.-zool. Ges. Liechtenstein-Sargans-Werdenberg* 15: 127-136, fig. 1.

1986d [Breeding habitat of Somatochlora alpestris at Amdenerhöhe, canton St Gallen, Switzerland]. *In*: P.S. Corbet, Current topics in dragonfly biology, pt 2, p. 3. *Soc. int. odonatol. rapid Comm.* (Suppl.), Vol. 6, Bilthoven.

1986e [A comment on dragonfly predation on insects associated with coniferous forests in the United States and the possibility of its beneficial bearing on forest conservation in the air-polluted areas]. *In*: P.S. Corbet, Current topics in dragonfly biology, pt 2, p. 6. *Soc. int. odonatol. rapid Comm.* (Suppl.), Vol. 6, Bilthoven.

1986f [A note on the problems of the field assessment of the odonate component in the structure of biotic communities]. *In*: P.S. Corbet, Current topics in dragonfly biology, pt 2, p. 7. *Soc. int. odonatol. rapid Comm.* (Suppl.), Vol. 6, Bilthoven.

1986g [A note on the odonate community structure in a Luzon fishpond, the Philippines, with references to the role of dragonflies as human food in Indonesia and as vectors of human diseases in Thailand]. *In*: P.S. Corbet, Current topics in dragonfly biology, pt 2, pp. 10-12. *Soc. int. odonatol. rapid Comm.* (Suppl.), Vol. 6, Bilthoven.

1986h [Effects of "acid rain" on the Leucorrhinia larval populations in Fenoscandia]. *In*: P.S. Corbet, Current topics in dragonfly biology, pt 2, p. 13. *Soc. int. odonatol. rapid Comm.* (Suppl.), Vol. 6, Bilthoven.

1986i [The modified life history of Lestes macrostigma in Dalmatia, as conditioned by the early summer dry season]. *In*: P.S. Corbet, Current topics in dragonfly biology, pt 2, p. 15. *Soc. int. odonatol. rapid Comm.* (Suppl.), Vol. 6, Bilthoven.

1986j [A statement on the variation in the numbers of dimorphic females in geographically distant populations of Boyeria irene]. *In*: P.S. Corbet, Current topics in dragonfly biology, pt 2, p. 16. *Soc. int. odonatol. rapid Comm.* (Suppl.), Vol. 6, Bilthoven.

1986k [A comment on the geographic variation in the number of larval instars in Ceriagrion coromandelianum, Cordulegaster and Epiophlebia]. *In*: P.S. Corbet, Current topics in dragonfly biology, pt 2, p. 24. *Soc. int. odonatol. rapid Comm.* (Suppl.), Vol. 6, Bilthoven.

1986l Aeshna caerulea (Ström, 1783) vs Aeshna coerulea (Anisoptera: Aeshnidae). *Notul. odonatol.* 2(8): 133.

1987 The first All-Union Symposium on Odonatology in the USSR. *Selysia* 16(1): 5-6.

1988a (Pinratana, A., – & M. Hämäläinen) *List of the Odonata of Thailand and annotated bibliography*. Viratham Press, Bangkok. iv+20 pp., 20 col. pls excl.

1988b [Migratory habits and vertical distribution of Pantala flavescens in the Himalayan region]. *In*: P.S. Corbet Current topics in dragonfly biology, pt 3, p. 5-6. *Soc. int. odonatol. rapid Comm.* (Suppl.), Vol. 8, Bilthoven.

1988c (– & M. Kiauta) The unusual recombination potential and its ecological implications in Coenagrion mercuriale (Charp.) from Liechtenstein (Zygoptera: Coenagrionidae). *Notul. odonatol.* 3(2): 34-35, fig. 1.

1988d The path towards the International Odonata Research Institute. *Anu. Rep. Int. Odonata Res. Inst.* 1: 1-5, fig. 1.

1990a (– & M. Kiauta) Early summer dragonflies of the Ryder Lake area, Chilliwack, British Columbia, Canada (Odonata). *Opusc. zool. flumin.* 58: 1-10, fig. 1.

1990b (– & M. Kiauta) Some dragonflies collected around Ryder Lake, Chilliwack district, in 1990. *Boreus* 10(2): 7.

1991 (– & M. Kiauta) Biogeographic considerations on Coenagrion hylas freyi (Bilek, 1954), based mainly on the karyotypic features of a population from North Tyrol, Austria (Zygoptera: Coenagrionidae). *Odonatologica* 20(4): 417-431, figs 1-2.

1992 (– & M. Kiauta) Aeshna caerulea (Ström), a new dragonfly for the fauna of Slovenia (Anisoptera: Aeshnidae). *Notul. odonatol.* 3(10): 168.

1993a (– & M. Kiauta) In memory of Janny Margaretha van Brink (5 May 1923 – 28 March 1993). *Odonatologica* 22(2): 129-131, portrait. – [Appreciation of her work in the SIO]

1993b (– & C. Soler) Die rätoromanischen Namen für die Libellen Graubündens: ein Diskussionsvorschlag (Odonata). *Opusc. zool. flumin.* 110: 1-8.

1994a Two photographs relative to the history of odonatology in Slovenia. *Notul. odonatol.* 4(4): 69-70, figs 1-2.

1994b (Thompson, D.J. & –) Odonatospeleology: dragonflies in caves, with a checklist of the known records (Odonata). *Opusc. zool. flumin.* 118: 1-10.

1994c (– & M. Kiauta) Dragonfly records from the districts of Ribnica and Kočevje, Lower Carniola, Slovenia (Odonata). *Opusc. zool. flumin.* 124: 1-8.

1994d (– & M. KIAUTA) A note on the dragonfly assemblage at the intermittent pond of Ledine, Julian Alps, NW Slovenia (Odonata). *Opusc. zool. flumin.* 124: 9-12, fig. 1.

1994e (– & M. Kiauta) On a small autumnal dragonfly collection from the Cariboo and southwestern British Columbia, Canada (Odonata). *Opusc. zool. flumin.* 126: 1-8.

1994f A contribution towards the odonatological bibliography of Slovenia. *Exuviae* 1(1): 9-15.

1995a (– & M. Kiauta) Odonatological periodicals in-print. *Notul. odonatol.* 4(5): 77-80.

1995b (– & M. Kotarac) Two dragonfly records from karst caves in Bosnia-Herzegovina and Slovenia (Anisoptera: Aeshnidae, Corduliidae). *Notul. odonatol.* 4(6): 106-107.

1995c (– & M. Kiauta) The karyotypes of Somatochlora meridionalis Nielsen from Slovenia and S. metallica (Vander L.) from Switzerland, with a tentative note on the origin of the central European S. metallica (Odonata: Corduliidae). *Opusc. zool. flumin.* 137: 1-5, figs 1-6.

1998 An interesting old observation on dragonfly behaviour at the seaside, Indonesia. *Notul. odonatol.* 5(2): 21.

1999 (– & M. Kiauta) A note on dragonfly response during the 93.8% solar eclipse of 11 August 1999 in the Netherlands (Odonata). *Opusc. zool. flumin.* 172: 1-6.

2000 Ferdinand J. Schmidt (1791-1878). *Erjavecia* 9: 1-9, figs 1 -5, portrait. – [Slovenian; biography and appreciation of his odonatological work]

2001 Snake associations in the European "dragonfly" folk appellations: distributional patterns reflecting the Urnfield culture expansion during the first millenium B.C.? *Abstr. 1st Int. Conf. 'The Veneti within the ethnogenesis of the central European population'*, Ljubljana, pp. 37-38.

2002 A visit of Professor František Klapálek, the founder of Czech scientific entomology, to Slovenia – a century ago. *Erjavecia* 14: 1-4, portrait. – [Slovenian; biography and appreciation of odonatological work]

2003a (–, M. Kiauta & K. Inoue) To Professor Dr Norman Winfrid Moore for his 80th birthday. *Odonatologica* 32(1): 1-4, portrait.

2003b A note on the Odonata collection of the Rev. Father Gabriel Strobl (1846-1925) in the Benedictine Abbey at Admont, Austria. *Notul. odonatol.* 6(2): 14-19, figs 1-3, portrait.

2003c Rev. Father Gabriel Strobl (1846-1925) and the dragonflies of Slovenia. *Erjavecia* 15: 1-5, portrait. – [Slovenian]

2003d Dr Roman Puschnig, the father of Carinthian odonatology (1875-1962). *Erjavecia* 16: 1-6, fig. 1, portrait. – [Slovenian;

biography, appreciation of odonatological work and bibliography]

2004a On the 1834 assemblage of the odonate fauna of Ljubljana listed by Franz Wilhelm Lippich (1799-1847). *Erjavecia* 17: 1-7, figs 1-2, portrait. – [Slovenian]

2004b Odonate faunistics at the southwestern reaches of the Slovenian ethnic territory: incidental collectors of the Alessandro Carlo principe Della Torre e TAsso circle (1881-1937). *Erjavecia* 18: 1-8, figs 1-3, portrait. – [Slovenian]

2005 Professor Dr Ciril Ažman, insect collector from Novo mesto (1881-1952). *Erjavecia* 20: 1-7, portrait. – [Slovenian; biography and appreciation of odonatological work].

Odonata : Biology of Dragonflies
Ed.: B.K. Tyagi
Scientific Publishers (India), 2007, pp. 25-32

2

STUDIES ON NEOTROPICAL PROTONEURIDAE. 20. *NEONEURA KIAUTAI* SPEC. NOV. FROM SOUTHEASTERN BRAZIL (ZYGOPTERA, PROTONEURIDAE)

ANGELO B.M. MACHADO

Departamento de Zoologia, Instituto de Ciências Biológicas, Universidade Federal de Minas Gerais, Belo Horizonte, Minas Gerais, Brasil, Caixa Postal 486, 31.270-901.
e.mail: angelo@icb.ufmg.br

ABSTRACT

Neoneura kiautai, sp.n. is described and illustrated. It is close to *N. ethela* but differs mainly by the shape of the decumbent process of cercus and the colour of abdominal segments 7-10.

INTRODUCTION

In his review of the genus *Neoneura* Garrison (1999) mentions 23 species, most of them Neotropical, with only two species reaching the Nearctic region. Since then, four new species have been described (Machado, 2002, 2003, 2005), all from Brazil. We describe now an additional Brazilian species from the State of Minas Gerais, thus increasing the number of known species to 28. Out of these, 18 occur in Brazil as recorded by Selys (1860, 1886), Williamson (1917), Machado (1975, 1989, 2002, 2003, 2005) and Garrison (1999). Most species of *Neoneura* are lotic, and since very few Brazilian rivers and streams have been surveyed for their *Neoneura* fauna so far, the number of Brazilian species of *Neoneura* is likely to increase considerably. We describe now another Brazilian species of *Neoneura* that lives both in lotic and lentic habitats.

NEONEURA KIAUTAI SP. NOV.

Material. - Holotype, male, allotype female, Brazil, Minas Gerais, Açucena, II-1951, A.B.M. Machado leg. (collected in a pond). Paratypes: 5 male, same data; 2 males, Cataguazes, 15/2/1958, N.D. Santos leg. (Riacho Meia Pataca); 4 males, Ubá, 9/XII/1959. N.D. Santos leg. (on a stream tributary of the Xopotó River); 1 male, Espirito Santo, Santa Tereza, 24/IV/1998, De Marco leg. Holotype, allotype and 4 male paratypes deposited in the author´s collection (Belo Horizonte); 7 male paratypes deposited in the Museu Nacional, Rio de Janeiro; 1 male paratype deposited in the Department of Zoology, Universidade Federal de Minas Gerais, Belo Horizonte.

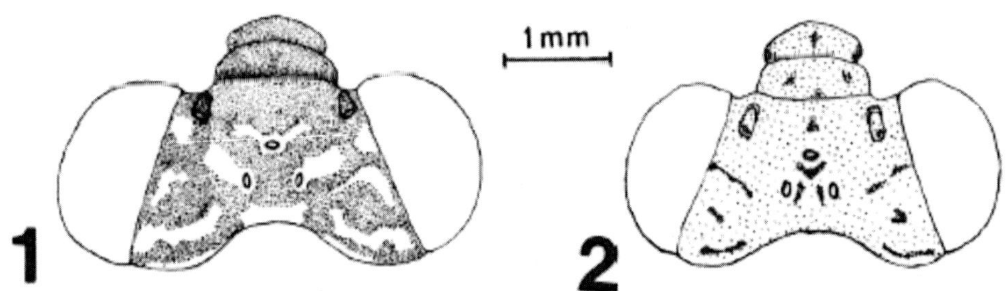

Figs. 1-2. *Neoneura kiautai* sp. nov., dorsum of head in the male holotype (1) and female allotype (2)

Etymology. - This species is named after my friend Prof. Dr. Bastiaan Kiauta on the occasion of his 70th birthday, in recognition for his enormous contribution to the development of odonatology throughout the world.

MALE (holotype). - *Head.* - labium yellow. Labrum dark brown bordered with yellow anteriorly; anteclypeus greyish yellow. Posclypeus black. Genae, base of mandibles and antefrons greyish yellow. Dorsal part of head black with greyish yellow markings as shown in Fig. 1. Rear of the head with a large central black area with irregular outlines, surrounded by yellow.

Thorax. (Fig. 3)- Prothorax light brown, with the anterior and posterior lobes black bordered with yellow, connected with each other by a black central band. Pterothorax: light brown on the posteriror part of mesepisternum and metepisternum and light yellow on the meso and metepimeron. Black areas as shown in Fig. 3. In one paratype the black areas were more extensive, occupying almost all the mesepisternum and a large part of the mesepimeron. Wings hyaline, slightly suffused with yellow in three paratypes, pterostigma yellowish brown. Legs: femora yellow on the flexor and dark brown on the extensor surfaces. Tibiae brown on the flexor and yellow on the extensor surfaces. Tarsi brown.

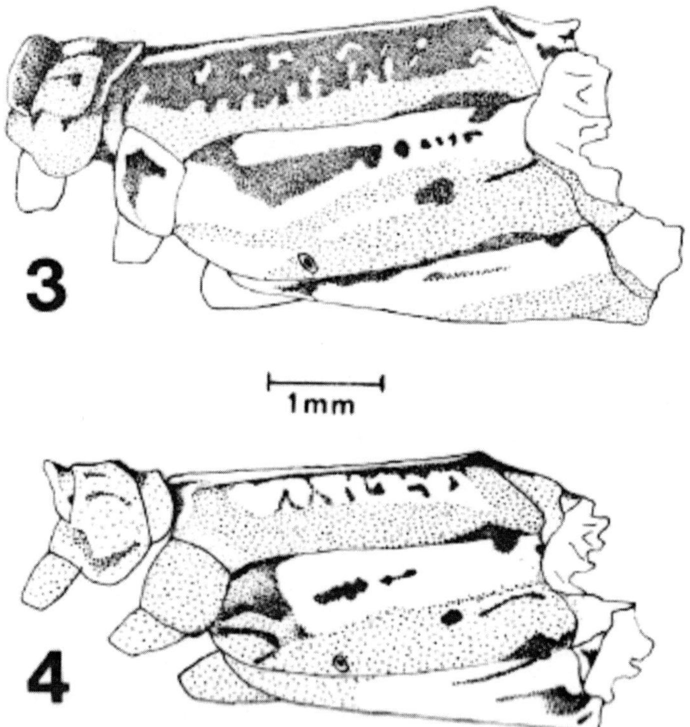

Figs. 3-4. *Neoneura kiautai* sp. nov., dorsolateral view of the thorax in the male holotype (3) and female allotype (4)

Venation: postnodals in forewing 8 (21.4%), 9 (21.4%), 10 (21.4%), 11 (35.7%); in hindwing 8 (35.7%), 9 (64.3%). R3 in forewing originating at the level of the 4th (64.3%) or 5 (35.7%) postnodal; in hindwing at the level of the 3rd (64.3%) or 4 (35.7%) postnodal. IR2 in forewing originating at the level of the 6th (11.9%), 7 (64.3%) or 8 (23.8%) postnodal; in hindwing at the level of the 6th (28.5%) or 7 (71.4%) postnodal. CUP in fore and hindwings terminating at about half the distance between the crossvein descending from the subnodus and that descending from the first postnodal (42.8%), slightly distal (38.0%) or distinctly distal from it (19.0%). Arculus in both wings distinctly distal to the second antenodal.

Abdomen - Segment 1 dorsally black, laterally light yellow with an oblique black spot. Segment 2 dorsally reddish orange, laterally light yellow with an elongated tapering dark brown stripe in the middle. Segments 3-6 reddish orange dorsally, ventrally light orange. Sternites 3-8 black. Segment 7 dorsally dark brown, laterally light brown with a black distal ring. Segments 8-10 black with orange red markings laterally. Distal half of 10 orange red. Cerci laterally orange red, dorsally black. In one paratype segments 7-10 are black throughout with no orange red area. In another paratype the extension of the orange red areas are greater. Paraprocts reddish orange.

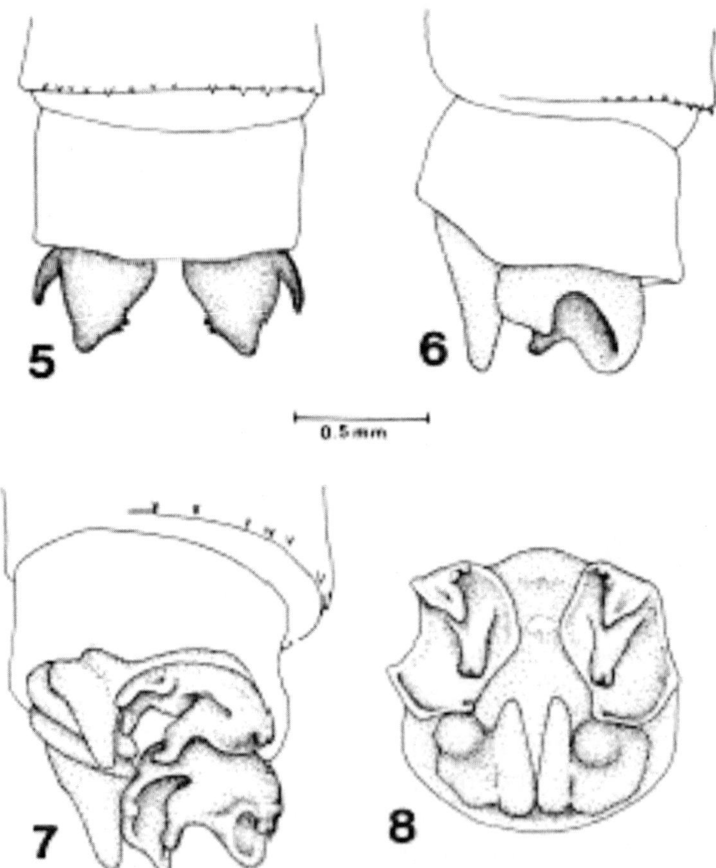

Figs. 5-8. *Neoneura kiautai* sp. nov., male holotype. Last abdominal segment with the anal appendages in dorsal (5), lateral (6), oblique posterior (7) and posterior (8) views

Structural characters (Figs. 5-8) - Posterior lobe of prothorax smoothly curved. Cercus shorter than segment 10, in dorsal view (Fig. 5), subtriangular with two teeth on the medial margin, the ventral branch projecting laterally. In lateral view (Fig. 6) subequal to the paraproct, the ventral branch much smaller than the dorsal one, armed with a blunt apical tooth. In oblique posterior view (Fig. 7) the medial surface of the dorsal branch of cerci shows at the distal third a raised area bordered anteriorly by an arcuated carina, connected above with the marginal teeth. In posterior view (Fig. 8) the decumbent processes of the dorsal branches are parallel and perpendicular to the transverse axis, armed with a distal, slightly bifid tooth.

Measurements (mm). - Abdomen 27.3-29.1 (mean 28.2); hindwings 18.4-20.0 (mean 19.4); pterostigma 09-10.

FEMALE (allotype). - *Head.* - Face yellow with a basomedial and a lateral black spot on the labrum and a black dot on either side of the postclypeus. Dorsal part of head greyish yellow with black markings as shown in Fig. 2. Rear of the head anteriorly black, posteriorly yellow.

Thorax. - Prothorax (Figs. 4, 9) greyish yellow. Anterior lobe with a large central triangular black spot and three small ones at each angle of the lobe. Posterior lobe with a transverse central black band as shown in Fig. 9. Pterothorax. - Middorsal carina, medial part of mesepisternum, meso and metepimeron greyish yellow. Remaining parts of pleuron light brown. Black markings as shown in Fig. 4. Wings hyaline, pterostigma yellowish.

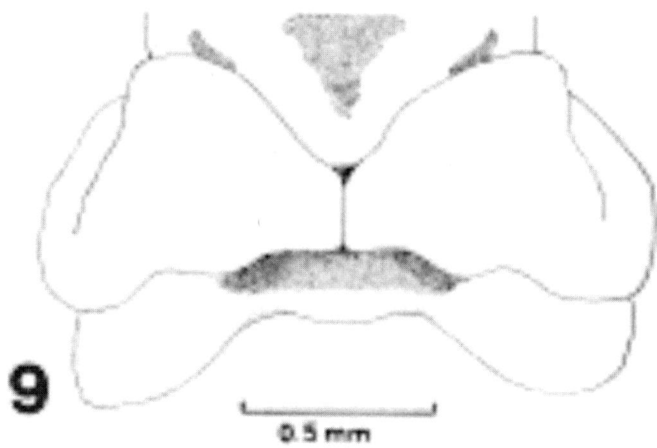

Fig. 9. *Neoneura kiautai* sp. nov. female allotype, prothorax in dorsal view

Venation. - Similar to that of the male.

Abdomen. - Light yellow. Anterior half of dorsal part of segment 1 black, anterolateral part dark brown. Segment 2 with a proximal, distal and lateral dark brown line. Dorsal parts of segments 2-7 with a narrow transverse dark preapical stripe.

Measurements (mm). - abdomen 26.5, hidwing 17.0; pterostigma 1.0.

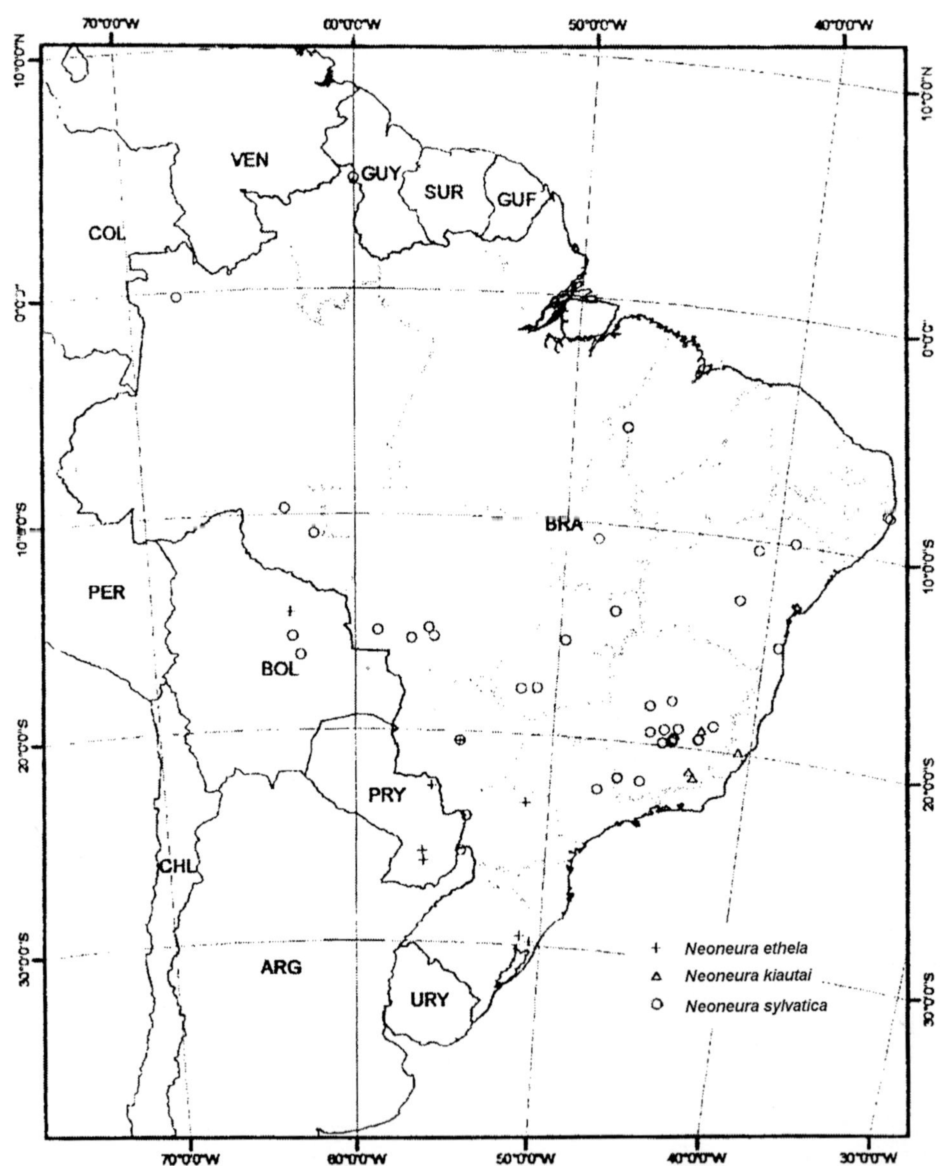

Fig. 10. Distribution map of *N. kiautai*, *N. ethela* and *N. sylvatica*

Table 1. Characters separating the species of neoneura whose males have a carina on the medioapical part of the dorsal branch of cerci.

CHARACTERS	*N. kiautai*	*N. ethela*	*N. sylvatica*
Cercus	with decumbent processes perpendicular to the transverse axis	Decumbent processes directed midventrally	Decumbent processes directed midventrally
Dominant color of head	black	black	sky blue
Dominant color of mesepisternum	black	black	sky blue
Abdomen color	Segments 1-6 orange red; 7 brown, 8-10 black with lateral reddish markings	Segments 1-10 red. 8-10 with lateral black markings	Segments 1-2 sky blue; 3-10 red
Geographical distribution (Fig. 10)	Southeastern Brazil	Southern Brazil, Paraguai and Bolivia	All over Brazil, Venezuela, Bolivia and Argentina

DISCUSSION

By having a large decumbent process of the cercus perpendicular to the transverse axis, when viewed posteriorly, *N. kiautai* would belong in the *N. fulvicollis* group of Garrison (1999), together with *N. cristina* Racenis, 1955 and *N. gaida* Racenis, 1953. However, it differs markedly from these species in color and mainly in the shape of the appendages that are completely different from those of the species of the *N. fulvicollis* group. Indeed, except for the perpendicular decumbent process, the cercus of *N. kiautai* is identical to those of *N. ethela* Williamson, 1917 and *N. sylvatica* Hagen in Selys, 1886, species belonging in the *N. rubriventris* group. The characters separating these 3 species are shown in Table 1. They are unique in the genus because, in oblique posterior view, the apical third of the dorsal branch of their cerci shows a raised area bordered anteriorly by a carina (Fig. 7). Confirming the close affinity of *N. kiautai* with *N. ethela* and *N. sylvatica*, the females of these three species are very similar. The only difference is in the color of the posterior prothoraxical lobe that has a black transverse band in *N. kiautai* (Fig. 9), reduced to two lateral spots in *N. ethela* and *N. sylvatica*. Although this character was present in several female specimens of *N. sylvatica* and *N. ethela*, we do not know its variability in *N. kiautai* because only a single female is known. *N. kiautai* occurs in Southeastern

Brazil whereas *N. ethela* is a Southern Brazil species (Fig. 10). This could be used to separate the females of these two species but not that of *N. sylvatica* since this species has a wide distribution in Brazil (Fig. 10). Most species of *Neoneura* occur only in streams or large rivers, being thus strictly lotic. *Neoneura kiautai*, however, has been found in both streams and ponds which means that it probably has adapted itself to live in lentic environments. The same occurred with *N. sylvatica* that can be found both in lotic and lentic habitats such as the Pampulha Lake in Belo Horizonte (Minas Gerais). Also *N. ethela*, which in Brazil had so far been found only in streams, was found in a pond in Bolivia (Garrison, 1999). Regarding its conservation status when evaluated under the IUCN (2001) criteria, *N. kiautai* can be categorized as Least Concern.

ACKNOWLEDGEMENTS

I thank Myrian Morato Duarte for the drawings illustrating this papers and Breno Reis V. De Mendonça for the map. I thank also Janira Martins Costa, for allowing me to study the specimens of Museu Nacional, Frederico Lencioni for providing the records of *Neoneura* in his private collection, and Lúcio Cadaval Bedê for critically reviewing this paper.

REFERENCES

Garrison, R.W. 1999. The genus Neoneura, with keys and description of a new species, *Neoneura jurzitzai* spec. nov. (Zygoptera: Protoneuridae). Odonatologica 28(4):343-375.

IUCN. 2001. IUCN Red List Categories and Criteria: Version 3.1. IUCN Species Survival Commission. IUCN, Gland, Switzerland and Cambridge, 26 pp.

Machado, A.B.M. 1975. *Neoneura schreiberi* nova espécie da região amazônica (Odonata: Protoneuridae). Ciência e Cultura, 27(7):764-766.

Machado, A.B.M. 1989. Studies on neotropical Protoneuridae. 8. *Neoneura desana* sp. nov. from the Amazonian region (Odonata, Zygoptera). Revta. bras. Ent. 33(2):225-228.

Machado, A.B.M. 2002. *Neoneura lucas* spec. nov. from Brazilian Pantanal (Zygoptera: Protoneuridae). Odonatologica 31(2):199-204.

Machado, A.B.M. 2003. *Neoneura moorei* spec. nov. from the Amazonian region of Brazil (Odonata: Protoneuridae). Odonatologica 32(1):89-93.

Odonata : Biology of Dragonflies
Ed.: B.K. Tyagi
Scientific Publishers (India), 2007, pp. 33-43

3

THREE NEW DAMSELFLY SPECIES FROM PAPUA NEW GUINEA (ZYGOPTERA : MEGAPODAGRIONIDAE, COENAGRIONIDAE)

G. THEISCHINGER[1] and S.J. RICHARDS[2]

[1]NSW Department of Environment and Conservation, PO Box 29, Lidcombe NSW 1825 Australia
e.mail: Gunther.Theischinger@environment.nsw.gov.au
[2]Vertebrates Department, South Australian Museum, North Terrace, Adelaide, S. A. 5000 Australia
e.mail: richards.steve@saugov.sa.gov.au

(Dedication : "To our friend Prof. Dr. Bastiaan Kiauta, one of Odonatology's Knights in Shining Armour.")

ABSTRACT

Three species of damselflies from Eastern Highland Province, Papua New Guinea are described as new. They are: *Argiolestes angulatus* sp. n. (Holotype ♂: Mamaifu, 21-XI-1997), *Argiolestes fornicatus* sp. n. (Holotype ♂: Mamaifu, 29-XI-1997) and *Austroagrion kiautai* sp. n. (Holotype ♂: montane lake behind Mamaifu, 27-XI-1997). Diagnostic characters of the adults are illustrated and the affinities of the species are discussed. All type material is deposited at the South Australian Museum, Adelaide.

INTRODUCTION

The Crater Mountain Wildlife Management Area (CMWMA) incorporates a large expanse of low to high-elevation rainforest in Eastern Highlands and Chimbu Provinces, Papua New Guinea. Recent studies indicate that the area has a remarkably diverse odonate fauna, and that a high proportion of species are undescribed (Oppel, 2005). Theischinger & Richards (2005, 2007) described a new species of *Drepanosticta* and a new species of *Teinobasis* respectively from Herowana Village at around 1400 m a.s.l. in the CMWMA. Here we describe three additional zygopteran species from the vicinity of Maimafu, a high-altitude village about 26 km from Herowana in the CMWMA. This is the sixth in a series of papers that aims to address taxonomic novelties in New Guinean Odonata collected by SJR between 1996 and 2001 (Theischinger & Richards, 2005, 2006a, 2006b, 2006c, 2007).

MATERIAL AND METHODS

Descriptive terminology largely follows Chao (1953) and Watson & O'Farrell (1991). Descriptions of coloration are based on preserved material and measurements are given in millimetres (mm). Illustrations were done with the aid of a camera lucida and are not to scale. Unless indicated otherwise material is deposited in the South Australian Museum (SAMA), North Terrace, Adelaide, South Australia.

ARGIOLESTES ANGULATUS SP. NOV.
Figures 1-3

Material – Holotype ♂ (Sama I 21778): Papua New Guinea, Eastern Highland Province, Mamaifu, (145°01'E, 6°30'S), 21-XI-1997, S.J. Richards.

Etymology - The specific name (angulatus = angular) refers to the shape of the superior anal appendages of the male.

MALE: *Head* – Labium black and brown, with part of mid-line narrowly yellow; labrum metallic greenish black; anteclypeus pale brownish yellow; postclypeus metallic black with a wide subtriagular patch each side adjacent to epistomal suture; remainder of head including antennae black with only an orange patch each side between base of postclypeus and eye-margin and an elongate brown patch each side between lateral ocellus and antennal base.

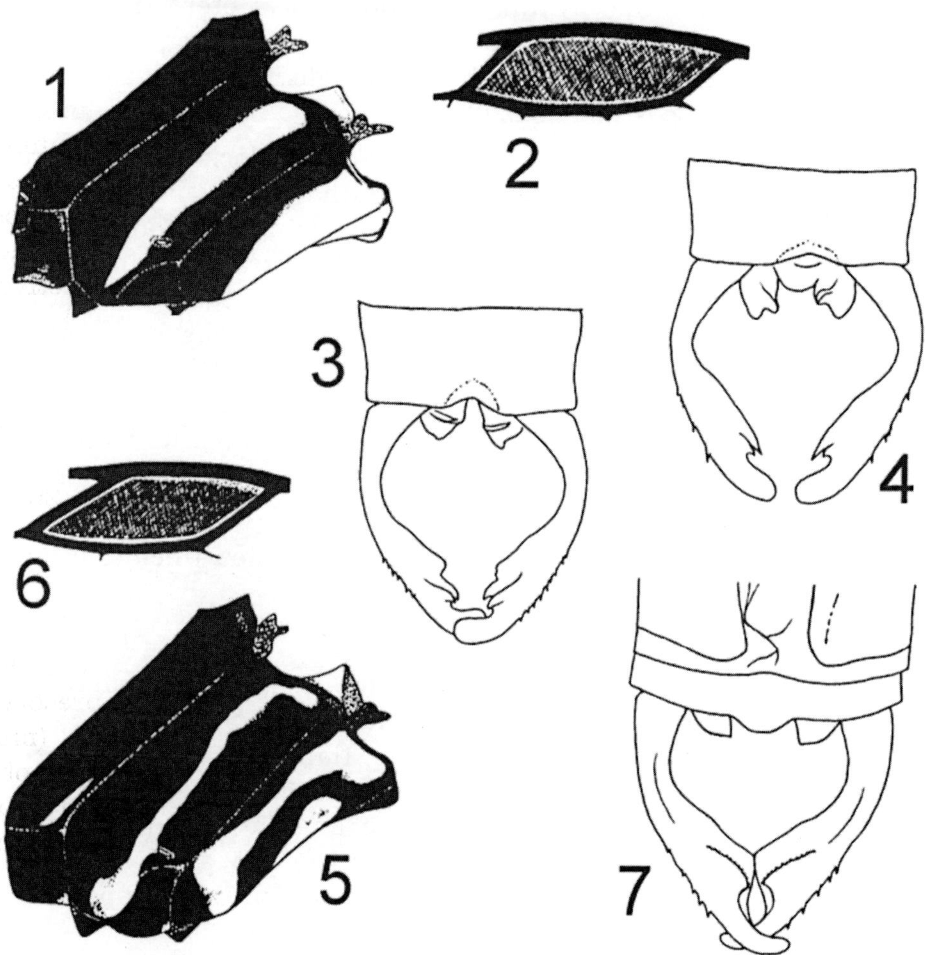

Figs 1-4. Male diagnostic characters of *Argiolestes angulatus* sp. n. (Figs 1-3) and *Argiolestes sidonia* Lieftinck (Fig. 4): (1) synthorax, lateral view; (2) pterostigma; (3, 4) anal appendages, dorsal view. **5-7.** *Argiolestes fornicatus* sp. n., male: (5) synthorax, lateral view; (6) pterostigma; (7) anal appendages, dorsal view.

Prothorax - Pronotum largely black with greyish patch each side on anterior, median and posterior lobe; propleura black; coxa blackish brown, trochanter blackish brown and dull yellow, remainder of leg pale to brownish yellow with outer surface of femur and tibia and apex of tarsus distinctly darkened/blackened and claws brown.

Synthorax – Largely black, most of mesepisternum and mesepimeron metallic; particularly anterior portion of metepisternum and posterior portion of metepimeron, and almost all of metapostepimeron and poststernum greyish yellow; legs much as in prothorax with coxae basally somewhat lighter.

Wings – Membrane hyaline; venation black; pterostigma greyish black, at least twice as long as wide, distinctly slanting, overlying 2 to 2 ½ cells; anal field containing mostly a single row of cells, only rarely some double cells; postnodals 22/20.

Abdomen – Terga 1 and 2 dorsally metallic black, laterally greyish yellow; terga 3-6 pale to dark brown with a small ill-defined basal portion and approximately apical 1/5 black; terga 7 and 8 largely black, only narrowly greyish yellow along ventral margin; tergum 9 and segment 10 black; posterodorsal margin of segment 10 laterally well sclerotised and straight, medially membranous and somewhat excised; sterna largely black. Anal appendages (Fig. 3): superiors black, forcipate with apex distinctly arched mediad, and with a narrow inner flange bearing two somewhat curved apical teeth of similar size and a distinct subapical angulation; inferiors largely brown and black, bilobed, inner margin ventrally membranous, whitish yellow.

Measurements – Hindwing 28.2 mm, abdomen (including appendages) 33.2 mm.

FEMALE: unknown.

DISCUSSION – *Argiolestes angulatus* sp. n. is most similar to and appears closely related to *A. sidonia* Martin. The male of *A. sidonia*, however, has the margin of the inner flange of the superior anal appendages smooth and its ventral tooth large and straight (Fig. 4), whereas this margin bears a tooth-like angulation and the ventral tooth is smaller and somewhat curved in the new species (Fig. 3). In addition the apical part of the superior appendages appears more strongly curved in *A. angulatus* than in *A. sidonia*.

ARGIOLESTES FORNICATUS SP. NOV.
Figures 5-7

Material – Holotype ♂ (Sama I 21779): Papua New Guinea, Eastern Highland Province, Mamaifu, (145°01'E, 6°30'S), 29-XI-1997, S.J. Richards.

Etymology - The specific name (fornicatus = arched) refers to the shape of the posterior dorsal margin of segment 10 of the male.

MALE: *Head* – Labium greyish brown; labrum metallic deep black; anteclypeus bright yellow; postclypeus black, narrowly greyish yellow along anterior and lateral margins and somewhat broader laterally at base; remainder of head including antennae brownish- to greyish black with only a small irregular greyish yellow patch each side between base of postclypeus and eye-margin.

Prothorax - Pronotum largely black with narrow anterior rim of anterior lobe yellow and lateral portions of median and posterior lobe greyish yellow; propleura black; coxa and trochanter brownish black, remainder of leg brown.

Synthorax – Mesanepisternum, mesepimeron and metepisternum black, slightly metallic with a thin dark yellow stripe anterior and close to, and along, ventral 1/4-1/3 of mesopleural suture, and a yellow patch on ventral portion of metanepisternum confluent or nearly so with a pale yellow stripe beginning at subalar ridge and running along intersegmental suture; metepimeron yellow, with irregular patches of variable brown to black along metapleural suture, along middle and part of posterior portion beginning from posterodorsal corner and in ventral corner; metapostepimeron brownish black; poststernum dark brown; legs much as in prothorax with coxae basally somewhat lighter

Wings – Membrane hyaline; venation black; pterostigma reddish to dark greyish brown, hardly 1.5 times longer than wide, strongly slanting, overlying 1½ to 2¼ cells; up to 3 rows of cells in anal field; postnodals 24-25/20-22.

Abdomen – Terga 1-6 largely brown; only tergum 2 metallic black and approximately apical 1/5 each of terga 3-6 black; terga 7-9 largely black, only mid-line of tergum 8 and hollowed out section of tergum 9 greyish brown; segment 10 black dorsally and laterally, brown ventrally; posterodorsal margin largely straight, medially rather narrowly produced into a smooth, nearly semicircular lobe; sterna largely black. Anal appendages (Fig. 7): superiors forcipate with narrow apex, subapical edge bearing between 12 and 15 small teeth, and with wide large triangular inner flange; inferiors short, plump, with well-developed apical corner directed mesad.

Measurements – Hindwing 32.0 mm, abdomen (including appendages) 37.3 mm.

FEMALE unknown.

DISCUSSION – Lieftinck (1956) and Michalski (manuscript) distinguish three morphological groups that contain the majority of Papuan *Argiolestes*: (a) species with posterior border of male segment 10 entire and approximately straight, (b) species with this border finely denticulate, at least so in the middle (on median tubercle), and (c) species with this border with a single (or bifid) median triangular tooth or spine. *Argiolestes fornicatus* sp.n. does not fit any of the three groups. It appears, however, that the evenly arched and non-serrated median projection of the posterior dorsal border of segment 10 is somewhat similar to the species of groups (b) and (c).

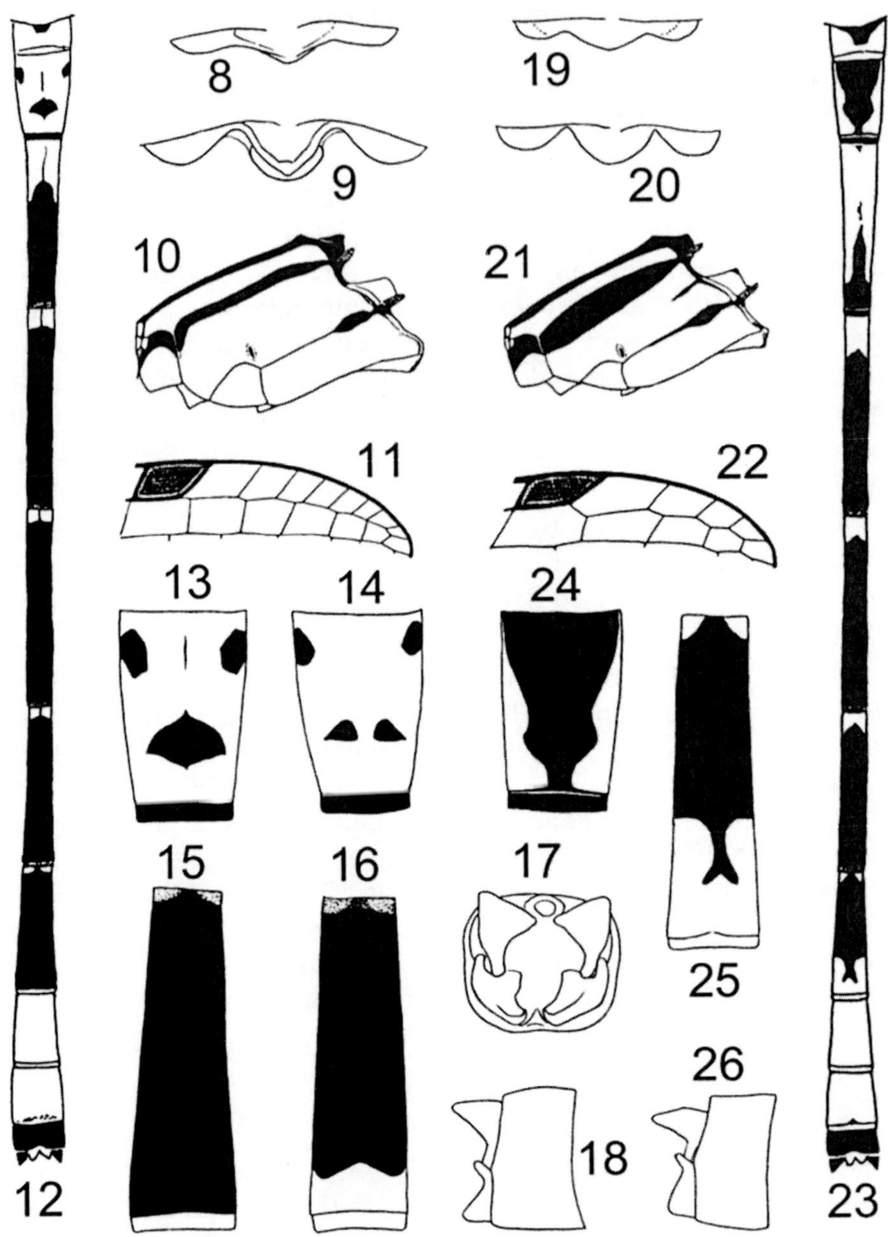

Figs 8-26. Diagnostic characters of *Austroagrion kiautai* sp. n. (Figs 8-18) and *Austroagrion cyane* (Selys) (Figs 19-26): (8, 9, 19, 20) prothorax, dorsal view: (8, 19) male; (9, 20) female; (10, 21) synthorax, male, lateral view; (11, 22) tip of hindwing, male; (12, 23) dorsal abdominal pattern, male; (13, 14, 24) tergum 2, male, dorsal view: (13) holotype ♂; (14) paratype ♂; (15, 16, 25) tergum 7, male: (15) holotype ♂; (16) paratype ♂; (17, 18, 26) segment 10 and anal appendages, male: (17) caudal view; (18, 26) lateral view.

AUSTROAGRION KIAUTAI SP. NOV.
Figures 8-18

Material – Holotype ♂ (Sama I 21780): Papua New Guinea, Eastern Highland Province, montane lake behind Mamaifu, (145°01'E, 6°30'S), 27-XI-1997, S.J. Richards; 1 paratype ♂ (Sama I 21781), 1 paratype ♀ (Sama I21782), same data as holotype.

Etymology – Dedicated to Prof. Dr. Bastiaan Kiauta in recognition of his great merits and service for odonatology.

MALE: *Head* – Labium, labrum except for mid-dorsal and lateral black spots, anteclypeus and frons greenish to bluish; postclypeus black with 2 small pale spots; dorsal and posterior surfaces of head black, except for 2 variable pale spots between lateral and median ocelli, and 2 large, transversely elongate postocular bluish green patches which merge with bluish green line on the postocciput; a narrow somewhat arched black mark on the frons, in front of median ocellus; antennae black; postgenae greenish.

Prothorax – Lateral cervical sclerites yellowish; pronotum mainly black, the anterior lobe and lateral portions of the posterior lobe greenish, the median lobe medially broadly black, laterally greenish yellow; posterior lobe with median section widely triangular, the apex thickened; episternum and epimeron bluish green; coxa, trochanter and inner surfaces of femur and tibia yellowish, the outer surfaces of femur and tibia black; tarsus brownish black, claws brown, tipped black.

Synthorax – Spiracular dorsum and collar black medially, greenish laterally; antealar ridge and sinus, dorsal carina and a narrow median portion of the front of synthorax black, its lateral portions broadly blue and joined to greenish spot on collar; humeral stripes (along mesopleural suture) narrow, black; rest of pleura bluish, except for black patch on dorsal 1/3 of metapleural suture; sterna and poststernum bluish; legs as in prothorax.

Wings – Membrane hyaline; venation black; pterostigma of both wings black, very narrowly translucent along bordering veins; 6-8 costal cells beyond pterostigma. Ac situated about midway between Ax1 and Ax2; anal vein leaving wing margin at Ac. Postnodals 11-12/10-11.

Abdomen – Tergum 1 blue with black mediodorsal mark in anterior half; tergum 2 largely blue anteriorly with 2 black laterodorsal spot that may be small, posteriorly with large black median mark which may be dissolved in 2 laterodorsal spots; terga 3-6 dorsally metallic green-black laterally blue, the blue continuous with narrow basal blue rings on each segment; tergum 7 black with either only intersegmental membrane or intersegmental membrane plus ca apical 1/5 blue; terga 8 and 9 blue, only sides of intersegmental membrane of 8 and posterior edge of 9 may be black; dorsal 2/3 of segment 10 black ventral 1/3 blue without clear definition; sterna largely brownish black. Anal appendages (Figs. 17, 18): superiors largely brownish black; inferiors largely greenish yellow; a wide angle between upper and lower branch of superior appendages.

Measurements – Hindwing 15.6-16.4 mm, abdomen (including appendages) 21.5-22.0 mm.

FEMALE: *Head* – Pattern much as in male; postclypeus lacking pale spots; scape blackish orange.

Prothorax – Pattern much as in male, but with pale orange instead of greenish; posterior lobe appearing definitely two-layered.

Synthorax – Pattern much as in male; front of synthorax laterally orange instead of greenish, and other pale areas lighter blue.

Wings – Much as in male; pterostigma greyish brown; 6-7 costal cells beyond pterostigma; postnodals 11-12/10.

Abdomen – All terga and Segment 10 dorsally largely black, laterally greyish to bluish-orange; anal appendages black; valves greyish yellow; sterna largely brownish black.

Measurements – Hindwing 17.1 mm, abdomen (including appendages) 21.6 mm.

DISCUSSION – *Austroagrion kiautai* sp.n. is similar and appears closely related to the species pair *A. cyane* (Selys) (Figs. 19-26) from South Australia and south-western Australia (Selys, 1876; Lieftinck 1982; Watson *et al.,* 1991; Theischinger & Hawking, 2006) and *A. watsoni* Lieftinck from eastern Australia (Lieftinck, 1982, Watson et al., 1991, Theischinger & Hawking, 2006). The males of only these three species have the superior anal appendages longer than the inferiors. The posterior lobe of the pronotum of both sexes, however, is more elaborate and the angle between upper and lower branch of the superior anal appendages of the male is markedly wider (as seen in lateral view) in the new species than in the other two. The dark pattern elements of the synthorax are much more reduced in the new species than in any of the hitherto described species of *Austroagrion* Tillyard, including *A. exclamationis* Campion from northern Australia and New Guinea (Campion, 1915; Lieftinck, 1933, 1949, 1951; Fraser, 1960; Watson, 1969; Watson *et al.,* 1991; Tsuda, 2000; Theischinger & Hawking, 2006) and *A. pindrina* Watson from north-western Australia (Watson, 1969; Watson *et al.,* 1991; Theischinger & Hawking, 2006).

The strongly reduced dark pattern of tergum 2, the much more extended dark element on tergum 3 and the more extensive and less complex dark element on tergum 7 of the male abdomen also clearly distinguish the new species from all its congeners. These distinct structural and colour characters, together with the larger size and richer venation (f.e. 5-8 costal cells beyond the pterostigma compared with 2-5 in the other species) of the new species and the information given by Michalski (manuscript) on *Austroagrion* from Papua New Guinea may even indicate that *A. kiautai* is the adelphotaxon of all other *Austroagrion* species as a group.

Habitat of the three new species. *Argiolestes angulatus* sp.n. and *A. fornicatus* sp.n. were found in heavily disturbed primary lower to mid-montane rainforest along a steep stream adjacent to Maimafu Village. They were active in exposed, sunny conditions created by gardens being constructed adjacent to the stream. *Austroagrion kiautai* sp.n. was found in primary forest around the verge of a large, montane lake several hours walk above Maimafu Village. The altitude of the lake is approximately 2000 m asl.

ACKNOWLEDGEMENTS

We wish to express our gratitude to : The PNG National Research Institute who assisted with SJR's Research Visas, and the PNG Department of Environment and Conservation for approving export of specimens. Fieldwork in PNG was supported by Conservation International, the Wildlife Conservation Society and the Research and Conservation Foundation (Goroka) as part of a USAID-funded field biology course for Papua New Guinean students. SJR is particularly grateful to Andrew Mack, Debra Wright and Robert Bino for their support. The people of Maimafu Village kindly hosted the training team and students on their land and gave approval for collection of specimens. We are also grateful to J. Michalski for making available a draft copy of the text for his unpublished book on the dragonflies of New Guinea (referred to as manuscript) and to Jan Forrest of the South Australian Museum for her ongoing assistance. GT is grateful for ongoing support by Tim Pritchard, manager at the NSW Department of Environment and Conservation.

REFERENCES

Campion, H. 1915. A new agrionine dragonfly from Northern Australia. *Am. Mag. nat. Hist.* (8)16: 105-108.

Chao, H.F. 1953. The external morphology of the dragonfly *Onychogo-mphus ardens* Needham. *Smithson. misc. Collns* 122(6): 1-56.

Fraser, F.C. 1960. *A Handbook of the Dragonflies of Australasia. With keys for the identification of all species.* – Royal Zool. Soc. N.S.W., Sydney.

Lieftinck, M.A. 1933. Odonaten aus Nordaustralien. *Rev. Suisse Zool.* 40: 409-439.

Lieftinck, M.A. 1949. The dragonflies (Odonata) of New Guinea and neighbouring islands Part VII. Results of the third Archbold Expedition 1938-1939 and of the Le Roux Expedition 1939 to Netherlands New Guinea (Zygoptera). *Nova Guinea* (N.S.) 5: 1-271.

Lieftinck, M.A. 1951. Results of the Archbold Expeditions. No. 64. Odonata of the 1948 Archbold Cape York Expedition, with a list of the dragonflies from the peninsula. *Am. Mus. Novit.* 1488: 1-48.

Lieftinck, M.A. 1956. Revision of the genus *Argiolestes* Selys (Odonata) in New Guinea and the Moluccas with notes on the larval forms of the family Megapodagrionidae. *Nova Guinea* (N.S.) 7 (1): 59-121.

Lieftinck, M.A. 1982. Settling up matters of taxonomy, nomenclature and synonymy in the genus *Austroagrion* Tillyard (Zygoptera: Coenagrionidae). *Odonatologica* 11(4): 287-295.

Oppel, S. 2005. Habitat associations of an Odonata community in a lower montane rainforest in Papua New Guinea. *Int. J. Odonat.* 8(2): 243-257.

Theischinger, G. & J. Hawking 2006. *The Complete Field Guide to Dragonflies of Australia.* CSIRO Publishing, Collingwood.

Theischinger, G. & S.J. Richards 2005. Two new species of *Drepanosticta* Laidlaw from Papua New Guinea (Zygoptera : Platystictidae). *Odonatologica* 34(3): 307-312.

Theischinger, G. & S.J. Richards 2006a. Two new species of *Nososticta* Hagen in Selys from Papua New Guinea (Zygoptera: Protoneuridae). *Odonatologica* 35(1): 67-71.

Theischinger, G. & S.J. Richards 2006b. Two new Zygoptera species from Papua New Guinea (Protoneuridae, Coenagrionidae). *Odonato-logica* 35(2): 199-204.

Theischinger, G. & S.J. Richards 2006c. A new species of *Argiolestes* Selys from Papua New Guinea (Odonata: Megapodoagrionidae). *Odonatologica* 35(4): 385-388.

Theischinger, G. & S.J. Richards 2007. A new species of *Teinobasis* Kirby from Papua New Guinea (Odonata : Coenagrionidae). *Odonato-logica* 36(1):

Tsuda, S. 2000. *A Distributional List of the World Odonata 2000.* Tsuda, Osaka.

Watson, J.A.L. 1969. Taxonomy, ecology, and zoogeography of dragonflies (Odonata) from the north-west of Western Australia. *Aust. J. Zool.*17: 65-112.

Watson, J.A.L. & F.A. O'Farrell 1991. Odonata (dragonflies and damselflies). *In*: CSIRO [Ed.]: *The Insects of Australia,* [2nd edn.], Melbourne Univ. Press, Melbourne.

Watson, J.A.L., G. Theischinger & H.M. Abbey 1991. *The Australian Dragonflies: A Guide to the Identification, distributions and habitats of Australian Odonata.* CSIRO, Canberra.

Odonata : Biology of Dragonflies
Ed.: B.K. Tyagi
Scientific Publishers (India), 2007, pp. 45-50.

4

DRAGONFLIES FROM THE OKAVANGO SWAMPS (BOTSWANA, SOUTHERN AFRICA) IN WINTER

HENRI J. DUMONT

*Animal Ecology, Ghent University, Ledegancktraat,
9000 Ghent, Belgium
e.mail: Henry.Dumont@ugent.be*

(Dedication : This paper is dedicated, on the occasion of his 70th birthday, to Bostjan Kiauta, a friend and colleague for the past 45 years. I have particularly fond memories of our joint expedition to Nepal in 1972, where altitudinal zonation of dragonflies provides a possible parallel with the temporal zonation observed in the Okavango.)

ABSTRACT

Geographic situation and altitude of the Okavango swamps combine to create a distinctly seasonal climate, with an outspoken cool season. The local dragonfly fauna in winter is distinctly less species-rich than in summer. Although low abundances (rare species) may slightly complicate the situation, it is probably fair to estimate the faunal impoverishment at a factor 2 to 4 (17 species were censed in July 2006, against *c.* 70 known from all seasons combined). The composition of the winter fauna is dominated by wide-ranging species, tolerant of strong variations in environmental conditions, but at least two *Pseudagrion* species (*P. deningi, P. coelestis*) appear to have taken advantage of the void created by the austral winter, and have their adult population peak in winter.

INTRODUCTION

The Okavango swamps, situated in southern Africa just north of the tropic of Capricorn, at an average altitude of *c.* 950 m a.s.l., are the *c.* 10,000 km^2 sized endorheic endpoint of the Okavango river, which springs as the Cubango in Angola's western highlands, crosses the Caprivi strip (here named the Kavango river), and finally enters the Botswana craton, where it spreads out in a fan-like manner. The centralmost branches of the delta are the Xudum and Boro, and all branches unite at the southeast of the swamps into the rivers Nghabe (a seasonal river) and Thalakamane (a permanent river). For a detailed account of the geography, geology, climate, and biology of the delta, and for a useful map, see Mepham (1987).

The water feeding the swamps is derived from seasonal rains in Angola (local precipitation contributes little), which extend from October to March-April. In the flat landscape of the swamps, the floods proceed slowly and Maun, the major city at the south end of the swamps, is only reached in July, creating the somewhat paradoxical situation that the high water levels here coincide with the peak of the dry (and cool) season. The swamps shrink back to a size of *c.* 9,000 km^2 during droughts, and may expand to a *c.* 16,000 km^2 during wet years. At such times the seasonal river Nghabe drains excess water to lake Ngami, a large temporary lake to the south-west of the swamps. Exceptionally, a link can even become established with the Zambezi basin, via the Khwai river to the east of the swamps.

The climate of the altitudinal Okavango is seasonal, in spite of being situated north of the tropic. October is the hottest month (mean maximum temperature 35°C, mean minimum 15°C), while the coldest ones are June and July (mean maximum 23°C, mean minimum 6°C). For typical warmth-loving insects like adult dragonflies, it may be expected that the mean minima will matter most, and so the swamps will display their highest species richness in summer. However, this also means the highest level of interspecific competition, and therefore there should be a selective pressure on species to alleviate competition by specializing on the lower temperature segment. The temperature range (10-11° across a typical year) could indeed be sufficient to allow a few species to specialize to its lower end. Provided the flying season of a species is not too long (Corbet, 1999), winter temperatures should select against true tropical species, and so the winter specialists, if any, should be taxa with a rather restricted, local distribution. Wide-ranging species are by definition generalists, not specialists, and can be ruled out.

Here, I briefly explore the dragonfly fauna present in the swamps at the height of the austral winter (July), and attempt to determine whether indeed winter specialists may be present, and what species are capable of surviving the cool season in the imaginal state.

MATERIAL AND METHODS

A census was made of any dragonfly species present, (1) along a 200 m stretch of the Thamalakane river near Maun on 18, 19 and 20 July (zone 1), (2) about 80 km of swamp along the Boro river from its origin to about halfway chief's island on 21-23 July (zone 2), starting around 8:30 a.m. and until 5-6 p.m. Additional observations were made on 18 July at Lake Ngami, at that time reduced to a residual puddle, greatly eutrophicated by large number of cattle and water birds (loc. 3), and the Khwai river (19° 14′ S, 23° 57′, E, loc. 4) on 24 July 2006. This inventory was compared to the findings of Pinhey (1976), whose collections were mainly made during several expeditions in summer (mostly in February).

List of species recorded

A letter S preceding the species name signifies a singleton (species recorded only in a single specimen; an asterisk denotes a wide-ranging species, known to be tolerant of strong environmental fluctuations).

1. **Lestes pallidus* Rambur: 2 males, zone 1.
2. *Lestes pinheyi* Fraser: 7 males, zone 1.
3. **Ischnura senegalensis*: rather common in both sexes, zone 1, zone 2, loc. 4
4. **Agriocnemis exilis* Selys: 1 male, 1 female, zone 1; 1 male, zone 2.
5. *Pseudagrion coelestis* Longfield: common, zone 1
6. *Pseudagrion deningi* Pinhey: rather common, zone 1; swarming, zone 2.

7. **Pseudagrion hamoni* Fraser: few territorial males, zone 1.
8. *Pseudagrion rufostigma*: few specimens, zone 2.
9. **Anax imperator* Leach: males patrolling stretches of river, zones 1 and 2; one female, ovipositing, loc. 3.
10. *Anax bangweuluensis* Kimmins: 1 male, zone 2.
11. **Crocothemis erythraea* (Brullé): males, females, zones 1 and 2, loc. 4.
12. **Diplacodes lefebvrei* (Rambur): males, females, loc. 4.
13. *Hemistigma albipuncta* (Rambur): 1 female, zone 2.
14. **Orthetrum chrysostigma* (Burmeister): 1 male, zone 1.
15. **Sympetrum fonscolombii* (Selys): males, females, zone 1; ovipositing tandems, loc. 3.
16. **Trithemis annulata* (P. de Beauvois): fairly common, zone 1 and loc 4. Many specimens looking quite aged.
17. **Trithemis arteriosa* (Burmeister): 1 male, loc. 4.

DISCUSSION

Pinhey (1976) lists about 70 species of dragonflies from the Okavango delta. The July 2006 list amounts to only one quarter of that, and all species recorded also appear in Pinhey's list. It would be tempting to jump to the conclusion that the winter season indeed reduces the number of dragonfly species to one fourth their summer's number, yet perhaps things are a bit more complicated. Not less than 5 out of 17 species recorded are indeed singletons, a strong indication that species has been overlooked because their abundances are so low that they are not easily observed. On the other hand, it remains possible that Pinhey's list is also incomplete, as it was based on the usual expeditions' hit-and-run approach. On balance, it seems safe to conclude that there is indeed a faunal impoverishment in winter, and that it amount to a factor 2 to 4 reduction in species richness.

Turning now to the nature of the winter fauna, we find that more than half (11 out of 17 species, marked with an asterisk in the species-list) is wide-ranging. These species occur over most of Africa, including the north and frequently spill over into the Iberian Peninsula and even western Europe in the West, and the Levant in the East. They are therefore certainly more tolerant to fluctuating environmental conditions than true tropical species, and may occur in a place like the Okavango year-round. A second fraction, composed of five species, is of more interest: it is composed of species that show restricted ranges in southern Africa. Three of these (*Anax bangweuluensis, Lestes pinheyi, Pseudagrion rufostigma)* were either not particularly common or even represented by singletons, and may therefore be either at the end or the beginning of their flight period, but the remaining two, *Pseudagrion deningi* and *P. coelestis* were common. Regarding the daily activity of the fauna, the first specimens to arrive at the water at the Thamalakane river were *Ischnura senegalensis* around 9 a.m., followed around 10 a.m. by *Pseudagrion coelestis*, of which the first tandems were observed around 11 a.m. By 10:30 a.m. the first Anisoptera appeared, with *Crocothemis* and *Trithemis annulata* most prominent, and *Anax imperator* starting patrols around 11 a.m. Inside the swamps, the first *Pseudagrion deningi* did not appear until 10:30 a.m. but by noon they had become extremely abundant, literally swarming over all patches of open water, either as solitary males, or as tandems, or as ovipositing couples. *Pseudagrion deningi* is morphologically atypical, perhaps not even a *Pseudagrion* (Pinhey, 1961, 1967), and with a restricted range. It was originally described from Lake Bangweulu in Northern Zambia, closer to the equator than the Okavango swamps, but at an even higher elevation (*c.* 1100 m) than the Okavango (Compère & Symoens, 1987). Beside Zambia and perhaps Katanga, it is only known with certainty from the Okavango swamps (Pinhey, 1984). It indeed appears to be a swamp species, and is our prime candidate for a winter-adapted taxon across its limited range. In second line, *Pseudagrion coelestis* also qualifies for the same status. It is more widely distributed in southern Africa, was not as common as the preceding, and seemed to avoid the depths of the swamps, but it was common and in full reproduction on the Thamalakane river. I predict that both, but especially the former species, will predominantly or exclusively occur during the cooler period of the year.

CONCLUSION

The circumstantial evidence gathered here suggests that in the Okavango swamps, where geographic location combined with altitude creates a marked seasonality, the winter season is characterized by a strong reduction in the abundance and species richness of the odonate fauna. Most surviving taxa belong to hardy, widespread taxa. Yet, as predicted by ecological theory, a small fraction of endemics to southern Africa seems to have adapted to the colder winter temperatures, and exploits the winter season's reduced interspecific competition to build up large local populations during that season.

ACKNOWLEDGEMENTS

I thank Dr. Rob Hart (Pietermaritzbug, South Africa) and Mr. Donald Hart for making my trip to the Okavango possible. The Oppenheimer Research Institute of the University of Botswana is thanked for making available the speedboat that allowed me to explore a significant stretch of swamp, and for allowing our party to use their camp facilities on Chief's island. My travel companion Jo Vermeir assisted with collecting dragonflies, catching, as usual, a few species that I never even saw.

REFERENCES

Compère, P. & J.J. Symoens, 1987. Lac Bangweulu et zones humides associées. In M.J. Burgis & J.J. Symoens (eds.), African wetlands and shallow water bodies. Directory. Orstom, Paris, pp. 404-412.

Corbet, P.S. 1999. Dragonflies. Behaviour and Ecology of Odonata. Harley Books, Martins, 829 pp.

Mepham, J.S., 1987. The Okavango delta and the Makgadikgadi pans. In M.J. Burgis & J.J. Symoens (eds.), African wetlands and shallow water bodies. Directory. Orstom, Paris, pp. 486-503.

Pinhey, E.C.G. 1961. Dragonflies (Odonata) of Central Africa. Occ. Pap. Rhodes-Livinstone Mus. 14: 1-97.

Pinhey, E.C.G. 1967. Odonata Zygoptera. In J.J. Symoens (ed.), Hydrobiological survey of the Lake Bangweulu Luapala river basin 14 (3): 43 pp.

Pinhey, E.C.G. 1976. Dragonflies (Odonata) from Botswana, with ecological notes. Occ. Pap. Natn. Mus. Monum. Rhod. B 5 (10): 524-601.

Pinhey, E.C.G. 1984. A check-list of the Odonata of Zimbabwe and Zambia. Smithersia 3: 1-64.

Odonata : Biology of Dragonflies
Ed.: B.K. Tyagi
Scientific Publishers (India), 2007, pp. 51-61

5

CORDULEGASTER INSIGNIS (SCHNEIDER, 1845) IN BULGARIA WITH NOTES ON ITS BIOLOGY AND ECOLOGY

MILEN MARINOV[1], BURKHARD GREBE[2] & YORDAN KUTSAROV[3]

[1]*P.O.Box 134, BG-Sofia, Bulgaria*
e.mail: mg_marinov@yahoo.com
[2]*Oberdorfallee 7a, D-53909 Zülpich, Germany*
e.mail: Burkhard.Grebe@t-online.de
[3]*18 "P. Kiselov" Str., BG-7600 Tutrakan, Bulgaria*
e.mail: kalimok@gmail.com

ABSTRACT

Bulgarian *Cordulegaster insignis* finding places are summarized and mapped. Special attention on its biotope is given with emphasis on species' biology and ecology. New records from Bulgaria enlarge *C. insignis* distribution to the west. Its possible existence in even westernmost areas, like Serbia & Montenegro, is briefly discussed.

INTRODUCTION

Nedelkov (1923) reports *Cordulegaster insignis* for the first time for Bulgaria near the town of Kjustendil, W Bulgaria (Fig. 1). He stated "Possibly it came here from Macedonia along the Struma river valley". Subsequent data from E Bulgaria (Beschovski, 1964, 1993) led to the conclusion that the species is typical for this part of the country (Beschovski, 1993).

Beschovski (1964) ascribed the Bulgarian members of the *C. insignis* group to the nominal subspecies. Dumont (1976) and following him Askew (2004) however ascribed them to *C. i. charpentieri* (Kolenati, 1846). They stated the distribution of this subspecies all along the Pontic Black Sea coast up to Georgia, with an additional remark that the species had also been found in W Bulgaria. Beschovski (1993, 1994) followed Dumont (1976) and Askew (2004) and gave detailed description of the local *C. insignis*, regarded as *C. i. charpentieri*, the latter being considered as a Pontic element. However, the assignation of the Bulgarian *C. insignis* to *C. i. charpentieri* seems now to be doubtful, the occurrence or lack of yellow spots on S9-10 as well as the presence or absence of a dark stripe on the frons, being too much variable in this group to allow any subspecific distinction. Currently *C. i. charpentieri* is only known with certainty in the eastern half of Turkey, Armenia and Georgia (Shengelia, 1975; Dumont, 1977; Demirsoy, 1982; Lohmann, 1993; Verschuren, 1989; van Pelt, 2004; Tailly & Ananian, 2004; Tailly *et al.,* 2004). So the subspecific rank of the Bulgarian *C. insignis* remains unclear and has to be solved including the Romanian records of the species described as an own subspecies *C. i. montandoni* (St Quentin, 1971) respectively *C. montandoni* (Lohmann, 1993). Its finding place Comana is only 20 km from Danube river and the Bulgarian border. Consequently in this paper the Bulgarian specimen are assigned to *C. insignis* without any subspecific determination.

Material recently collected by the authors brought new localities for the country. Studies on *C. insignis'* habitat and its behaviour were done as well. Here we report new information on some important topics related to its biology and ecology.

MATERIAL AND METHODS

All *Cordulegaster insignis* records from Bulgaria from published information are summarized, including specimen in the collections of the National Museum of Natural History, Sofia (NMNH), Institute of Zoology, Sofia (IZ) and Faculty of Biology by the Sofia University (SU) as well as original author's information. *C. insignis'* biotope, biology and ecology were studied along a small springbrook – locality 5. Material was collected in four successive years (1998-2001) during several short visits. Mainly exuviae were collected, and larvae and adults were observed. Behavioural notes were taken and compared with information from other places within the country – N Black Sea

Figure 2. Barzartsi River by the village of Archar, NW Bulgaria.

Figure 3. Springbrook passing through the village of Dunavets, NE Bulgaria.

Figure 4. Springbrook the village of Vetren, NE Bulgaria.

Figure 5. The mouth of the springbrook near the village of Vetren, NE Bulgaria.

Figure 6. *Cordulegaster insignis* male at perching position.

Figure 9. Patrolling area of *Cordulegaster insignis*.

Figure 10. Teneral *Cordulegaster insignis*.

coast and NW Bulgaria. Greater attention was paid to the breeding place itself. Some measurements of the physical components of the water in the localities along the Danube River are given, too.

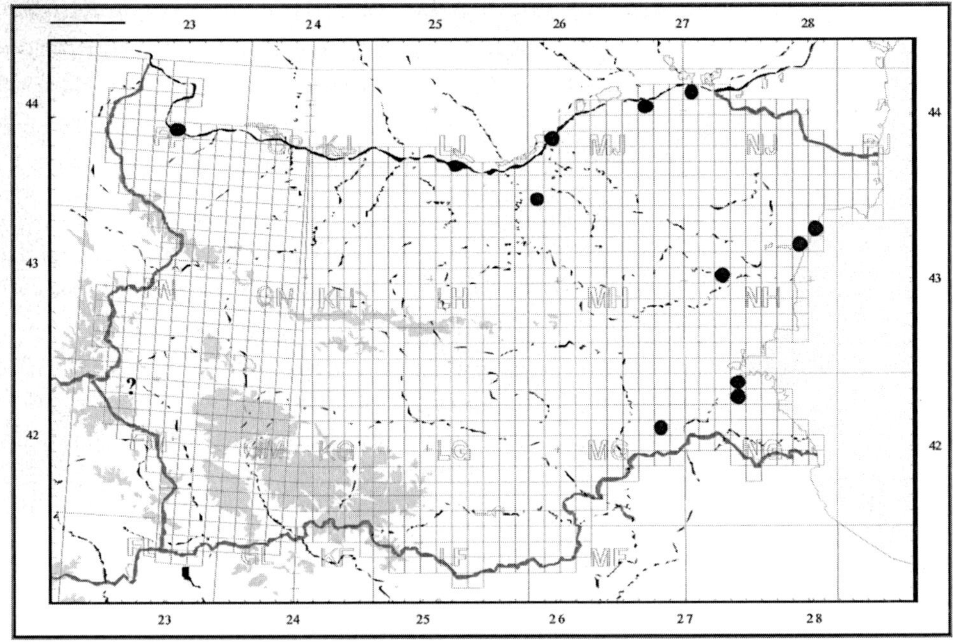

Figure 1. Horizontal distribution of *Cordulegaster insignis* in Bulgaria.

RESULTS

Cordulegaster insignis' records from Bulgaria

Figure 1 represents all currently known localities for Bulgaria. These are given with more details below:

1. Barzartsi River by the village of Archar (43°48'N; 22°55'E), ca. 51 m a.s.l.: 17.06.2000, Lazarov leg.; 26.06.2005, Marinov & Lazarov leg. (Fig. 2).
2. Swamps by the "Obraztsov chiflik" suburb of the town of Russe (43°50'N; 25°57'E), ca. 20 m a.s.l.: 07.06.1937, 16.06.1937, Buresh leg., NMNH, van Pelt det.
3. Small forest S from the village of Volovo and N from the junction to the town of Borovo (43°29'N; 25°48'E), ca. 225 m a.s.l.: 12.06.2006, Marinov leg.
4. Springbrook passing through the village of Dunavets (44°03'N; 26°44'E), ca. 30 m a.s.l.: 27.05.2006, Kutsarov leg. (Fig. 3.).

5. Springbrook by the Danube River bank near the village of Vetren (44°08'N; 27°00'E), ca. 20 m a.s.l.: 04.06.1992, Georgiev leg.; 06.06.1998, Marinov leg.; 28.06.1999, Marinov, Grebe leg.; 28.06.2000, Marinov leg.; 02.-05.07.2001, Marinov leg. (Fig. 4).
6. Village of Krasimir (43°01'N; 27°17'E), ca. 71 m a.s.l.: 07.1940, Zlatarski leg. NMNH-Sofia, van Pelt det.
7. Springbrook near the village of Frangata (43°15'N; 27°55'E), ca. 50 m a.s.l.: 11.06.1961, Beschovski (1964).
8. Springbrook within the Varna Botanical Garden (43°15'N; 27°55'E), ca. 10 m a.s.l.: 18.07.1992, Marinov leg.; 22.07.1993, 13.06.2001, 05.07.2001, 07.07.2001, 23.07.2001, 27.06.2002, SU, van Pelt det.
9. Aladja monastery N from the town of Varna (43°17'N; 28°01'E), ca. 50 m a.s.l.: 15.05.1960, Beschovski (1964); 26.06. 1978, Beschovski leg., IZ, van Pelt det.; 25.07.1993, SU, van Pelt det.
10. Town of Kyustendil (42°17'N; 22°41'E), ca. 519 m a.s.l.: July, Nedelkov (1923).
11. Popovska River above the town of Bolyarovo (42°10'N; 26°49'E), ca. 200 m a.s.l.: 21.06.1985, Beschovski (1993).
12. Forest W from the village of Krushovets (42°16'N; 27°29'E), ca. 250 m a.s.l.: 18.06.1985, Beschovski leg.; Bevshovski, van Pelt det.
13. Canal near the "Varli Bryag" mine, W from the town of Burgas (42°26'N; 27°24'E), ca. 50 m a.s.l.: 31.05.1998, Marinov leg.

Biotope description

The study site (locality 5, Fig. 4) was a springbrook situated exactly at Danube riverbank. It was formed by the seepage water flowing from the hill, which marks the riverbank. During spring highflow, the Danube floods the mouth of the springbrook. This place was used by the larvae as some exuviae were found there. The springbrook was of variable wideness 0.50-1.50 m. Its depth was ranging from about 0.10 to 0.50 m. It is flowing through a deep shady place covered by surrounding trees and bushes: *Fraxinus exelsior* L., *Salix alba* L., *Alnus glutinosa* (L.) Gaertner, *Ulmus minor* Miller, *Ulmus glabra* Huds., *Sambucus nigra* L., *Cornus sanguinea* L. Their branches were entwined by *Humulus lupulus* L. The springbrook entered the river at an open area (Fig. 5), where it formed very shallow (0.01-0.10 m) floods in the nearby vegetation consisting of

Holcus lanatus L., *Glyceria aquatica* Wahl., *Chelidonium majus* L., *Ranunculus repens* L., *Equisetum telmateia* Ehrh., *Equisetum arvense* L., *Myosoton aquaticum* (L.) Moench, *Galium spurium* L., *Sparganium erectum* L., *Phragmites communis* (Cav.). At that part the water surface itself was covered completely by *Berula erecta* (Huds.) Coville.

Physical parameters of the water are given in the Table 1. They are shown together with situation measured at another breeding locality found along the Bulgarian Danube River stretch.

Notes on biology and ecology

At the breeding place (locality 5) *C. insignis* males started flying actively around noon. They moved very slowly above the water surface. Usually they hovered for several seconds and very often changed their flight direction returning back continuously. The place where they flew above was of very small size (about 1 m in diameter). Males often perched on the plant stems (*Holcus lanatus*, *Phragmites communis*, *Sparganium erectum*) holding their abdomen angled to the surface beneath (Fig. 6). Before perching at a chosen place they checked it very carefully, up to six times. Males showed site attachment – after being frightened they flew around and perched exactly at the same place. The duration of the interval between two perches varied usually, decreasing from 3 to 5 min (at the initial stage after males' arrival) down to 0.18 to 0.50 min (after four subsequent flights followed by perches) (n = 2). Due to their slow flight and frequent perching individuals could be caught even by hand. Males flew away from the water, too.

Table 1. Physical components of the waters in some of the Bulgarian *Cordulegaster insignis'* localities

Physical component	Locality 4	Locality 5
Dissolved oxygen (mg/l)	7.52	8.22
pH	8.31	7.90
TDS (mg/l)	1346	1152
Water current (m/s):		
- measured at the bank	0.2	0.116
- measured at the middle section	0.4	0.4
Temperature (°C)	19.1	22.5

Observations of copulation and oviposition were scarce. One pair was caught at noon flying high between tree branches about ten meters from the water. A single female was seen ovipositing low above the water surface at a very densely vegetated place. It was noticed first by the sound of its wings beating the surrounding plants (*Glyceria aquatica* Wahl.). After finishing oviposition it was unable to fly directly because of the plant stems and stayed perching for several seconds. The oviposition place was 0.40 m wide and 1 cm deep. The time it spent for ovipositing was about 9 min. During 3.17 min it dipped 322 times to the water.

Emergence was observed before (10:00 am) or around noon. Individuals preferred green stems mainly. Fifty four exuviae were collected within an area of about 5 m^2 along the springbrook. All of them were found in different positions: on plant stems, on the lower surfaces of leaves, parallel or perpendicular to the surface beneath. The height at which emergence occurred varied from 0.12 m (open place – on *Galium spurium* L.) to 0.48 m (shady place – on *Scutellaria gallericulata* L.). All exuviae were on plants above the water or floods around, not anyone on the ground. One teneral female was caught resting in the grass at a shady place about 5 m from the water.

Other dragonfly species seen at same locality were Calopteryx splendens, Lestes parvidens, Platycnemis pennipes, Erythromma viridulum, Ischnura elegans, Aeshna mixta, A. affinis, Gomphus flavipes, Sympetrum sanguineum and S. meridionale.

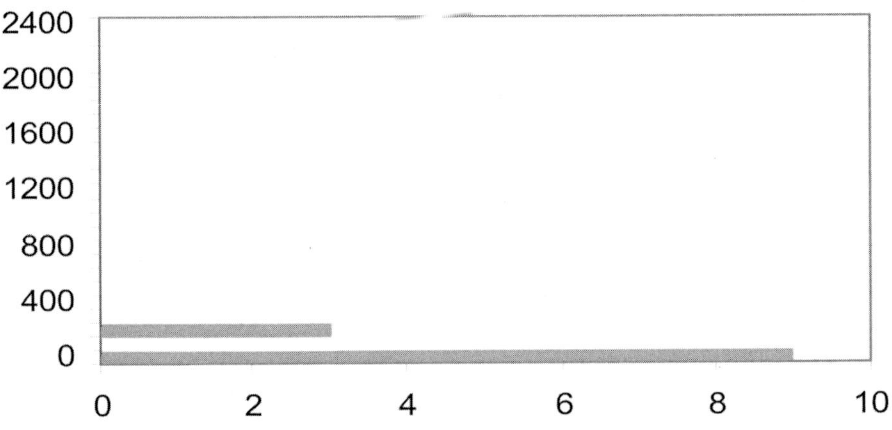

Figure 7. Vertical distribution of *Cordulegaster insignis* in Bulgaria (Abscissa presents the number of observations per 100 meters, but the ordinate - the intervals of 100 meters in vertical direction).

DISCUSSION

In Bulgaria *C. insignis* is distributed mainly at lowland areas up to 100 m a.s.l. (Fig. 7). Single records at higher altitude (around 200 m a.s.l.) are from SE part of the country and a male from NE found about 20 km away from the possibly breeding area. In the southern part of its range the species is also found at much higher altitude, *i.e.* up to 2000 m a.s.l. (Belyshev, 1958; Blom, 1982; Seidenbusch, 1995). It occurs (as *C. i. coronatus*) in the Central Asian high mountain zones of East Kazakhstan (Belyshev, 1961), where it reaches 3150 m a.s.l. and is regarded as an obligatory mountain element (Borisov, 1989).

In Bulgaria, *C. insignis* seems to be a spring-summer flyer with a comparatively long flying period of more than two months (Fig. 8). The earliest observation date is 15.05.1960 (Beschovski, 1964), the latest 25.07.1993 (deposited at Sofia University), both at the same place (locality 9). The peak of activity around the beginning of July is related to the observations done at the breeding place described here. Those figures surely vary from place to place and fall within the overall situation reported for the species. Extremes dates found in the literature are on 18.04.1949 (Kimmins, 1950) for Afganistan and latest on 10.08.1997 (van Pelt, 2004) for Turkey.

Figure 8. Flying period of *Cordulegaster insignis* in Bulgaria (Abscissa presents the months, but the ordinate the number of records per ten-days periods of the month).

Beschovski (1994) gives *Cordulegaster insignis* (as *C. i. charpentieri*) as a Pontic zoogeographic element in the Bulgarian fauna, being mainly distributed along the whole Bulgarian Black Sea coast. The newly collected information from some places along the Bulgarian Danube river shore enlarges *C. insignis* distribution to the west. This information is in accordance with Nedelkov (1923) and proves *C. insignis* as an inhabitant of W Bulgaria. However, it cannot confirm his statement about its possible origin (see the introduction section). It seems to be more probable that *C. insignis* is extending its area to the west using the Danube river's shore. Biotopes similar to the one we described here are typical for many places along the Danube River in its Bulgarian stretch. In its north-westernmost place known, *C. insignis* inhabits nearly the same type of biotope – similar in structure, but bigger in size. Given the well known selectivity of *C. insignis*, we may suppose that the species reproduces in other localities along this river. Because the species is not easy to be found, more detailed investigations are necessary in this area in order to establish its true status in Bulgaria. Possibly, similar places may be found even westernmost in Serbia and Montenegro along the Danube River. This is supported by the fact that the Archar's population is well developed.

D'aguilar *et al.* (1998) identify the *C. insignis* habitat as: flowing waters: rivulets, brooklets, seepage waters with small receptacles filled with water on slopes, embankments and peaty areas, from sea level to at last 1700 m a.s.l. Species ecology is insufficiently studied. Lopau (1995, 1999) gave a short overview basing on its own experience in Greek islands. Leipelt (2005) searched for an explanation of the low effectiveness of this species to withstand strong hydraulic stress and for possible reasons of its attachment to springs and springbrooks. Investigations from Bulgaria are in accordance with some of their conclusions. We confirm the species preference to develop in small streams including springs. However, our studies did not fully match with the habitat description in Lopau (1999), as none of the known Bulgarian breeding places has sandy bottom. All are muddy with deeply shaded banks. The shading plants, both grasses and bushes, are used as substrate during emergence. Green stems are preferred, which is also proved for Greek islands (Lopau, 1999).

New data on species behaviour are reported here. *C. insignis* performs a very specific pattern above the water surface. Morning hours males spend hunting away from the stream in some occasions up to several hundred meters from the breeding site (locality 8). Early morning flying activity in an open dry treeless area was observed by Reinhardt (1992) in Georgia. In Bulgarian localities individuals also keep over open areas. They use the trees and other bank vegetation as perching sites and sometime stay at a height of about 2.50 m (locality 1, Fig. 9). All individuals are observed at places exposed to the sun only and show high site attachment to the perching substrate. Green stems and leaves are preferred, but some individuals perch on dead twigs as well. In the typical posture they keep their abdomen angled towards the surface beneath. Perpendicular abdomen is characterized mainly for the tenerals (Fig. 10). Males appear near or over the water around noon. The time before that they fly rapidly over the springbrooks without stops. In some occasions we spent hours in the morning at the breeding place seeing not any signs of *C. insignis*. Once at the water, males search for perching places at sunny areas again low in the grass just above the water surface. They chose that areas after careful investigations by hovering continuously. They are not seen perching at shady parts of the water. Usually, those places are passed in flight. Probably same areas are used as maturity places only as one teneral was found resting low in the grass shaded by the surrounding trees.

In spite of the number of localities given above *C. insignis* is proposed to be included in the Bulgarian Red Data Book as Critically Endangered. In most of the so far known finding places only single individuals were observed. Surely breeding areas are exposed to a high human pressure by tourists, fishermen, local farmers. Danube River water fluctuations also may cause some negative influence, changing the physical conditions of the habitat dramatically, but this influence is not well studied. Current investigations bring new information about *C. insignis* ecology and biology in Bulgaria, which reaches the limits of its range on the territory of Bulgaria.

ACKNOWLEDGMENTS

The authors thank all scientists that contributed to this article: Ch. Gussev, D. Dimitrov, D. Uzunov, G.J. van Pelt, H. Dumont, I. Georgiev, J.P. Boudot, K.G . Leipelt, M. Schorr, M. Tailly, N. Hacet, N. Matushkina, R. Seidenbusch, S. Lazarov and W. Lopau. Investigations were supported financially by International Dragonfly Fund, PANDION-D tour-operator firm and National Museum of Natural History, Sofia.

REFERENCES

Askew, R., 2004. The dragonflies of Europe. Harley Books.

Aguilar, J., J.-L. Dommanget & R. Prechag, 1998. A field guide to the dragonflies of Britain, Europe and North Africa. Collins.

Belyshev, B.F., 1958. To the Odonata fauna of Middle Asia. Fragmenta Faunistica, Polska Akademia Nauk, Institut Zoologiczny 8, No 5: 97-109.

Belyshev, B.F., 1961. To the dragonfly fauna (Odonata) of the Eastern Kazakhstan. Faunistica, Polska Akademia Nauk, Institut Zoologiczny 9, No 4: 27-42.

Beschovski, V., 1964. Dragonflies (Odonata) from the Bulgarian Black Sea coast. Bull. Inst. Zool. Mus. Sofia, XV: 115-129.

Beschovski, V., 1993. A critical notes on some Odonata species (Insecta, Odonata), reported by Bulgarian authors for the territories of Bulgaria, Greece and Macedonia. Acta zool. Bulg., 46: 39-43.

Beschovski, V., 1994. Fauna of Bulgaria. T.23. Insecta, Odonata. Publ. Bulg. Acad. Science.

Blom, W.L., 1982. List of Odonata collected during various lepidopterological trips in Iran (1971-1974). Notulae Odonatologicae 1(9): 150-151.

Borissov, S.N., 1989. On the vertical distribution of the dragonflies in the Pamir-Alay. Proceedings of the Scientific conference "Topics in the ecology of the mountain regions" (section Odonatology), Dushanbe, 9-13 October 1989: 30-38.

Demirsoy, A., 1982. Türkiye faunasi. Odonata. Seri 8, Bölüm 4, Sayi 8, Tübitak mat. Ankara.

Dumont, H., 1976. *Aeshna charpentieri* Kolenati, 1964, a synonym of *Cordulegaster insignis* Schneider, 1845 and on the correct status of *Cordulegaster charpentieri* auctorum (Anisoptera, Cordulegastridae). Odonatologica 5, No4: 313-321.

Dumont, H., 1977. A review of the dragonfly fauna of Turkey and adjacent Mediterranean islands (Insecta Odonata). Bulletin et Annales de la Société royale entomologique de Belgique 113: 119-171.

Kimmins, D.E., 1950. The 3rd Danish Expedition to Central Asia. Zool. Res. 4, Odonata (Vidensk. Medd. Dansk naturh. Foren.) 112: 235-241.

Leipelt, K.G., 2005. Behavioral differences in response to current: implications for the longitudinal distribution of stream odonates. Arch. Hydrobiol. 163: 81-100.

Lohmann, H., 1993. Revision der Cordulegastridae. 2. Beschreibung neuer Arten in der Gattungen Cordulegaster, Anotogaster, Neallogaster und Sonjagaster (Anisoptera). Odonatologica, 22 (3): 273-294.

Lopau, W., 1995. Beitrag zur kenntnis der Odonatenfauna der griechischen Inseln Rhodos, Kos, Samos und Chios; Libellenbeobachtungen aus den Sommern 1992, 1993 und 1994 und andere naturkundliche Notizen (Libellen, Lurche, Kriechtiere). Naturkdl. Reiseber. 4: 1-60.

Lopau, W., 1999. Die Libellenfauna der griechischen Inseln Thassos, Samothraki und Limnos. Libellula Supplement 2: 43-61.

Nedelkov, N., 1923. Eight contribution to the entomological fauna of Bulgaria. Magazine of Bulg. Acad. Science, Nature-Mathematical branch, XXV, No 12: 45-52.

Reinhardt, K., 1992. Beobachtungen von Libellen an Fliegewässern in Georgien (Odonata). Entomologische nachrichten und Berichte, 36: 243-248.

Shengelia, E.S., 1975. Dragonflies of Georgia. In: "Materialy k faune Gruzii". vol. 5, Mezniereba, Tbilisi: 61-81.

Seidenbusch, R., 1995. Description of the last instar larva of *Sonjagaster insignis* Schneider 1845 (Odonata : Cordulegastridae). Sulzbach-Rosenberg Libellenrundbriefe 1: 3 pp.

St Quentin, D., 1971. Zum Vorkommen von *Cordulegaster insignis* Schneider in Rumänien. Studii communicari Sibiu Muz. Brukenthal 16, 205-208.

Tailly, M. & V. Ananian, 2004. Preliminary checklist and atlas of Armenian dragonflies. <http://users.pandora.be/tailly/armenodon/armenian-dragonflies.htm>

Tailly, M., V. Ananian & H. Dumont, 2004. Recent dragonfly observations in Armenia, with an updated checklist. Zoology in the Middle East, 31: 93-102.

van Pelt, G.J., 2004. New records of dragonflies from Turkey (Odonata). Libellula Supplement 5: 3-38.

Verschuren, D. 1989. Revision of the larvae of West-Palaearctic Cordulegaster Leach, 1815 (Odonata, Cordulegastridae), with a key to the considered taxa and a discussion on their affinity. Bull. Annls. Soc. r. belge Ent., 125: 5-35.

Odonata : Biology of Dragonflies
Ed.: B.K. Tyagi
Scientific Publishers (India), 2007, pp. 63-72

6

THE DRAGONFLY FAUNA OF THE SHIVAPURI HILLS, NEPAL (ODONATA: ZYGOPTERA, ANISOZYGOPTERA, ANISOPTERA)

T. BROCKHAUS[1], S.G. BUTLER[2], R.G. KEMP[3] & G.S. VICK[4]

[1] *An der Morgensonne 5, D-09387 Jahnsdorf/Erzgebirge*
e.mail: t.brockhaus@t-online.de
[2] *'Red Willow' AllStretton, Shropshire.*
[3] *5 Maltings Close, Ackleton, Wolverhampton, WV6 7WB.*
[4] *'Crossfields' Little London Basingstoke, Hampshire.*

ABSTRACT

26 species of Odonata are placed on record from the Shivapuri mountains, Kathmandu, Nepal.

(Die Libellenfauna der Shivapuri-Berge, Nepal (Odonata: Zygoptera, Anisozygoptera, Anisoptera): 26 Libellenarten werden für die Shivapuri-Berge, nördlich von Kathmandu, Nepal, gemeldet).

INTRODUCTION

The check list of the Nepalese Odonata (Vick, 1989) is now more than 17 years old. Probably, due to civil war over the last 10 years, little information about dragonflies from this country has been published since then (e.g. Mahato, 2000; Kemp & Butler, 2001; Brockhaus, 2003).

The following article provides a list of Odonata recorded from the Shivapuri mountains, an area which is well known because of the relict species *Epiophlebia laidlawii* (Tani & Miyatake, 1979; Asahina, 1982a; Saville, 1990; Mahato, 1993). Other odonate species from this area have been reported by Vick (1985), Mahato (1993) and Brockhaus (2001) (Fig. 1).

SURVEY AREA

The Shivapuri Hills are situated in the northern part of the Kathmandu Valley approx. 15 km north of the capital of Nepal, Kathmandu. The highest point is the Shivapuri peak at 2732 metres. The hills form part of the Mahabharat region with mountains reaching 3000 m. About 50% of the area is covered with forest with both subtropical and temperate vegetation types. In the uninhabited areas, oaks, pines and rhododendron predominate. The lower levels surrounding the mountains have been mainly converted into terraced agricultural land. About 15000 people in the region derive their living from agriculture.

The hottest month is August averaging 25°C. However, even in winter daytime temperatures can reach up to 20°C. During the monsoon, between May to September, approx. 80% of the annual rainfall occurs (3200 mm a year).

Table 1. Four recent visits to the Shivapuri mountains undertaken by the first author

Date	Route	Participants
8 May 2000	climb steeply from the mountains in the south to a height of about 2500 m.	Thomas Brockhaus, Ronald Peusche l, Andreas Trautmann
25 May 2000	climb steeply from the mountains in the south to a height of about 2300 m.	Thomas Brockhaus
14 – 16 February 2002	Bhotechaur – Chisapani - northern brink of the hills between 2000 and 2400 m, back to Bhotechaur	Renate Brockhaus, Thomas Brockhaus, Andreas Hirsch, Ilona Hirsch
10 – 12 April 2005	Sundarijal – Chisapani - northern brink - Shivapuri peak - Naghy monastery - Budhanilkantha	Tanja Brockhaus, Thomas Brockhaus, Frank Kramer

The Shivapuri Hills have been protected since 1975 and were initially refered to as the "Shivapuri-Watershed" but since 1985 have been known as the "Shivapuri Watershed and Wildlife Reserve". The hills cover an area of about 144 square kilometres. They are the most important drinking water source for the capital Kathmandu (Chaudhary 1998).

There are several different access routes from Kathmandu to the Shivapuris. One can go by bus from Kathmandu to Budhanilkantha and take a trekking path on the south side the mountains. Another way is by car to Bhotechaur and from there trek to Chisapani and take the northern edge of the hills at altitudes between 2000 and 2400 m. The last possibility is by car to Sundarijal and trek from south to north crossing the mountains to Chisapani.

SPECIES RECORDED WITH OBSERVATIONS AND COMMENTS

Zygoptera

Caliphaea confusa (Hagen in Selys, 1859)

During May, 2000, in the southern part of the Shivapuris, we found both sexes of this species inhabiting vegetation close to montane streams. The males appear to be territorial (Fig. 2).

Determination : Fraser (1934).

Habitat : small hill streams in woods.

Altitude : 730-2350 m a.s.l. (Vick, 1989).

Distribution: Bhutan, Burma, China, India, Laos, Nepal, Thailand (Tsuda, 1991).

Indolestes cyaneus Selys, 1862

The first record was an individual male, seen in February 2002, on the northern edge of the hills. On 12 April 2005, several males and females were observed close to a cattle pool in the central part of the Shivapuris, situated on a grassy, treeless plateau used for cattle grazing, at approx. 2500 m.

Determination: Fraser (1933, with the synonym *Ceylonlestes cyanea*)

Habitat: grass land, watering place.

Altitude: 670-3870 m a. s.l. (St. Quentin ,1970; Vick, 1989).

Distribution: Bhutan, India, Nepal, Taiwan (Tsuda, 1991).

Anisoptera

Davidius aberrans senchalensis **Fraser, 1926**

In May 2000, the emergence of *Davidius aberrans* was observed at a altitude of 2400 m. The larva was positioned on a large, moss-covered, stone, approx. 30 cm about the water surface. The habitat of the larvae is lithotelms in steep montane streams. With the help of Kiyoshi Inoue the dragonfly was determined by Kazuma Kitagawa as *Davidius aberrans senchalensis*. This subspecies is new for Nepal and very rare, having been only recorded previously from Darjeeling, India.

Determination: Fraser (1934).

Habitat: small strong running streams, Lithotelms, in woods.

Altitude: 2400-2800 m a.s.l.

Distribution: India, Darjeeling (Fraser, 1934; Asahina, 1961), new for Nepal.

Anotogaster nipalensis **Sélys, 1854**

In May 2000 we observed a mating pair of *Anotogaster nipalensis* in the southern part of the Shivapuris. They were perched in a broad-leaved tree at a height of approximately 3 metres.

Determination: Fraser (1936).

Habitat: small streams.

Altitude: 450-3250 m a.s.l. (Vick, 1989).

Distribution: India, Nepal, China (Tsuda, 1991).

Neallogaster ornata Asahina, 1982

In May 2000, a female of *Neallogaster* was *seen* in the southern part of the mountains. It flew very low, approx. 30 cm, above the mountain path during moderate rain. In Brockhaus (2001) this species was wrongly determined as *Neallogaster hermionae*. Asahina described *Neallogaster ornata as* new in 1982. It differs from the other species of *Neallogaster* by having yellow spots on the second abdominal segment (Fig. 3).

A teneral specimen was observed in April, 2005 on the northern edge of the mountains.

Determination: Asahina (1982b, 1991).

Habitat: fast-flowing streams, pools.

Altitude: 1550-2710 m a.s.l. (Vick, 1989).

Distribution: India, Nepal, only in the western part of the Himalayas (Asahina, 1982b).

***Chlorogomphus atkinsoni* (Sélys, 1878)**

In 2005, at three sites on the northern edge of the mountains, we recorded both teneral adults and larvae of *Chlorogomphus atkinsoni.* All were associated with steep, mountain streams (Fig. 4).

Determination: Fraser (1936).

Habitat: fast-flowing streams, little pools with detritus.

Altitude: 620–2230 m a.s.l. (Vick, 1989).

Distribution: India (West Bengal, Darjeeling, Sikkim), Nepal (Kumar & Prasad, 1981).

***Orthetrum glaucum* (Brauer, 1865)**

In April 2000, an aged specimen was taken near the Shivapuri peak. In February 2002 we observed 1 male and 1 female of *Orthetrum glaucum* on a small ox-bow formed from a mountain stream on the east side of the Shivapuris.

Determination: Fraser (1936).

Habitat: little ox-bows of hill streams.

Altitude: 240-2600 m a.s.l. (Vick, 1989).

Distribution: everywhere in South East Asia.

***Pantala flavescens* Fabricius, 1798**

Dozens of teneral individuals flew near the trekking path in the northern part of the Shivapuris in April 2005. It is unknown, whether the species breed here.

Determination: Askew (1988).

Habitat: ponds, pools.

Altitude: 130-6300 m a. s.l. (Vick, 1989).

Distribution: circum-equatorial (Belyshev & Haritonov, 1983).

***Sympetrum decoloratum* Sélys, 1884**

We observed several individuals in February 2002 on a pond which irrigated rice fields. This species is new for Nepal this record is a confirmation of the entry in Vick, 1989 which is listed with a question mark.

Determination: Fraser (1936), Askew (1988).

Habitat: little ponds near the terrace fields.

Altitude: no informations.

Distribution: North Africa, Asia Minor, Middle Asia, India new for Nepal.

Table 2. Additional Odonata records from the Sivapuri Hills

Species	Date	Altitude m above sea-level	Stage, number	Source
Zygoptera				
Calopterygidae				
Caliphaea confusa	14.7.1984	1800	1xmale	Vick (1985)
(Hagen in Selys, 1859)	3.6.1987	2350	1xmale	C. Smith (Vick det.)
	10.-13.6.87	2100	3xmale	C. Smith (Vick det.)
	8.5.2000	2150	males, females, abundant	Brockhaus (2001)
	-	-	-	Bahadur B. (2005)
Chlorocyphidae	10.7.1984	1800	1xmale	Vick (1985)
Rhinocypha unimaculata Selys, 1853	-	-	-	Bahadur B. (2005)
	1985	-	-	C. Smith (Vick det.)
Amphipterygidae	15.2.2002	2600	1xmale	Brockhaus
Philoganga montana (Hagen in Selys, 1859)	12.4.2005	2400	males, females, abundant	Brockhaus
	14.7.1984	1800	1xmale	Vick (1985)
Lestidae	28.5.2000	2000	1xmale	Kemp & Butler
Indolestes cyaneus Selys, 1862				
	13.6.1987	1950	2xmale 2xfemale	C. Smith (Vick det.)
Synlestidae	3.5.1979	2732	1xfemale	Tani & Miyatake
Megalestes major Selys,	3.5.1979	2400	1xfemale	(1979)
1862	3.5.1979	2520	1xexuviae	Ibidem
	3.5.1979	2300	several males	Ibidem
	-	-	several larvae	Asahina (1982a)
	-	-	-	Bahadur B. (2005)
Platycnemididae	6./7.12.1988	2240	"abundant"	Saville (1990)
Calicnemia pulverulans	13.12.1988	2070	"three found"	Ibidem
Selys 1886		1940	"abundant"	Ibidem
Anisozygoptera		1860	"abundant"	Ibidem
Epiophlebiidae	14.12.1988	2380	"two found	Ibidem
Epiophlebia laidlawi	28.5.2000	2000	all larvae	Butler & Kemp
	30.5.2000		larvae	

Fig. 1. Larvae of *Epiophlebia laidlawi* (…,Photo: BOB KEMP)

Fig. 2. *Caliphaea confusa,* male (8.5.2000, Photo: T. Brockhaus)

Fig. 3. *Neallogaster ornate,* male (11.4.2005, Photo: T. Brockhaus)

Fig. 4. *Chlorogomphus atkinsoni,* young male (11.4.2005, Photo: T. Brockhaus)

Tillyard 1921	11.6.1987	1555	3xmale	C. Smith (Vick det.)
Anisoptera	30.5.2000	2100	4xmale	Kemp & Butler
Gomphidae	10.7.1984	1800	1xmale	Vick (1985)
Anisogomphus bivittatus Selys, 1854	8.5.2000	2400	1 exuviae 1 male hatched	Brockhaus (2001)
Anisogomphus occipitalis Selys, 1854	11.6.1987	1555	1xmale	C. Smith (Vick det.)
	8.5.2000	2200	1xmale 1 x female mating	Brockhaus (2001)
Davidius abberans senchalensis Fraser 1926	30.5.2000	2100	3 x male exuviae emergence seen	Kemp & Butler
Cordulegasteridae	14.5.1980	2732	1xmale	Asahina (1982b)
Anotogaster nipalensis Selys 1854	13.6.1987	2710	1xmale	C. Smith (Vick det.)
Neallogaster hermionae Fraser, 1927	25.5.2000	2200	1xfemale	Brockhaus (2001) (wrong determined as *N. hermionae*)
Neallogaster ornata Asahina, 1982	30.5.2000	2100	1xmale	Kemp & Butler
	11.4.2005	2350	several males and females, all teneral	Brockhaus
Corduliidae	1986	1620	1xmale	C. Smith (Vick det.)
Macromia moorei Selys, 1874	13.6.1987	1980	1xmale 1 x female	C. Smith (Vick det.)
Idionyx stevensi Fraser, 1924	28.5.2000	2100	numerous males 1xfemale larvae/ emergence	Kemp & Butler
Chlorogomphidae				
Chlorogomphus atkinsoni (Selys, 1878)	30.5.2000	2100	1xmale	Kemp & Butler
Chlorogomphus selysi Fraser, 1929	11., 12.4.2005	2350	several males, all teneral, larvae	Brockhaus
Libellulidae				
Crocothemis servilia (Drury, 1770)	1987	-	1xmale	C. Smith (Vick det.)
Orthetrum glaucum (Brauer 1865)	10.7.1984	1480	1xmale	Vick (1985)
Orthetrum sabina (Drury, 1770)	8.5.2000	2732	1xmale, teneral	Brockhaus (2001)
	14.2.2002	1600	1xmale 1x female	Brockhaus
Orthetrum taeniolatum (Schneider, 1845)	10.7.1984	1480	1xmale	Vick (1985)
Orthetrum triangulare (Selys, 1878)	14.7.1984	1800	1xmale	Vick (1985)
Neurothemis fulvia (Drury, 1773)	10.7.1984	1480	1xmale	Vick (1985)
Pantala flavescens	14.7.2000	1480	1xmale	Vick (1985)
	11., 12.4.2005	2200	Several males	Brockhaus

Fabricius 1798 *Sympetrum decoloratum* Selys 1884	14.2.2002	1700	Several males and females, all teneral	Brockhaus
Tritemis festiva (Rambur, 1842)	10.7.1984	1480	1xmale	Vick (1985)

SOME NOTES ON LARVAL COLLECTIONS BY S.G. BUTLER

1992

Larvae of *C. atkinsoni* were found in coarse sandy stretches of fast-flowing streams, these showed markedly different colouration from larvae collected elsewhere and from the description in Fraser (1936).

The larva of *I. stevensi* is typical of the genus, but more robust than other species so far (description in preparation by S.G. Butler).

1996 and 2000

The larvae of *E. laidlawi* were found in a stream which passed under the forest track at approx 1800m on the south side above the Kathmandu valley. Other similar streams were searched, but no larvae were found, probably due to human pollution (Subodh Sharma *pers. comm.*). A larva of *Davidius* sp. (det G.S. Vick) was also observed emerging on the same stream as *E. laidlawi* (1.6.96. S.G.B. collection)

Aeshnid larvae were collected in 1996 and exuviae in 2000 of two species which unfortunately died before emergence. Both species show affinities to Brachytronini, one of which appears close to *Cephalaeschna.*

DISCUSSION

Currently, 26 odonates are known from the Shivapuri hills and they include several scarce and taxonomically important species. The most notable species is *Epiophlebia laidlawi*, an insect considered threatened (IUCN, 2006, Red Data Book) and endemic to the Himalayan foothills. Other notable species include *Davidius aberrans senchalensis*, *Neallogaster ornata* and *Chlorogomphus atkinsoni* which appear to be rare with restricted ranges.

Apart from *Indolestes cyaneus* and the Libellulids recorded, all are specialists, typical of fast-flowing, montane streams. Their ecology and behaviour remain largely unknown and would well repay further investigation.

REFERENCES

Asahina, S. (1961): Description of some dragonfly larvae from Darjeeling. Kontyu 29: 240-246.

Asahina, S. (1982a): Survey of the relict dragonfly Epiophlebia laidlawi Tillyard in Nepal, Mai 1981. Rep. Odon. Specialist Group Int. Un. Conserv. Nat. 1: 1-6.

Asahina, S. (1982b): A Revision of the Himalayan Dragonflies of the genus Neallogaster (Odonata, Cordulegasteridae). Bull. Natn. Sci. Mus. Tokio, Ser. A 8: 153-171.

Asahina, S. (1991): Notes on some problematical dragonfly specimens from Nepal. TOMBO 34: 23-26.

Askew, R.R. (1988): The Dragonflies of Europe. Harley Books. Colchester.

Bahadur Budha, P. (2005): Dragonfly biodiversity: Important management tool for providing safe drinking water to Kathmanduties, Nepal. 2nd Symposium on Biodiversity of the Himalayans – interactive report. www.zin.ru/Animalia/ coleoptera/eng/05i.htm.

Belyshev, B.F. & A.J. Haritonov (1983): Geografia strekos (Odonata) meridionalnowo faunistitscheskowo Tscharstwa. Nauka, Novosibirsk. [Russ.].

Brockhaus, T. (2001): Beobachtungen zur Libellenfauna der Shivapuri Berge, Nepal (Odonata). Entomologische Nachrichten und Berichte 45: 221-223. [German].

Brockhaus, T. (2003): Pantala flavescens (Fabricius) in Khumbu Himal, Nepal (Anisoptera: Libellulidae). Notulae Odonatologicae 6: 2-3.

Chaudhary, Ram, P. (1998): Biodiversity in Nepal. Know Nepal, Series No. 17, S. Devi, Saharanpur (U.P.), India & Tecpress Books Bangkok, Thailand.

Fraser, F.C. (1933): The fauna of British India including Ceylon and Burma. Odonata. Vol. I. Taylor and Francis, London.

Fraser, F.C. (1934): The fauna of British India including Ceylon and Burma. Odonata. Vol. II. Taylor and Francis, London.

Fraser, F.C. (1936): The fauna of British India including Ceylon and Burma. Odonata. Vol. III. Taylor and Francis, London.

IUCN (2006): 2006 IUCN Red List of Threatened Species. <www.iucnredlist.org>. Downloaded on 25 August 2006.

Kemp, R.G. & S.G. Butler (2001): Some dragonfly records from Phewa Tal, Nepal with notes on Philoganga montana (Selys) (Zygoptera: Amphipterygidae). Notulae odonatologicae 5: 88-91.

Kumar, A. & M. Prasad (1981): Field ecology, zoogeography and taxonomy of the Odonata of Western Himalaya, India. Rec. Zoo. Surv. India, Occ. Paper 20: 1-118.

Mahato, M. (1993): Epiophlebia laidlawi – a living ghost. Selysia 22: 2.

Mahato, M. (2000): Resource partitioning among larvae of six coexisting odonate species of the Khali Gandaki River, central Nepal (Anisoptrea). Odonatologica 29: 209-223.

Saville, N. (1990): Survey of the status of Epiophlebia laidlawi (Tillyard) in the Shivapuri Watershed and

Wildlife Reserve, 1988. In: Report of the Cambridge. Entomological Expedition to Nepal 1988: 31-35.

St. Quentin, D. (1970): Odonata aus Nepal. Khumbu Himal 3: 389-411. [German]

Tani, K. & Y. Miyatake (1979): The discovery of Epiophlebia laidlawi Tillyard, 1921 in the Kathmandu Valley, Nepal (Anisozygoptera: Epiophlebiidae). Odonatologica 8: 329-332.

Tsuda, S. (1991): A distributional list of World Odonata. Osaka.

Vick, G.S. (1985): Odonata collected by Shiplake College Trekking Society expedition to Nepal in 1984. Notulae Odonatologicae 2(5) : 80-82

Vick, G.S. (1989): List of the dragonflies recorded from Nepal, with a summary of their altitudinal distribution (Odonata). Opuscula Zoologica Fluminensis 43: 1-21.

Odonata : Biology of Dragonflies
Ed.: B.K. Tyagi
Scientific Publishers (India), 2007, pp. 73 - 87

7

THE COMPOSITION AND HISTORY OF SIBERIAN ODONATE FAUNA

A. YU. HARITONOV

Institute of Animal Systematics and Ecology,
Siberian Division of Russian
Academy of Sciences, Frunze str. 11, Novosibirsk 630091, Russia
e.mail: pc@eco.nsc.ru

ABSTRACT

Based on literature analysis and the latest special researches the more precise list of dragonflies taxons of Asian part of Russia, counted 129 species, is presented. The species are distributed among 7 geographical regions: Ural and Transural, West-Siberian Plain, Altai and Sayan, North of East Siberia, South of East Siberia, North of Far East, South of Far East. The ancient autochthonous odonate fauna of south part of Russian Far East is the most rich of species and is a member of Subholarctic faunistic region. The rest of territory is a member of Holarctic faunistic region and to divide into two parts: Euro-Siberian and Siberian subregions. Their odonate fauna is not rich, young and allochthonous.

INTRODUCTION

The word "Siberia" is generally used to denote the Asian part of Russia stretching from the Urals in the west to the Pacific in the east and occupying the area of 12 million sq km which is considerably larger than the area of the European continent. The first records of Siberian dragonflies appeared in scientific literature 150 years ago in publications of H. Hagen (1856 a,b) who listed 17 species for the outskirts of Irkutsk, 8 species for Kamchatka, 4 species for Okhotsk coast, and a few species for some other areas. Some later studies undertaken by H. Hagen (1858), E. De. Selys Longchamps (1872, 1887), F. Trybom (1889) included short lists of dragonflies mainly from the eastern part of Siberia. Thus the first studies of Siberian odonate fauna were first initiated by German, Belgian, English, Swedish and other foreign scientists in the 19th century but it was not until the 20th century that Siberia became the study object of Russian odonatologists. Great contribution to the studies of Siberian odonate fauna was made by A.N. Bartenev in the first half of the 20th century. It is quite symbolic that his first student paper (Bartenev, 1908) was devoted to the outlying areas of Western Siberia – Lake Uvildy and some other reservoirs of the eastern macroslope of the Urals, and his final posthumous work (Bartenev, 1956) embraced the odonate fauna of the Far East, *i.e.* the eastern outlying areas of Siberia. On the whole A.N. Bartenev's 24 papers (1910, 1911, 1912, 1915, 1930 *etc.*) contain different materials on Siberian dragonflies, although most of them did not result from the analysis of his personal collections but those belonging to different collectors.

The greatest part of the 2nd part of the 20th century is associated with the name of B.F. Belyshev. His first paper devoted to *Epitheca bimaculata* biology and systematics in Kulunda Steppe (South of the Western Siberia) was published in 1951. B.F. Belyshev launched a number of expeditions to collect his material on dragonflies in such regions as Altai and South of the Western Siberia, outskirts of Lake Baikal, the lower reach of the Lena River, the southern part of the Russian Far East. Not only did he collect the faunistic material but he also described ecological and biological characteristics of Siberian dragonflies he used to observe in nature. Special attention was paid to the analysis of distribution of dragonflies. Studying Siberia B.F. Belyshev created his own system of zoogeographical demarcation of the world earth based on distribution of dragonflies. His 87 papers (Belyshev, 1951, 1964, 1968; Belyshev & Haritonov, 1981; Belyshev *et al.,* 1989, etc.) contain different materials on Siberian dragonflies generalized in a Doctoral thesis and a three-volume monograph "Dragonflies of Siberia" (1973, 1974) which outlines about 119 dragonfly species. Thanks to B.F. Belyshev Novosibirsk Institute of Animal Systematics and Ecology of the Russian Academy of Sciences has become a core of the Russian odonatology. B.F. Belyshev died in

1993. His latest works were published in collaboration with the author of the present paper, his follower A.Yu. Haritonov who continued studying Siberian dragonflies in Ural and Trans-Ural, Yamal and Taimyr Peninsulas, the whole southern part of the Western Siberia, Altai-Sayan highlands, the southern part of the Far East and Kamchatka (Haritonov, 1972, 1974, 1975a,b, 1976, 1978, 1986, 1990, *etc.*). Under the supervision of A.Yu. Haritonov a number of post-graduate students wrote their Ph.D. theses on different Siberian regions (Sukhacheva, 1989; Malikova, 1995; Popova, 1999; Dronzikova, 2000; Yanybaeva, 2002; Perepelov, 2003; Chaplina, 2004; Yurchenko, 2004; Belevich, 2005). Several tens of papers on Siberian dragonflies have been recently published among which it is worth while mentioning Kolosov (1914; 1916a,b; 1927), Vnukovsky (1928), Kosterin (1992, 1996, 1999, 2004, etc.). All of the research projects made it possible to gain a more precise concept of the composition and geographical distribution of Siberian odonate fauna allowing analyzing the zoogeography and interpreting the history of Siberian odonate fauna.

RESULTS

The present list of Siberian dragonflies includes a fair amount of new faunistic data and recent taxonomic and nomenclature modifications (Table 1). A number of taxa have been recently excluded from the list of Siberian odonate fauna and are considered as synonyms. The species *Libellula relicta* Belyshev & Kiauta, 1968 is placed in synonymy with *Libellula quadrimaculata ab. praenubila* Newman (Ketenchiev & Haritonov, 1999). The subspecies *Leucorrhinia albifrons obensis* Belyshev, 1973 described due to the predominance of females with yellow scurf on wings in a small population in the South of Western Siberia, does not exist, in fact, because this character is quite variable and may be revealed in dragonflies of this species in diverse places at any time. The study of series of palaearctic species of the genus *Sympetrum* (Popova, 1999) has shown strong modification variability, especially cline variability. As a result, the subspecies *S.p. intermedium* Belyshev, 1956 and *S.p. kurentzovi* Belyshev, 1956 of the species *S. pedemontanum* are placed in synonymy. It has been found that this species has two subspecies: a nominal subspecies on the continent and *S.p. elatum* Selys, 1883 on Far Eastern islands. The subspecies *Sympetrum striolatum dochidordzi* Belyshev, 1958 (Schmidt, 1961) and *S. croceolum fuscoatrum* Belyshev, 1964 are also placed in synonymy to a nominal form.

Table 1. Siberian Dragonfly species & their geographical disrtibution

	Species of dragonflies	West Siberia			East Siberia		Far East	
		Ural & Trans-ural	West-Sib-erian Plain	Altai & Sayan	North of East Siberia	South of East Siberia	North of Far East	South of Far East
1	2	3	4	5	6	7	8	9
	ANISOPTERA							
1	*Pantala flavescens* (Fabricius, 1798)							+
2	*Libellula depressa* Linnaeus, 1758	+	+					
3	*L. quadrimaculata quadrimaculata* Linnaeus, 1758	+	+	+	+	+	+	
	L.q. orientalis Belyshev, 1956							+
4	*Leucorrhinia albifrons* (Burmeister, 1839)	+	+	+				
5	*L. caudalis* (Charpentier, 1840)	+	+	+				
6	*L. dubia* (Vander Linden, 1825)	+	+	+				
7	*L. orientalis* Selys, 1887 (=*L. ussuriensis* Bartenev, 1914)	+1			+	+	+	+
8	*L. pectoralis* (Charpentier, 1825)	+	+	+				
9	*L. intermedia* Bartenev, 1910				+	+	+	+
10	*L. rubicunda* (Linnaeus, 1758)	+	+	+				
11	*Sympetrum baccha matutinum* Ris, 1911							+
12	*S. cordulegaster* (Selys, 1883)							+
13	*S. croceolum fuscoatrum* Belyshev, 1964		+2	+2				+
14	*S. danae* (Sulzer, 1776)	+	+	+	+	+	+	+
15	*S. depressiusculum* (Selys, 1841)							+
16	*S. eroticum eroticum* (Selys, 1883)							+
17	*S. flaveolum flaveolum* (Linnaeus, 1758)	+	+	+	+	+	+	+
18	*S. frequens* (Selys, 1883)							+
19	*S. infuscatum* (Selys, 1883)							+
20	*S. imitans* (Selys, 1886)					+		+
21	*S. kunckeli* (Selys, 1884)							+
22	*S. meridionale* (Selys, 1841)	+2						
23	*S. parvulum* (Bartenev, 1912)							+
24	*S. pedemontanum pedemontanum* (Mueller, 1766)	+	+	+		+		+
	S.p. elatum (Selys, 1872)							+3
25	*S. risi* Bartenev, 1914							+
26	*S. sanguineum sanguineum* (Mueller, 1764)	+	+	+				

1	2	3	4	5	6	7	8	9
27	*S. striolatum striolatum* (Charpentier, 1840)			+4				
	S.s. imitoides Bartenev, 1919							+
	S.s. kurilis Belyshev, 1976							+3
28	*S. vulgatum vulgatum* (Linnaeus, 1758)	+	+	+	+	+		
29	*S. uniforme* (Selys, 1883)							+
30	*Lyriothemis pachygastra* (Selys, 1878)							+
31	*Neurothemis fluctuans* (Fabricius, 1793) (= *Palliata palliata* Rambur, 1842)							+
32	*Deielia phaon* (Selys, 1883)							+
33	*Orthetrum cancellatum cancellatum* (Linnaeus, 1758)	+	+					
	O.c. orientale Belyshev, 1958			+				
34	*O. brunneum brunneum* Fonscolombe, (1837)					+5		
35	*O. albistylum albistylum* (Selys, 1848)					+		
	O.a. speciosum (Uhler, 1858)					+		+
36	*O. melania (=triangulare)* (Selys, 1883)							+
37	*Selysiothemis nigra* (Vander Linden, 1825)	+6						
38	*Epitheca bimaculata bimaculata* (Charpentier, 1825)	+	+		+			
	E.b. altaica Belyshev, 1951			+		+		
	E.b. sibirica Selys, 1887						+	+
39	*Somatochlora alpestris* (Selys, 1840)	+	+	+	+	+	+	+
40	*S. arctica* (Zetterstedt, 1840)	+1	+	+	+	+	+	+
41	*S.exuberata* Bartenev, 1910					+	+	+
42	*S. flavomaculata* (Vander Linden, 1825)	+	+	+				
43	*Somatochlora graeseri graeseri* Selys, 1887	+		+	+	+	+	+
	S.g.aureola Oguma, 1915							+3
44	*S. metallica metallica* (Vander Linden, 1825)	+	+	+	+	+		
	S.m. abacanica Belyshev, 1955			+				
45	*S. sahlbergi* Trybom, 1889	+1		+	+	+	+	
46	*S. uchidai* Foerster, 1909							+3
47	*S. viridiaenea* (Uhler, 1858)							+

1	2	3	4	5	6	7	8	9
48	*Cordulia aenea aenea* (Linnaeus, 1758)	+	+	+	+	+		
	C.a. amurensis Selys, 1887						+	+
49	*Macromia amphigena fraenata* Martin, 1907	+6		+		+		+
50	*M. manchurica* Asahina, 1964							+
51	*Epophthalmia elegans* (Brauer, 1865)							+
52	*Aeshna affinis* Vander Linden, 1820	+	+					
53	*Ae. caerulea* (Strom, 1783)	+	+	+	+	+	+	+
54	*Ae. crenata crenata* Hagen, 1856	+	+	+	+	+	+	+
	Ae.c. nigroflava Martin, 1908							+3
55	*Ae. cyanea* (Mueller, 1764)	+						
56	*Ae. grandis* (Linnaeus, 1758)	+	+	+		+		
57	*Ae. juncea* (Linnaeus, 1758)	+	+	+	+	+	+	+
58	*Ae. mixta mixta* Latreille, 1805	+	+	+		+		+
59	*Ae. serrata* Hagen, 1856	+	+	+	+	+	+	
60	*Ae. subarctica subarctica* Walker, 1908	+	+	+	+	+	+	+
61	*Ae. viridis* Eversmann, 1836	+	+	+		+		+
62	*Anax imperator imperator* Leach, 1815	+						
63	*A. junius* Drury, 1770						+7	
64	*A.parthenope parthenope* Selys, 1839	+		+				
	A.p. julius Brauer,1865							+
65	*Brachytron pratense* Mueller, 1764	+						
66	*Aeschnophlebia kolthoffi* Sjostedt, 1925							+
67	*Ae. longistigma* Selys, 1883							+
68	*Cordulegaster boltonii boltonii* (Donovan, 1807)	+6						
69	*Anotogaster sieboldii* (Selys, 1854)							+3
70	*Anisogomphus maacki* (Selys, 1872)							+
71	*Asiagomphus melanopsoides* (Doi, 1943)							+
72	*Gomphidia confluens* Selys, 1878							+
73	*Gomphus schmidti* Asahina, 1956							+
74	*G. vulgatissimus* (Linnaeus, 1758)	+	+	+				
75	*Davidius (=Hologomphus) lunatus* (Bartenev, 1914)							+
76	*Nihonogomphus ruptus* (Selys, 1858)			+		+		+
77	*Onychogomphus forcipatus forcipatus* (Linnaeus, 1758)	+						
1	2	3	4	5	6	7	8	9

78	*Ophiogomphus cecilia* Fourcroy, 1785	+	+	+				
79	*O. obscurus* Bartenev, 1909			+		+		+
80	*O. reductus* Calvert, 1898					+8		
81	*O. spinicornis* Selys, 1878					+8		
82	*Shaogomphus postocularis epophthalmus* Selys, 1872			+		+		+
83	*Sieboldius albardae* Selys, 1886							+
84	*Sinictinogomphus(=Ictinus) clavatus clavatus* (Fabricius, 1775)							+
85	*Stylurus (=Davidius) annulatus* (Djakonov, 1926)							+
86	*S. (=Gomphus) flavipes flavipes* (Charpentier, 1825)	+	+	+		+		+
	S.f. sibirica (Bartenev, 1910)		+	+				
87	*Trigomphus anormolobatus* Bartenev, 1911							+
88	*T. citimus citimus* (Needham, 1931)							+
89	*T. nigripes* (Selys, 1887)							+
	ZYGOPTERA							
90	*Calopteryx atrata* Selys,1853							+
91	*C. japonica japonica* Selys, 1869			+		+		+
	C.j. altaica Belyshev, 1955			+				
92	*C. splendens splendens* (Harris, 1782)	+	+	+		+		
93	*C. virgo virgo* (Linnaeus, 1758)	+	+					
94	*Mnais pruinosa costalis* Selys, 1869							+3
95	*Lestes barbarus* (Fabricius, 1798)	+	+	+				
96	*L. dryas* Kirby, 1890	+	+	+	+	+	+	+
97	*L. japonicus* Selys, 1883							+
98	*L. macrostigma* (Eversmann, 1836)	+	+	+				
99	*L. temporalis* Selys, 1883							+
100	*L. sponsa* (Hansemann, 1823)	+	+	+	+	+	+	+
101	*L. virens vestalis* Rambur, 1842	+	+	+				
102	*Sympecma fusca* (Vander Linden, 1820)	+9				+		+
103	*S. paedisca braueri* Bianki, 1905	+	+	+		+		+
104	*Cercion calamorum calamorum* (Ris, 1916)							+
105	*C. hieroglyphicum* (Brauer, 1865)							+
106	*C. plagiosum* (Needham, 1930)							+
107	*C. (=Coenagrion) v-nigrum* (Needham, 1930)							+
1	2	3	4	5	6	7	8	9

108	*Coenagrion armatum* (Charpentier, 1840)	+	+	+	+	+	+	+
109	*C. johanssoni johanssoni* Wallengren, 1859 (=*C. concinnum* Johanssen, 1859)	+	+	+	+	+	+	
	C.j. bartenevi Belyshev, 1955							+
	C.j. convalescens Bartenev, 1914							+
110	*C. ecornutum* (Selys, 1872)	+		+		+		+
111	*C. glaciale glaciale* (Selys, 1872) (=*C. tugur* Bartenev, 1956)				+	+		
	C.g. orientale (Belyshev, 1973)						+	+
112	*C. hastulatum* (Charpentier, 1825)	+	+	+	+	+	+	+
113	*C. hylas hylas* (Trybom, 1889)	+1		+	+	+		
	C.h. ussuriense (Belyshev, 1956)						+	+
114	*C. lanceolatum* (Selys, 1872)				+	+	+	+
115	*C. lunulatum* (Charpentier, 1840) (=*C. vernale* Hagen, 1839)	+	+	+	+	+	+	+
116	*C. puella puella* (Linnaeus, 1758)	+	+	+				
117	*C. pulchellum* (Vander Linden, 1823)	+	+	+				
118	*Erythromma humerale* Selys, 1887					+		+
119	*E. najas* (Hansemann, 1823)	+	+	+	+	+		
120	*Enallagma cyathigerum cyathigerum* (Charpent, 1840)	+	+	+	+	+	+	+
	E.c. risi Schmidt, 1961	+8	+8	+8		+		
121	*E. boreale circulatum* (Selys, 1883)							+3
122	*Ischnura aralensis* Haritonov, 1979	+						
123	*I. asiatica* (Brauer, 1865) (=*I. lobata* Needham, 1930)							+
124	*I. elegans elegans* (Vander Linden, 1823)	+	+	+		+		+
125	*I. pumilio* (Charpentier, 1825)	+		+				
126	*Nehalennia speciosa* (Charpentier, 1840)	+	+	+		+		+
127	*Copera tokyoensis* Asahina, 1948 (=*Denticnemis bicolor* Bartenev, 1956)							+
128	*Platycnemis pennipes* (Pallas, 1771)	+	+	+		+		+
129	*P. phyllopoda* Djakonov, 1926							+

1 – only Polar Ural; 2 – extremely local; 3 – only Far East islands; 4 – only South-West Altai; 5 – data of Hagen, 1856; 6 – single find on South Urals; 7 – old data for Kamchatka; 8 – south of region; 9 – data of Boev *et al.*, 1989.

The species *Macromia sibirica* Djakonov, 1926 and *Macromia bartenevi* Belyshev, 1973 are placed in synonymy with *Macromia amphigena fraenata* Martin, 1907 (Malikova, 1995). We place the subspecies *Somatochlora flavomaculata minor* Belyshev, 1955 and *S.sahlbergi relicta* Belyshev, 1971 in synonymy to nominal forms because they were described on the basis of a few specimens without taking into account their individual variability.

The taxa of the genus *Aeshna* have been greatly varied. The variability of all 15 species of the genus known on the territory of Siberia and the neighboring regions has been studied. The morphological variability of more than 3000 imago specimens, 3500 exuviae and 2000 larvae of the genus *Aeshna* has been estimated according to the collections of the Institute of Animal Systematics and Ecology of the Russian Academy of Sciences and some other Russian, European and American museums. As a result, the most prevailing species *Aeshna juncea* has been found to have a great individual variability of both colour and structure characteristics resulting in inaccurate descriptions of its six subspecies in the Palaearctic region. It has been found appropriate to place these species as well as *Aeshna undulata* Bartenev, 1930 and *Ae. baicalensis* Belyshev, 1964 in synonymy with *Ae. juncea* (Belevich, 2005). It has also been shown that the subspecies *Aeshna crenata wnukovskii* Belyshev, 1973 is baseless and should be placed in synonymy with *Aeshna crenata crenata* Hagen, 1856. *Aeshna nigroflava* Martin, 1908 is to be considered as a subspecies of *Aeshna crenata* inhabiting Japan and other Far Eastern islands. The subspecies *Aeshna serrata osiliensis* Mierzejwski, 1913 is considered as a synonym of the nominal form *Aeshna serrata serrata* Hagen, 1856 (Belevich, 2005). The species *Aeshna palmata* Hagen, 1856 has also been excluded from the list of Siberian odonate fauna because it has proved to be fallacious to recognize this North American species in Kamchatka (Gloyd, 1941).

The species *Aeschnophlebia zygoptera* Belyshev, 1965 is placed in synonymy with *Aeschnophlebia kolthoffi* Sjostedt, 1925 (Malikova, 1995). The species *Temnogomphus amurensis* Bartenev, 1930 is placed in synonymy with *Anisogomphus maacki* (Selys, 1872). The species *Gomphus epophthalmus* Selys, 1872 is considered as a synonym of *Shaogomphus postocularis epophthalmus* Selys, 1872 according to the system recognized by many authors (Kosterin, 2004). The species *Gomphus chankae* (=chancae) Bartenev, 1956 is placed in synonymy with *G. schmidti* Asahina, 1956 (Malikova, 1995).

The species *Coenagrion striatum* Bartenev, 1956 mentioned by Belyshev (1974) is to be considered as a synonym of *Cercion plagiosum* (Needham, 1930) and *Coenagrion brevicauda* Bartenev, 1956 should to be considered as a synonym of *Cercion v-nigrum* (Needham, 1930) (Malikova, 1995).

The Palaearctic taxa of *Enallagma* are viewed quite inconsistently. To solve this problem series of characteristics of more than 1500 *Enallagma* specimens have been studied. It has been concluded that the only species *Enallagma cyathigerum* Charpentier, 1840 inhabits the continental part of the Palaearctic whose strong morphological variability has led to a wrong description of a number of species and subspecies (Yurchenko, 2004). According to Yurchenko *Enallagma deserti* Selys, 1871; *E. risi* Schmidt, 1961; *E. strouhali* Quentin, 1962; E. *nigrolineata* Belyshev et Haritonov, 1975 are to be placed in synonymy with the polymorphic *E. cyathigerum*. *E. risi* may be considered as a subspecies of *Enallagma cyathigerum risi* Schmidt, 1961 because of cline variability of characteristics from northern to southern populations. Though the most northern and southern populations are well differentiated, the characteristics of subspecies from the temperate zone of Russia are mixed up and are difficult and even impossible to differentiate. The taxon *E. circulatum* Selys, 1883 is to be considered as an island Asian subspecies of the North-American species *E. boreale* Selys, 1875. *E. belyshevi* Haritonov, 1975 is to be placed in synonymy with the subspecies *E. boreale circulatum*.

Thus the more precise list of dragonflies taxa of Asian part of Russia counted 129 species, 16 from them have in Siberia on 2 3 sub species. The species are distributed among 7 geographical regions which have difference at climate and natural conditions: Ural and Transural, West-Siberian Plain, Altai and Sayan, North of East Siberia, South of East Siberia, North of Far East, Sourh of Far East. The ancient autochthonous odonate fauna of south part of Russian Far East is the most rich of species and is a member of Subholarctic faunistic region. The rest of territory is a member of Holarctic faunistic region and to divide into two parts: Euro-Siberian and Siberian subregions. Their odonate fauna is not rich, young and allochthonous, because it was formed only in postglacial time thanks to migrations from outsiberian territories. Against a background of poor Holarctic fauna as a whole the odonate fauna of South Ural and Altai-Sayan highlands look some more rich because it can be refugiums in glacial period and save the elements of before glacial fauna till present time.

ACKNOWLEDGEMENT

The research was supported by the Grants from Russian Foundation of Fundamental Investigations (04-04-48778a and 06-04-63088).

REFERENCES

Bartenev, A.N. 1908. Kollektstiya strekoz iz okrestnostei oz. Uvil'dy Ekaterinburgskogo uezda Permskoi Gubernii [A collection of dragonflies of the environs of Lake Uvil'dy of Ekaterinburg District of Perm' Province]. Proc. Naturalist Soc. Kazan Univ. 41: 1-40. – [Russ.]

Bartenef, A.N. 1910. Data relating to Siberian dragonflies. Zool. Anz. 35: 270-278.

Bartenev, A.N. 1911. Materialy po strekozam palearkticheskoi Azii iz kollektsii Zoologicheskogo muzeya Imperskoi Akademii nauk, 1 [Material on dragonflies of the Palaearctic Asia from the collection of the Zoological Museum of the Emperor Academy of Sciences, 1]. Annls Mus.zool. Acad. imp. Sci. St Petersbourg 16: 409-448. – [Russ.]

Bartenev, A.N. 1912. Materialy po strekozam palearkticheskoi Azii iz kollektsii Zoologicheskogo muzeya Imperskoi Akademii nauk, 2 [Material on dragonflies of the Palaearctic Asia from the collection of the Zoological Museum of the Emperor Academy of Sciences, 2]. Annls Mus. zool. Acad. imp. Sci. St Petersbourg 17: 289-310. – [Russ.]

Bartenev [Bartenef], A.N., 1915. Insectes Pseudoneuroptères (Insecta Pseudoneuroptera). Volume I. Libellulidae. Livraison 1. Faune de la Russie et des pays limitrophes. N. V. Nasonov (ed.), Académie Imperiale des Sciences, Petrograd. – [Russ., with French title). 352 pp.

Bartenef, A.N. 1930. Noch einmal uber die Artehgruppe Aeschna juncea (Odonata, Aeschninae) in der Palaearktik. Zool. Anz. Bd.89, N 710. P. 229-245.

Bartenef, A.N. 1956. Materialy po odonatologicheskoi faune (Insecta, Odonata) Dal'nego Vostoka [Materials on the odonate fauna (Insecta, Odonata) of the Fae East]. Trudy dal'nevost. Fil. Akad. Nauk SSSR 3: 201-283. – [Russ.]

Belevich, O.E., 2005. Strekozy roda *Aeshna* Palearktiki [The dragonflies of *Aeshna* genus of Palaearctic region]. Dissertation for the scientific degree of candidate of biological sciences (Ph. D. Thesis). Novosibirsk, 385 pp. – [Russ.]

Belyshev, B.F., 1951. K biologii i sistematike Epitheca bimaculata Charp. (Odonata) v Kulundinskoi stepi [On biology and sistematics Epitheca bimaculata Charp. (Odonata) in Kulunda steppe]. Entom. Revue 31: 504-505. – [Russ.]

Belyshev, B.F., 1964. Contributions to the knowledge of the fauna of dragonflies (Odonata) of Siberia. I. Fragmenta faunistica 11: 53-74. Warszawa. – [Russ., with Engl. title]

Belyshev, B.F., 1968. Contributions to the knowledge of the fauna of dragonflies (Odonata) of Siberia. IV. Geography of the dragonflies of Siberia. Fragmenta faunistica 14: 407-536. Warszawa. – [Russ., with Engl. title]

Belyshev, B.F., 1973. Dragonflies (Odonata) of Siberia. Vol. 1. Parts 1-2. Nauka, Siberian Division, Novosibirsk, 620 p. – [Russ.]

Belyshev, B.F. 1974. The Dragonflies (Odonata) of Siberia. Vol. II, Part. 3. Nauka, Siberian Division, Novosibirsk, 351 p. – [Russ.]

Belyshev, B.F. & A.Y. Haritonov, 1981. Geografiya strekoz (Odonata) Boreal'nogo faunisticheskogo tsarstva [Geography of Dragonflies (Odonata) of the Boreal Faunistic Kingdom]. Nauka, Siberian Division, Novosibirsk, 278 pp. – [Russ.]

Belyshev, B.F., A.Y. Haritonov, S.N. Borisov, Z.D. Spuris, G.A. Mazokhin-Porshnyakov, P.A. Mokrushov, R.S. Pavlyuk, L.N. Pritykina, G.I. Ryazanova, E.S. Shalopenok, L.N. Pisanenko, G.A. Sukhacheva, I.N. Haritonova, V.V. Zaika & L.I. Frantsevich, 1989. *Fauna and Ecology of Dragonflies*. Nauka, Siberian Division, Novosibirsk, 206 pp. – [Russ.]

Boev, V.G., Bayanov M.G. & Karataeva N.L. 1989. Materialy k kadastru fauny strekoz Bashkirskoi ASSR [Data to cadastre of fauna of dragonflies of Bashkirian ASSR]. Abstr. of papers of All-Union konf. on problems of cadastre and register of animal world. Part 4. Ufa: 9-11.

Chaplina, I.A., 2004. Fauna i ekologiya strekoz Kazakhstana [Fauna and ecology of dragonflies of Kazakhstan]. Dissertation for the scientific degree of candidate of biological sciences (Ph. D. Thesis). Novosibirsk, 257 pp. – [Russ.]

Dronzikova, M.V., 2000. Strekozy basseina reki Tomi [Dragonflies of the basin of Tom' river]. Dissertation for the scientific degree of candidate of biological sciences (Ph. D. Thesis). Novosibirsk, 244 pp. – [Russ.]

Gloyd, L.K., 1941. The identity of three geographically misplaced species of Odonata. *Bull. Chicago Acad. Sci.* 6: 130-132.

Hagen, H., 1856a. Odonates de la Russie. Etud. entomol. 5: 52-59.

Hagen, H., 1856b. Die Odonaten-Fauna des russischen Reichs. Stett. Entom. Z. 17: 363-381.

Hagen, H., 1858. Beitrag zur Odonaten-Fauna des russischen Reichs. Stett. Entom. Z. 19: 96-101.

Haritonov, A. Yu. 1972. Rasprostranenie strekoz na territorii Chelyabinskoi oblasti [Distribution of dragonflies on territory of Chelyabinsk district]. Abstr. of papers of scienc. konf. of Chelyabinsk division USSR Geographic Society, Chelyabinsk: 51-52. – [Russ.]

Haritonov, A. Yu. 1974. Zayeniseiskie strekozy na Polyarnom Urale [Trans Yenisei dragonflies at the Polar Ural]. In: Voprosy entomologii Sibiri [Problems of entomology of Siberia]. Nauka, Novosibirsk: 68-69.

Haritonov, A. Yu. 1975 a. Strekozy Urala i Zaural'ya (fauna, ekologiya, zoogeografiya) [Dragonflies of Ural and Transuralia (Fauna, Ecology, Zoogeography)]. Ph. D. Thesis, Novosibirsk, 205 pp. – [Russ.]

Haritonov, A. Yu. 1975 b. Strekozy Il'menskogo zapovednika [Dragonflies of Il'menskii Nature Reserve]. In: Voprosy zoologii, vyp.2 [Problems of zoology, issue 2]. Chelyabinsk: 63-65. – [Russ.]

Haritonov, A. Yu. 1976. Fauna strekoz (Insecta, Odonata) Urala i Vostochnogo Priural'ya [The fauna of dragonflies of Ural and Eastern Priural]. In: Fauna gel'mintov i chlenistonogikh Sibiri. Trudy Biologicheskogo instituta SO AN SSSR, vyp. 18 [Fauna of Helmints and Artropods of Siberia. Proc. Biol. Inst. SB AS USSR, issue 18]. Nauka, Novosibirsk: 157-161. – [Russ.]

Haritonov, A. Yu. 1978. Zoogeographicheskoe raionirovanie Vostochnogo Priural'ya na osnovanii rasprostraneniya strekoz [Zoogeographic division of Eastern Priural based on dragonflies distribution]. In: Chlenistonogie Sibiri [Artropods of Siberia]. Nauka, Novosibirsk: 47-54. – [Russ.]

Haritonov, A. Y. 1986. Otryad Odonata – Strekozy [Order Odonata – Dragonflies]. In: Opredelitel' nasekomykh Dal'nego Vostoka SSSR v shesti tomakh. Tom. I. Pervichnobeskryye, drevnecrylye s nepolnym prevrashcheniem [Guide for the Insects of the Far East of the USSR, in six volumes. Volume I. Apterygota, Archaeoptera, Hemimetabola]. Dr. P. A. Ler, ed. Nauka, Leningrad, p. 142-162. – [Russ.]

Haritonov, A. Y., 1990. Boreal'naya odonatofauna i ekologicheskie faktory rasprostraneniya strekoz [The Boreal odonatofauna and the ecological factors of distributions of dragonflies]. Dissertation for the scientific degree of doctor of biological sciences [D. Sci. Thesis], Novosibirsk, 531.

Haritonov, A. Y. & E. I. Malikova, 1998. Odonata of the Russian Far East: a summary. Odonatologica 27: 375-381.

Ketenchiev, KH.A. & Haritonov, A. Y., 1999. Strekozy Sredizemn-omor'ya [The dragonflies of Mediterranea], Nal'chik, 116 pp. – [Russ.]

Kolosov, Yu. M., 1914. Zametki o nasekomykh Tobol'skoi gubernii [Notes on insects of Tobol province]. In: Zapiski Ural'skogo obshchestva lubitelei estestvoznaniya. T. 34 [Notes of Ural society of lovers of natural history. T. 34]. Ekaterinburg: 13-36. – [Russ.]

Kolosov, Yu. M., 1916a. Materialy k poznaniu entomofauny Srednego Urala. V. Somatochlora arctica Zett. v Permskoi gubernii [Materials to study of insects faune of Perm' province]. In: Zapiski Ural'skogo obshchestva lubitelei estestvoznaniya. T. 36 [Notes of Ural society of lovers of natural history. T. 36]. Ekaterinburg: 71-73. – [Russ.]

Kolosov, Yu. M., 1916b. Materialy k poznaniu entomofauny Srednego Urala. VI. Aeschna cyanea Mull. v Permskoi gubernii [Materials to study of insects faune of Perm' province]. In: Zapiski Ural'skogo obshchestva lubitelei estestvoznaniya. T. 36 [Notes of Ural society of lovers of natural history. T. 36]. Ekaterinburg: 73-78. – [Russ.]

Kolosov, Yu. M., 1927. Zametki o strekozakh Chelyabinskogo okruga [Notes on dragonflies of Chelyabinsk district]. In: Sbornik materialov po izucheniyu Chelyabinskogo okruga, vyp. 1 [Collection of materials on study of Chelyabinsk district, issue 1]. Chelyabinsk: 7-13. – [Russ.]

Kosterin, O.E., 1992. New distribution records *of Somatochlora sahlbergi* Trybom (Odonata, Corduliidae). Acta Hydroentomologica Latvica 2:22-26.

Kosterin, O.E., 1996. Dragonflies (Odonata) of the city of Omsk. Acta hydroent. Latvica 3: 10-21.

Kosterin, O.E., 1999. Fauna of dragonflies (Odonata) of the Daurskii State Nature Reserve and its surroundings. In: Nasekomye Daurii i sopredel'nykh territorii (Sbornik nauchnykh trudov). Vypusk II [Insects of Dauria and adjacent territories (a collected paper edition). Issue II]. Novosibirsk: 5-40. – [Russ.]

Kosterin, O. E., 2004. Odonata of the Daurskii State Nature Reserve area, Transbaikalia, Russia. *Odonatologica* 33: 41-71.

Malikova, E. I., 1995. Strekozy (Odonata, Insecta) Dal'nego Vostoka Rossii [Dragonflies (Odonata, Insecta) of the Far East of Russia]. Dissertation for the scientific degree of candidate of biological sciences (Ph.D. Thesis). Novosibirsk, 233 pp. – [Russ.]

Perepelov, E.A., 2003. Evoluciya kariotipov strekoz (Odonata, Insecta) Severnoi Palearktiki [The evolution of karyotypes of dragonflies (Odonata, Insecta) of the North Palaearctic region]. Dissertation for the scientific degree of candidate of biological sciences (Ph.D. Thesis). Novosibirsk, 144 pp. – [Russ.]

Popova, O.N., 1999. Strekozy roda Sympetrum [Dragonflies of the Sympetrum genus]. Dissertation for the scientific degree of candidate of biological sciences (Ph. D. Thesis). Novosibirsk, 260 pp. – [Russ.]

Selys-Longchamps, E.DE., 1872. Materiaux pour une faune Nevropterologique de l'Asie septentrionale. Premiere partie. Odonates. Annls. Soc. ent. Belg. 15: 25-45.

Selys-Longchamps, E.DE., 1887. Odonates de l' Asie Mineure, et revision de ceux des autres parties de faune palearctique (dite europeenne). Annls. Soc. ent. Belg. 31: 1-85.

Sukhacheva (Smirnova), G.A., 1989. Strekozy Zapadno-Sibirskoi lesostepi i ih trophicheskie svyazi [The dragonflies of West-Siberian forest-steppe and their trophic connections]. Dissertation for the scientific degree of candidate of biological sciences (Ph.D. Thesis). Novosibirsk, 204 pp. – [Russ.]

Trybom, F., 1889. Trollslander (Odonater) insamlade under Svenska expeditionen till Jenisei, 1876. Bih. Svenska Vetensk. Akad. Handl. 15 (4): 1-21.

Yurchenko, Y.A., 2004. Sistematika i ekologiya palearkticheskikh strekoz roda *Enallagma* [Systematics and ecology of palaearctic dragonflies of *Enallagma* genus]. Dissertation for the scientific degree of candidate of biological sciences (Ph. D. Thesis). Novosibirsk, 246 pp. – [Russ.]

Vnukovskii, V.V., 1928. K faune strekoz (Odonata) Zapadnoi Sibiri [On odonate fauna (Odonata) of Western Siberia]. Trudy Sibirskogo instituta sel'skogo khozyaistva i lesovodstva. T. 10 [Ann. Siberian Institute agriculture and forestry. T. 10] Omsk: 3-9. – [Russ.]

Yanybaeva, V.A., 2001. Fauna i ekologiya strekoz Uzhnogo Urala [Fauna and ecology of dragonflies of South Ural]. Dissertation for the scientific degree of candidate of biological sciences (Ph. D. Thesis). Novosibirsk, 221 pp. – [Russ.]

Odonata : Biology of Dragonflies
Ed.: B.K. Tyagi
Scientific Publishers (India), 2007, pp. 89 - 104

8

THE DRAGONFLIES OF FOREST-STEPPE IN WEST SIBERIA : FAUNA, ECOLOGY AND BIOLOGY

O.N. POPOVA

Laboratory of insect ecology, Institute of Animal Systematics & Ecology, Siberian Branch, Russian Academy of Sciences, 630091, Frunze, 11, Novosibirsk, Russia
e.mail: pc@eco.nsc.ru

ABSTRACT

Results of studying fauna, zoogeography, ecology and biology of Odonata in forest-steppe zone of West Siberia are discussed. In forest-steppe 68 species of dragonflies have been found, which is the richest in species landscape in Siberia. The highest species diversity is noted in the remote parts – the Urals and Prisalair ones. The number of species common for the whole area is 44. Stationary study (in the Baraba part of West Siberian forest-steppe) of local dragonfly population made possible to obtain data on its species composition, number, biotopical distribution and structure of larvae and imago dominance and to ascertain life cycles of several mass species.

INTRODUCTION

The study of Odonata in the forest-steppe zone of southern part of West Siberia has been carried out for 30 years (the author has been participating in it from 1992 to 2006). The study has been performed in two main directions.

The first direction concerns stationary study of seasonal and interyear dynamics of the composition and structure of local dragonfly populations and of biology and ecology of mass species. In the period of the last 5 years a complex approach has been undertaken when equal attention was paid both to larval and imago phases, which has not been common practice before. Such annual ecological examination is performed on the territory of Barabinsk part of West Siberian forest-steppe, in Chany lake-river system on the base of the station of Institute of Animal Systematics and Ecology Siberian Branch Russian Academy of Sciences, set up in the point of confluence of the Chulym and Kargat rivers into Small Chany lake (lat. 54° 36' 55"N; lon. 78° 13' 03"E; 120 m a.s.l.).

The second direction refers to performing radial trips, regarding the base, throughout West Siberian forest-steppe to collect fauna material for the study of species composition and Odonata fauna structure. The abundant material collected in different places of WS forest steppe made it possible to carry out fauna-geographical analyses of the area examined (Popova & Haritonov, 2004). The data acquired during the following 2 years considerably supplemented and broadened that analyses: the species composition, distribution and the number of separate species have been defined more accurately.

MATERIAL AND METHODS

While studying the dragonfly population attention was paid equally to both larvae and imago.

The composition and structure of *Zygoptera* imago population were studied by means of 2 basic count methods: (1) timed catching with an air net, and (2) "mowing" with an entomological net. At timed catching (15 minutes each) all flying and sitting damselflies were being caught within a site of 200 m^2. "Mowing" was performed in four recurrences of 25 net sweeps each (100 net sweeps one count). When making these counts the following weather factors were registered: the ambient air and water temperature, cloudiness, the direction and strength of the wind. After the specimen's species, sex and age have been defined, most of them were let free.

The composition and structure of *Anisoptera* imago population were studied by means of 3 basic count methods: (1) visual counting of dragonflies on a route of random length and the band 2 m wide, (2) catching of 100 specimens on a 500 m^2 site, and (3) capture-recapture method when 300 specimens were marked on a 500 m^2 site and the next day about the same quantity of specimens were caught in similar weather.

On the water bodies nearest to the count sites dragonfly larvae were collected. Hydrobiological samples were collected with a water biocenometre and a water net. Biocenometre is a metal tube of square section 25 x 25 cm and 70-90 cm high. A water net was used to "mow"-drag by the net through water plants; the length and depth of the run, the distance from the bank, the diameter of the net were taken into account.

Stationary work was performed at Chany lake-river system at the mouth of the Kargat river, on Small Chany and Fadikha lakes. An average field season corresponds to 100 km of rout count (made on foot), about 100 counts of imago and 500 hydrobiological samples.

Dragonfly larvae were found both in lake and river biotopes, they occurred in about 85 % of the samples. Other hydrobiontes were taken from samples, but not zooplankton. In the aggregate dragonfly larvae comprise about 17 % of the number of registered hydrobiontes and as to mass – 34 %.

FEATURES OF THE TERRITORY EXAMINED

West Siberia covers the middle part of Russia and is the largest plain of the planet, framed by Altai-Sayany mountain province on the south-east and the Urals on the west. West Siberia lowland is second to none as far as water supplies and swamped surface are concerned. Its important hydrobiological feature is slow surface run-off and weak natural drainage of subsoil waters which caused the appearance of numerous swamps and lakes. The total number of lakes there amounts to several hundreds of thousands (Rikhter, 1957). The river system belongs mainly to the Ob' river basin.

In West Siberia the spread of warm and moisture distinctly correspond to the zones, which consequently defines the latitudious type of natural zones sequencing: tundra, forest-tundra, forest zone, forest-steppe and steppe.

The forest-steppe zone of West Siberia presents in itself a wide band stretching between taiga and steppe from the Urals to the Altai and Kuznetsk Ala Tau. Its length is about 1500 km and average width is 270 km (from 150 to 400 km), with the total area – 500,000 km^2. As to absolute marks, the zone is not uniform. The parts close to the Urals and to the Ob' are higher and have better drainage. The swamped territory of the forest-steppe lies in its central lower part and comprises 100,000 km^2, i.e. 20% of the whole area.

West Siberian forest-steppe (WS f-s) has the following distinctive features: (1) presence of two main relief forms – "*griva*" (low stretched hill) and *lowland* (hollow, fall – "zapadina"), (2) corresponding presence of flora associations of two basic types – steppish meadows on the "griva" and reedbeds in lowland, (3) island-type of aspen-birch groves, (4) strong water supplied and swamped surface, (5) heavy saline soils, and (6) temperate-continental climate.

Forest-steppe zone has lakes of the following types: (1) flood-lands lakes, (2) erosion lakes of ancient drain bottom, mostly salty, typical for Kulunda and Baraba, (3) suffusion lakes – of a rounded form and little surface (up to 2 km^2) and depth (up to 4 m); the banks are low and sloping here and covered with mire vegetation in some parts; no run off, in arid years many of these lakes dry up; the level of mineralization is high, and (4) end lakes are the largest ones (Chany, Kulundinskoye, Ubinscoye, Sartlan); shallow with silt-covered floor, sloping shores grown with thick reeds; no run off; slightly mineralized.

Barabinsk forest-steppe (Baraba) is the area of long-term stationary examination. Baraba is the country between the Ob' and Irtysh rivers, sretching between 54 and 57 degrees North within most part of the Novosibirsk region and eastern part of the Omsk region. As to the relief, the eastern and western parts of Baraba are more raised, while the central part is lowered (100-200 m above the sea level) and that is where Chany's and Sartlan's basins and the middle stretch of the Om' river lie. Most part of Baraba forest-steppe is swamped and salinized and is a reedbed-meadow-saline forest-steppe.

Subsoil waters come close to the surface and are often saline (1-12 g/l). Sites located nearby the rivers have fresh water. There the Om' river and its tributaries flow as well as the Kargat, Chulym, Bagan and Karasuk rivers. River valleys are mildly outlined, the depth of there cut is small (10-15 m), that is why in spring their waters interflow in some places. The current is slow (0,2-0,5 m/second in its lowest period and 0,6-1,0 m/second in flood-time). In especially arid years they dry up. Baraba lakes are mostly shallow (2-10 m) and small (up to 5 km^2). The lakes of the most northern part of Baraba have fresh water and of the most southern part - mineralized water (mostly soda, sulfate and chloride). Big lakes (Chany with its surface of 3,300 km^2, Tandovo, Sartlan) are less saline in comparison with little ones.

The climate of Baraba forest-steppe is predetermined by its location in east of West Siberian plain and the depression where in winter cold air stagnation occurs. Within the central depression in

Baraba with a large Chany lake (so called Chany depression) the transition of the temperature over 0° is in the beginning of the third decade of April and in the second decade of October; over 15° - in the first decade of June and in the third decade of August. Vegetation period lasts 150-155 days. From June, 1 late frost stop on the most part of Baraba and return on September, 10-15. The frost-free period is 110-120 days. In the area of Chany lake and in the Karasuk depression it lasts 80 days.

The average temperature in July is 18-19°, while the absolute maximum is 38°. The total sum of temperatures during the period of above 10° amounts to 1800-2100°. The precipitation total during summer period is 150-200 mm. Autumn is dry, but frost occurs.

Winter lasts from the beginning of November to the end of March. It is usually cold with the average temperature in January –20°, and the absolute minimum temperature is –54°. Snow cover appears in the first decade of November and disappears in the middle of April; it lies for 150-165 days. Snow cover is 30-50 cm thick. In the beginning of winter and in the transient spring season strong winds often blow. In March the average wind speed is 6.3 m/sec. In the beginning of summer it reaches more than 15-20 m/sec. Snowstorms are frequent in winter. In spring snow in Baraba melts away very fast (during 7-8 days).

RESULTS

Odonata fauna of forest-steppe in west siberia

According to division into districts of Northern Asia on the grounds of dragonfly spread, WS f-s is included into forest-steppe province of European-Siberian subregion of Golarctic region of Boreal faunistic kingdom (Belyshev & Haritonov, 1981). Taking into consideration the practice of odonatogeographical division of land, landscape peculiarities and our own recent foundings we isolated three subprovinces within WS f-s province: Urals, Ishim-Baraba and Prisalair. Ishim-Baraba subprovince might be divided into two sites: Ishim (the Ishim river flows almost through the middle of it) and Baraba (corresponds to Baraba forest-steppe); their differences as to the dragonfly fauna are insignificant – on the Ishim site *Calopteryx virgo* and *Libellula depressa* were caught, which do not populate Baraba (P. Zhakkar factor – 96%). We allotted right bank areas along Ob' river including Kuznetsk hollow into Prisalair subprovince. As to the landscape Ishim-Baraba subprovince is typical plain forest-steppe and Urals and Prisalair subprovinces – typical foothills forest-steppes.

The total number of dragonfly species in WS f-s is 68, being indicative of its fauna richness (Table 1). To make the comparison the following data can be presented: Russia – 148 (Haritonov, 1997), Siberia (except Far East) – 74 (Haritonov, 2002), the Caucasus – 83 (Popova, 1997), the Southern Urals – 64 (O.N. Popova & A. Yu. Haritonov, *unpublished*). Compared to other natural zones of Siberia WS f-s is the richest in the dragonfly species. This fact testifies to the opinion, which is not popular yet, that forest-steppe enjoys the role of a centre of species abundance and faunagenesis in West Siberia. Widely accepted opinion holds that only forest and steppe are such centres, while forest-steppe is merely a site where forest and steppe faunas are mechanically mixed up.

In the area of WS forest-steppe province *Zygoptera* suborder amounts to 25 species and *Anisoptera* – 43. The most representative as to species genera (up to 5-7 species) are *Coenagrion, Lestes, Aeshna, Somatochlora, Leucorrhinia, Sympetrum.*

When odonata faunas of different subprovinces were compared in pairs, Urals and Prisalair ones appeared to be the least similar (P. Zhakkar factor – 65%), though they refer to the same landscape – foothills forest-steppe. Odonata faunas of Urals and Ishim-Baraba subprovinces provide to be more alike (P. Zhakkar factor – 83 %), though they refer to different landscapes – foothills and plain forest-steppes. Prisalair subprovince odonata fauna appeared to be closer to Ishim-Baraba one (plain forest-steppe) (P. Zhakkar factor – 75 %).

To make a more detailed analyses of these similarity and divergence of dragonfly faunas their structure was examined taking into account such feature as abundance. Three levels of species abundance were esteemed on the bases of integral expert evaluation: mass, common and rare ones.

Table 1. List of dragonfly species of Asian part of forest-steppe province (West Siberia forest-steppe)

No.	SPECIES	Subprovinces			Abundance of species in province as a whole
		Urals	Ishim-Baraba	Prisalair	
	1	2	3	4	5
	Suborder Zygoptera				
1	*Calopteryx japonica* Selys, 1869	-	-	+	C
2	*C. splendens* (Harris, 1782)	+	+	+	M
3	*C. virgo* (L., 1758)	+	+	+	R

4	*Coenagrion armatum* (Charpentier, 1840)	+	+	+	M
5	*C. ecornutum* (Selys, 1872)	+	-	+	R
6	*C. hastulatum* (Charpentier, 1825)	+	+	+	M
7	*C. johanssoni* (Wallengren, 1894)	+	+	+	R
8	*C. lanceolatum* (Selys, 1872)	-	-	+	R
9	*C. puella* (L., 1758)	+	+	+	C
10	*C. pulchellum* (Vander Linden, 1823)	+	+	+	M
11	*C. lunulatum* (Charpentier, 1840)	+	+	+	M
12	*Enallagma cyathigerum* (Charpentier, 1840)	+	+	+	M
13	*Erythromma najas* (Hansemann, 1823)	+	+	+	C
14	*Er. viridulum* (Charpentier, 1840)	+	-	-	R
15	*Ischnura aralensis* Haritonov, 1979	+	-	-	R
16	*I. elegans* (Vander Linden, 1820)	+	+	+	R
17	*I. pumilio* (Charpentier, 1825)	-	-	+	R
18	*Nehalennia speciosa* (Charpentier, 1840)	+	+	+	R
19	*Lestes barbarus* (Fabricius, 1798)	+	+	+	C
20	*L. dryas* Kirby, 1890	+	+	+	M
21	*L. macrostigma* (Eversmann, 1836)	+	+	+	R
22	*L. sponsa* (Hansemann, 1823)	+	+	+	M
23	*L. virens* (Charpentier, 1825)	+	+	+	C
24	*Sympecma paedisca* (Brauer, 1882)	+	+	+	M
25	*Platycnemis pennipes* (Pallas, 1771)	+	+	+	C
	Suborder Anisoptera				
26	*Aeshna affinis* Vander Linden, 1820	+	+	+	C
27	*Ae. caerulea* (Stroem, 1783)	-	-	+	R
28	*Ae. crenata* Hagen, 1856	+	+	+	C
29	*Ae. cyanea* (Mueller, 1764)	+	-	-	R
31	*Ae. juncea* (L., 1758)	+	+	+	C
32	*Ae. mixta* Latreille, 1805	+	+	+	C
33	*Ae. serrata* Hagen, 1856	+	+	+	C
34	*Ae. subarctica* Walker, 1908	-	-	+	R
35	*Ae. viridis* Eversmann, 1836	+	+	+	C
36	*Anax imperator* Leach, 1815	+	-	-	R
37	*An. parthenope* Selys, 1839	+	-	+	R
38	*Anaciaschna isoceles* Mueller, 1767	+	-	-	R
39	*Brachytron pratense* (Mueller, 1764)	+	-	-	R
40	*Cordulia aenea* (L., 1758)	+	+	+	M
41	*Epitheca bimaculata* (Charpentier, 1825)	+	+	+	R
42	*Macromia amphigena* Selys, 1871	-	-	+	R
43	*Somatochlora alpestris* (Selys, 1840)	+	-	+	R
44	*S. arctica* (Zetterstedt, 1840)	-	+	+	R
45	*S. flavomaculata* (Vander Linden, 1825)	+	+	+	R

46	*S. metallica* (Vander Linden, 1825)	+	+	+	M
47	*Gomphus vulgatissimus* (L., 1758)	-	+	+	R
48	*Nihonogomphus ruptus* (Selys, 1858)	-	-	+	R
49	*Onychogomphus forcipatus* (L., 1758)	+	-	-	R
50	*Ophiogomphus cecilia* (Fourcroy, 1785)	+	+	+	R
51	*Oph. obscurus* Bartenef, 1909	-	-	+	R
52	*Shaogomphus postocularis epophthalmus* Selys, 1872	-	-	+	C
53	*Stylurus flavipes* (Charpentier, 1825)	+	+	+	C
54	*Leucorrhinia albifrons* (Burmeister, 1839)	+	+	+	R
55	*L. caudalis* (Charpentier, 1840)	+	+	+	R
56	*L. dubia* (Vander Linden, 1825)	+	+	+	C
57	*L. orientalis* Selys, 1887	-	-	+	R
58	*L. pectoralis* (Charpentier, 1825)	+	+	+	C
59	*L. rubicunda* (L., 1758)	+	+	+	M
60	*Libellula depressa* L., 1758	+	+	-	R
61	*L. quadrimaculata* L., 1758	+	+	+	M
62	*Orthetrum cancellatum* (L., 1758)	+	+	+	R
63	*Sympetrum croceolum* (Selys, 1883)	-	-	+	R
64	*S. danae* (Sulzer, 1776)	+	+	+	M
65	*S. flaveolum* (L., 1758)	+	+	+	M
66	*S. pedemontanum* (Mueller, 1766)	+	+	+	C
67	*S. sanguineum* (Mueller, 1764)	+	+	+	M
68	*S. vulgatum* (L., 1758)	+	+	+	M
	68	55	47	60	

+ presence of species, – absence of species; M – mass species, C – common species, R – rare species.

Out of 68 species of WS f-s province mass ones amount to 18 (26% of the total number), common – 17 (25%), rare – 33 (49%); that is the number of mass and common species taken together is commensurable with the number of rare ones. We registered 55 species in Urals subprovince (18 mass, 15 common, 22 rare), in Ishim-Baraba subprovince – 47 (18, 15, 14, correspondingly), in Prisalair subprovince – 60 (18,17,25, correspondingly). The number of the species common for the three subprovinces is 44 (65% of the total number of the species) which testifies to a significant likeness of their odonata faunas.

The highest species diversity is noted in the boundary subprovinces – the Urals and Prisalair ones. Only in Urals subprovince 7 species inhabit and all of them are rare: *Ischnura aralensis, Erythromma viridulum, Aeshna cyanea, Anax imperator, Anaciaschna isosceles, Brachytron pratense, Onychogomphus forcipatus*. Only in Prisalair subprovince 11 species inhabit and all of them are rare also: *Calopteryx japonica, Coenagrion lanceolatum, Ischnura pumilio,*

Aeshna caerulea, Ae. subarctica, Macromia amphi-gena, Nichonogo-mphus ruptus, Ophiogomphus obscurus, Shaogo-mphus postocularis, Leucorrhinia orientalis, Sympetrum croceolum. There no species specific for only Ishim-Baraba subprovince.

Thus there are good reasons to refer to WS f-s odonata fauna as a relatively uniform one and ecological data on Baraba forest-steppe can be extrapolated to the most part of the WS f-s area as a whole.

LOCAL ODONATA POPULATION

Annual stationary ecological research is carried out in Baraba forest-steppe, namely on the expeditionary base of Institute of Animal Systematics and Ecology Siberian Branch Russian Academy of Sciences, set up in the point of confluence of the Chulym and Kargat rivers into Small Chany lake and near Fadikha lake. The territory of the base belongs to Chany lake-river system and latter's is a part of so-called Chany-Baraba lake district. The latter's total area is 117,000 km^2 and the 2,500 lakes located there amount to total 4,900 km^2, that is 4,2% which is the richest in lakes area in the south of West Siberia. The Chany lake is the largest water body of WS f-s located in the central part of Baraba lowland. Its basin comprises several lakes: Small Chany, Large Chany, Yarkul', Uryum and Sargul'.

The district is characterized by so-called "*grivas*" (rows of low stretched hills divided by narrow valleys), rare birch-aspen small groves and shelter-belts, growing meadows of all types in the open space.

As a result of long-term work in Chany lake-river system the data were obtained on the seasonal and interyear dynamics of species composition, number, biotopical distribution and structure of dragonfly larvae and imago dominance.

IMAGO

Species composition and number

In the Chany basin 39 dragonfly species were found referring to 6 families (Calopterygidae, Coenagrionidae, Lestidae, Aeshnidae, Corduliidae, Libellulidae) and 16 genera (*Calopteryx, Coenagrion, Enallagma, Erythromma, Nehalennia, Lestes, Sympecma, Aeshna, Anax, Cordulia, Epitheca, Somatochlora, Leucorrhinia, Libellula, Orthetrum, Sympetrum*) (Popova, 2004). Sixteen species belong to suborder *Zygoptera*, 23 – to suborder *Anisoptera;* amongst which **Mass (dominant) species** are 15: *Coenagrion armatum* (Charpentier,

1840); *C.pulchellum* (Vander Linden, 1823); *C.lunulatum* (Charpentier, 1840); *Enallagma cyathigerum* (Charpentier, 1840); *Erythromma najas* (Hansemann, 1823); *Lestes sponsa* (Hansemann, 1823); *Sympecma paedisca* Brauer, 1839; *Aeshna mixta* Latreille, 1805; *Ae. serrata* Hagen, 1856; *Leucorrhinia pectoralis* (Charpentier, 1825); *L. rubicunda* (Linnaeus, 1758); *Libellula quadrimaculata* Linnaeus, 1758; *Sympetrum danae* (Sulzer, 1776, 1811); *S. flaveolum* (Linnaeus, 1758); *S. vulgatum* (Linnaeus, 1758); **Common species** have 7: *Lestes barbarus* (Fabricius, 1798); *L. dryas* Kirby, 1890; *L. virens* (Charpentier, 1825); *Aeshna grandis* (Linnaeus, 1758); *Cordulia aenea* (Linnaeus, 1758); *Somatochlora metallica* (Vander Linden, 1825); *Sympetrum sanguineum* (Mueller, 1764); **Rare species** have 11: *Coenagrion concinnum* Iohanssen, 1859; *C. hastulatum* (Charpentier, 1825); *C. puella* (Linnaeus, 1758); *Nehalennia speciosa* (Charpentier, 1840); *Lestes macrostigma* (Eversmann, 1836); *Aeshna affinis* Vander Linden, 1825; *Ae. juncea* (Linnaeus, 1758); *Ae. viridis* Eversmann, 1836; *Epitheca bimaculata* (Charpentier, 1825); *S. flavomaculata* (Vander Linden, 1825); *Leucorrhinia dubia* (Vander Linden, 1825); **Singular species** have 6: *Calopteryx splendens* (Harris,1782); *Aeshna crenata* Hagen, 1856; *Anax parthenope* Selys, 1839; *Somatochlora arctica* (Zetterstedt, 1840); *Orthetrum cancellatum* (Linnaeus, 1758); *Sympetrum pedemontanum* (Mueller, 1766).

Out of 6 singular species 5 were in imago phase and 1 – in larva phase (*Anax parthenope*). The last finding is especially peculiar. According to our counts and observations the distribution of imago in the air space is mainly spontaneous and only in some cases is aggregated (reproduction, overnight, evening flight). The distribution of larvae in water bodies, on contrary, is mainly aggregated. In such a case the probability of finding an imago is much higher than finding larvae, especially for rare and single species. Nevertheless in 2005 we found an *Anax parthenope* larva of the final age, while nobody either caught or saw an imago of this species for the period of 25 years of observation on the Chany part of Baraba forest-steppe.

Dominance structure in mass species

For all dominant species a very high level of number is noted which is provided by regular counts of imago carried out using various methods including the marking with repeated capture of insects: *Libellula quadrimaculata* up to 3.9 imago per 1 m^2 (the beginning of June), *Leucorrhinia* (*rubicunda* and *pectoralis*) – 5.6 (the beginning of June), *Sympetrum flaveolum* – 10-15 (the end of June – July), *S. vulgatum* – 8.5 (July – the beginning of August), *Aeshna mixta* – 1.4

(August); *Enallagma cyathigerum* – 3.4 (July). The total imago population density at optimal feeding stations can amount to 30 specimens per 1 m^2 being indicating of high biocenotic significance of these insects.

Dominant species are relatively unstable and they change one another during both flying season and from season to season. As an example species from *Leucorrhinia* and *Sympetrum* genera can be examined.

Thus in the season of 2006 at June, 5 we had the count – 1 *Leucorrhinia pectoralis* and 60 *L. rubicunda.* And at June, 13 we had the count – 1 *L. rubicunda* and 5 *L. pectoralis*, that is the change of dominance took place inside one dragonfly genus for one week only. According to long-term observations, as a rule, *L. rubicunda* dominates over *L. pectoralis* during their whole flying period (the 3rd decade of May – the 1st decade of July).

Inside the genus *Sympetrum* mass species are *S. flaveolum*, *S. vulgatum* and *S. danae.* The first to open the flight is *S. flaveolum* in the middle of June, the second flies *S. vulgatum* in the beginning of July, the third is *S. danae* from the middle of July, then they fly together and finish their seasonal flight almost simultaneously during the end of September – the beginning of October. For these species 2 dominance schemes are observed, changing each other from season to season. Thus according to one scheme during their flying period the change of their dominance takes place in turns, the sequence being defined by the time of their emerging and is of advancing nature. But such predetermined rotation does not take place in *Sympetrum* genus every season and to the other scheme the dominance of one species is occasionally observed through the whole flight time. The role of that dominant species is more often gained by *S. flaveolum*. The fact can probably be explained by its being everybiotic and everytopic, *i.e.*, its larvae populate water bodies of any type and imago – in any landscapes. As to other two species, *S. vulgatum* tends to fly in the open space – in the steppe, while *S. danae* prefers to fly in close space – in the forest and brushwood (Popova, 1999).

Seven dominant *Zygoptera* species form four distinct seasonal groups. The first, spring group of dominants comprises species hibernated in imago phase, *e.g.*, *Sympecma paedisca* and two species of *Coenagrion* genus - *lunulatum* and *armatum*, for which a mass highly synchronous emerging in the second half of May is typical. In the beginning of June in early summer *Coenagrion pulchellum* joins them. By the end of the first decade of July all these species practically disappear and are replaced by summer dominant group –

Enallagma cyathigerum, *Erythromma najas* and *Lestes sponsa*. In the first half of August new specimens of *S. paedisca* appears on the background of sharp decline in the number of the summer group, comprising a single-species hibernating autumn group.

LARVAE

Seasonal dynamics of Zygoptera number

22 dragonfly species were discovered on Fadikha lake (56% of 39 species), 10 of which belong to the Zygoptera suborder, 12 – to Anisoptera suborder. The Fig. 1 illustrates seasonal changes of number of larvae populations of 5 damselfly species in 2003 on this lake. Two tendencies are clearly traced in these changes. In some populations the change in the number of larvae is non-uniform, spasmodic and pulsed (e.g., *Enallagma cyathigerum* and *Erythromma najas*). In other populations this change is more advancing and regular (e.g., *Coenagrion armatum* and *C. lunulatum*). Each of these tendencies are probably linked with peculiarities of imago flight period. This link is clearly traced when seasonal changes in the age structure of larvae populations are analysed.

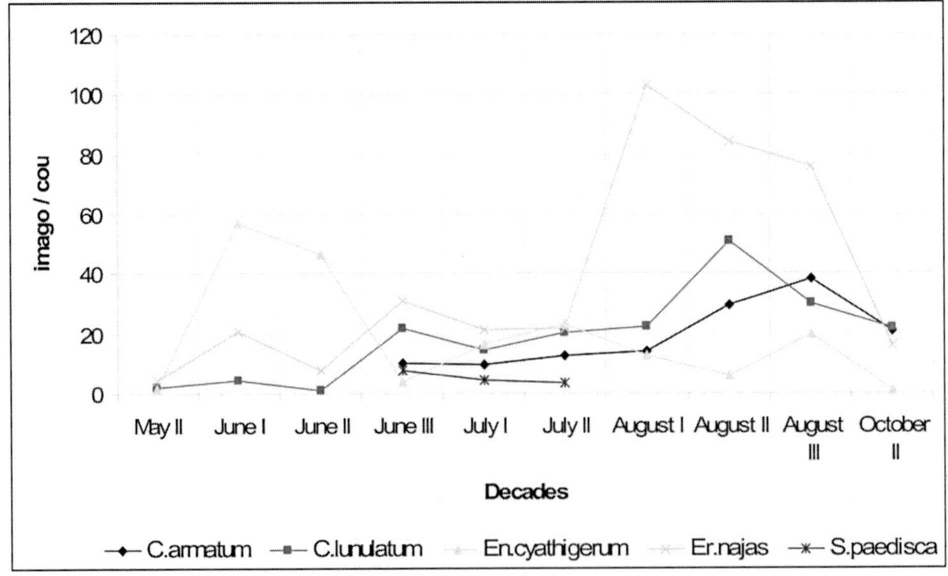

Fig. 1. Seasonal dynamics of number of damselfly larval populations on Fadikha lake.

C. lunulatum is an example of a regular change in the number of larva population. In the samples beginning from the middle of May up to the middle of June mainly older-age larvae are present, having synchronous mass emergence of imago, which takes place basically in the begining of June. Such emergence promotes a series of sequencing synchronous events. First of all the imago enter into reproductive period simultaneously, finishing it and dying likewise simultaneously. Correspondingly, all eggs, laid in water, develop more or less simultaneously and within one time interval (during June). Then mass hatching of first larvae *C. lunulatum* comes in the end of June. They feed, grow and by the middle of July middle-age larvae appear, and by the middle of August – older ages. Middle-age and older-age larvae hibernate.

Er. najas is an example of pulsed change in the number of larva population. From the middle of May up to the middle of June the samples contain larvae of different ages (young, middle, older), which keep emerging from water body gradually, in due time as far as they grow up without any peaks in imago emergence. Such uniform-stretched emergence promotes uniform prolonged laying the eggs by *Er. najas* as far as the end of their flight period in the end of July. All their eggs laid in water are correspondingly at different stages of their development which is displayed in the various age of the larvae as far as the beginning of hibernating period.

Life cycle of Zygoptera

For the first time life cycles of *Coenagrion armatum*, *C. pulchellum*, *C. lunulatum*, *Enallagma cyathigerum*, *Erythromma najas* were studied in the south of West Siberia. The larva phase in their development includes 11-12 ages, all species are monovoltine. A widely ranged and everybiotic species, *En. cyathigerum* is of special interest in this aspect. As a result of age analyses of hydrobiological samples it was found out that one-year long life cycle of *En. cyathigerum* is realized in two ways – with hibernation and without hibernation of eggs. *En. cyathigerum* emerging begins from the first days in June and lasts to the middle of July. The last meetings of imago take place in the end August – the beginning of September. Reproductive behaviour is observed throughout the whole summer period. Eggs laid by females in August evidently hibernate, because the samples taken in May contain only middle and older-age larvae but not younger ones. In the end of May – the beginning of June mass hatching from hibernated eggs takes place (which were laid in water bodies in August of the previous summer). These larvae approach final stage in their development and emerge in small quantities by the middle of July (in average 1 larva per 1 m^2). Single imago of *En. cyathigerum* flying in the end of July – in August belong to this very generation of larvae (who hibernated in eggs).From eggs laid in June – the beginning of July larvae hatch in the same summer season. By autumn larvae grow to the 7-9 ages (the middle age group) and hibernate in these ages. Next summer these larvae give mass emergence *En. cyathigerum* imago in the first-second decades of June (in average 61.7 larvae per 1 m^2. Up to the middle summer the multi-age larva population exists in water body, which hatched from both hibernated and "fresh" eggs. From the middle of summer up to its end the young and middle-age larva population exists in water body, which mostly hatched from "fresh" eggs. *En. cyathigerum* is the only species flying on the examined territory all the summer through. All above mentioned peculiarities of this species' ontogenesis appear to explain its circumboreal spread, everybiontic life style, high morphological variability and important biocenotic role, correspondingly.

TOPICAL GROUPS IN ZYGOPTERA POPULATIONS

Another interesting fact from the life of West Siberian dragonflies is connected with peculiarity of forest-steppe water bodies. Thus, there is an opinion that just after emergence young imagos fly to the land – to biotopes close to water bodies, where they feed, find shelter from predators and bad weather, overnight and having attained maturity begin to fly to their "native" water body to breed there. This scheme of dragonfly shifts in the space is characteristic of most water bodies where these insects live: river water-meadows, shallow lakes, swamps, ponds. The near-water zone of these water bodies can be grown by macrophytes (sedge, reed, mace reed, club-rush); but macrophytes do not grow into impassable thickets and are rather thinned out.

According to the type of vegetation in the near-water zone there are other water bodies typical for forest-steppe landscapes, where macrophytes, first of all reeds, make impassable thickets stretched into the water for several tens and even hundreds meters. These reedbeds are typical of Baraba forest-steppe and give it uniqueness.

On water bodies growing by reedbeds another scheme of dragonfly shifts is formed, observed by the author on Fadikha, Small Chany lakes and Kargat river. To analyze this phenomenon examination of Fadikha lake is enough. Massive reedbeds (further referred to as *reeds*) stretches into the water for 60-80 m, then there is "plyos" with wide of 100-200 m. "Plyos" are open water parts with small islands of *thin reeds* ("*redelya*"). Six Zygoptera species were found there: *Coenagrion armatum, C. pulchellum, C. lunulatum, Enallagma cyathigerum, Erythromma najas* and *Sympecma paedisca*. Larvae inhabits both "plyos" and reeds. Besides dragonfly larvae a complex of hydrobiontes was found in the samples taken in the reeds which resembles that of the lake "plyos": lower crustacean, flies, water bugs, ephemerae, water beetles, shellfish, leeches and water ticks. Almost all of them are food for dragonfly larvae. Thus reeds are an independent hydrobiocenosis and serve as a natural barrier between two biotopes – a water one ("plyos") and a land one (water-meadow). This peculiarity in the structural organization of water – near-water space leads to intrapopulation differentiation.

The emerging of Zygoptera starts usually at the end of the first decade of May when the average water temperature in the daytime is not lower than 12-13°C. During the whole season the water in "plyos" is on average 5°C higher than in reeds. By the moment of emergence the following larva age structure is observed in the lake: in the reeds – all ages, and in the "plyos" – only middle and older, no younger ones. The first to emerge are Zygoptera from "plyos" – spring *Coenagrion armatum* and *C. lunulatum*. The emerging young imago spend both days and nights on "plyos" and do not fly to the shore. In a few days, having attained reproductive age they start breeding there, on "plyos". The proof of this fact is total absence of Zygoptera imago in land, near-water biotopes. By the end of the second decade of May *C. armatum* and *C. lunulatum* begin to emerge in the reeds and fly to the shore. This emergence coincides with the beginning of the first young specimens regisrations in daily land counts. Mating period is spent on land where in comparison with the situation in the beginning of May there is enough food and places to shelter and spend nights. To breed they fly to the place where they emerged – to the reeds. Finally for the spring examined species (*C. armatum* and *C. lunulatum*) it is correct to divide their populations into two topical groups ("plyos" and reeds ones) having their own peculiarities in emergence and migrations caused by not typical space-structured organization of water ecosystem.

ACKNOWLEDGEMENTS

The author wishes to express her heartfelt gratitude to her husband, Dr A.Yu. Haritonov (Institute of Animal Systematics and Ecology SB RAS, Novosibirsk, Russia) for working discussions and valuable advices.

The research was supported by the Grants from Russian Foundation of Fundamental Investigations (04-04-48778a & 06-04-63088).

REFERENCES

Belyshev, B.F. & A.Yu. Haritonov, 1981. *Geografiya strekoz (Odonata) Boreal'nogo faunisticheskogo tsarstva.* Nauka, Novosibirsk. – [Russ.]

Haritonov, A., 1997. *Opredelitel' presnovodnykh bespozvonochnykh Rossii.* Tom 3. S. 221-246, 330-363: Otryad Odonata – Strekozy. ZIN RAN, Sankt-Peterburg. – [Russ.]

Haritonov, A., 2002. The dragonfly fauna of the Asian part of the former USSR. *Abstr. Papers I Symposium of the SIOROEA (Daejeon, Korea, July 26-29. 2002).* P. 23-24. Daejeon. – [Engl.]

Popova O.N., 1997. Dragonflies of the Caucasus. *Abstr. papers XIV International Symposium of Odonatology (Maribor, Slovenia, July 12-23, 1997).* P. 31. Maribor, 1997. – [Engl.]

Popova, O.N., 1999. *Strekozy roda Sympetrum.* Ph.D. thesis. Novosibirsk. – [Russ.]

Popova O.N., 2004. Struktura naseleniya strekoz (Insecta, Odonata) ozyorno-rechnoy sistemy. *Materialy mezhdunarodnoy sibirskoy zoologicheskoy konferentsii (15-22 sentyabrya 2004).* S. 69. Novosibirsk. – [Russ.]

Popova, O.N. & A.Yu. Haritonov, 2004. Fauna strekoz (Insecta, Odonata) zapadnosibir-skoy lesostepi. *Vestnik Tomskogo Gos. Universiteta* 11: 75-79. – [Russ.]

Rikhter G.D. (editor-in-chief), 1963. *Zapadnaya Sibir'.* AN SSSR, Moskva. 488 s. – [Russ.]

Odonata : Biology of Dragonflies
Ed.: B.K. Tyagi
Scientific Publishers (India), 2007, pp. 105 - 136

9

ODONATA OF MEXICO REVISITED

E. GONZÁLEZ SORIANO[1] & R. NOVELO GUTIÉRREZ[2]

[1]*Instituto de Biología, Departamento de Zoología, UNAM. Apartado Postal 70-153, C.P. 04510 México, D.F.*
e.mail: esoriano@ibiologia.unam.mx
[2]*Instituto de Ecología, A.C., Departamento de Entomología. Km 2.5 carr. Antigua a Coatepec # 351, Congregación El Haya, 91070 Xalapa, Veracruz.*
e.mail: rodolfo.novelo@inecol.edu.mx

(Dedication : This paper is dedicated to our mutual friend Dr. Bastiaan Kiauta on his 70th birth anniversary. Thanks a lot Bastiaan for your never ending friendship and support).

ABSTRACT

Odonata of Mexico, comprising one of the greatest biodiversity regions in the world, is discussed with as many as 19 species, including also one of the hitherto unrecorded genus *Ophiogomphus*, enlisted since 1996. A large number of taxonomic, geographic and other features associated with many of these species in different ecosystems are also elaborated.

INTRODUCTION

Since our last paper published in 1996 dealing with Mexican Odonata (González & Novelo, 1996), several new taxa have been added, new records have been published, some taxonomical re-arrangements have been made (e.g. Garrison 1994a, 1994b, 1996, 1999; Needham *et al.,* 2000; Westfall & May, 1996; Förster, 2001; von Ellenrieder, 2003, von Ellenrieder & Garrison, 2003) and specially many larvae have been described, keyed or both. In the 1996 paper we erroneously cited a total of 330 Mexican species, but in fact there were only 327 species.

In this chapter we update the current knowledge of the Odonata of Mexico until August 2006, considering both larval and adult stages. Although the taxonomical arrangement of the list is almost as the one published by Paulson & González (1994, last actualization in March 2006) some minor discrepancies with that paper are considered here.

NEW TAXA ADDED IN THE PERIOD, INCLUDING NEW COUNTRY RECORDS AND DESCRIPTION OF NEW LARVAL STAGES

Adults

A total of 19 new taxa have been added since 1996 (Table 1). It is important to notice that, some of the species listed here were in fact described as new in 1994. These taxa were not mentioned in our 1996 work because we indeed wrote such paper in 1994 but it was published until 1996. The genus *Argia* contributed with six (31.6%) of the new species described, followed by the genus *Erpetogomphus* with five (26.5%). The remaining genera contribute with only two or one species.

Table 1. New taxa for Mexico added since 1996

TAXA	REFERENCES
AESHNIDAE	
Coryphaeschna apeora Paulson 1994	Paulson, 1994
C. diapyra Paulson 1994	Paulson, 1994
Oplonaeschna magna González & Novelo 1998	González & Novelo, 1998
GOMPHIDAE	
Erpetogomphus agkistrodon 1994	Garrison, 1994
E. bothrops Garrison 1994	Garrison, 1994
E. erici Novelo 1999	Novelo & Garrison, 1999
E. heterodon 1994	Garrison, 1994
E. liopeltis 1994	Garrison, 1994

LIBELLULIDAE	
Brechmorhoga latialata González 1999	González, 1999
Elasmothemis aliciae González & Novelo 2006	González & Novelo, 2006

One of the highlights of the new species described was the discovery of a new species of *Ophiogomphus,* a previously unrecorded genus for Mexico (see below). Considering all the Mexican states, we had on average an increment in knowledge of 19.8% for each state, with a range of 2 to 54% (we did not consider the state of Tlaxcala, which had one species eliminated; the state of Aguascalientes with no increment at all; and Zacatecas which is the poorest known state with only one species recorded [see Appendix 1]). One important aspect that need to be outlined here is that in nine states (Chiapas, Hidalgo, Jalisco, Morelos, Nayarit, Oaxaca, SLP, Tamps. and Veracruz) there are more than 100 species (in the 1996 paper only four states surpassed this number). Veracruz still being the richest state with 220 species, followed by Chiapas with 170 species.

Table 2. Endemic species for Mexico. Those with an asterisk represent species with narrow distribution.

TAXA	DISTRIBUTIONAL NOTES
AMPHYPTERYGIDAE	
Amphypteryx longicaudata	Cloud forest, states of Hidalgo and Veracruz
LESTIDAE	
A. regalis	Cloud forest, states of SLP, Ver, Hgo, Qro.
*Lestes alfonsoi**	Cloud forest, state of Veracruz
MEGAPODAGRIONIDAE	
*Heteragrion azulum**	Rain forest at Los Tuxtlas region, state of Veracruz
*Paraphlebia hyalina**	Cloud forest, state of Veracruz
P. zoe	Cloud forest, states of SLP, Ver. and Hgo.
PLATYSTICTIDAE	
Palaemnema paulicaxa	Cloud forest, states of Puebla and Veracruz
P. paulicoba	Rain and cloud forests of Veracruz
P. paulitaba	Rain forests of states of Chiapas and Veracruz
PROTONEURIDAE	
Protoneura rojiza	Transition between pine-oak and tropical dry forests of states Oaxaca, Jalisco and Colima
COENAGRIONIDAE	
Argia deami	Mexican High-Plateu
A. carlcooki	Tropical dry and humid forests along Pacific coast

A. garrisoni	Humid forests of states of SLP, Tamps and NL
A. harknessi	Tropical dry forest along Pacific coast
A. percellulata	Tropical and cloud forests of states of Chiapas and Veracruz
A. westfalli	Tropical and subtropical forests, Chis,Ver, NL.
*Enallagma eiseni**	Baja California deserts
Ischnura posita atezca	Pine-oak forest state of Hidalgo
Neoerythromma gladiolatum	Ponds and lakes at lower altitudes along Pacific coast
*Telebasis incolumis**	Desert spring and oases in Baja California
AESHNIDAE	
*Rhionaeshna manni**	Desert spring and oases in Baja California
R. vazquezae	Streams in transition between temperate and tropical forests, states of Guerrero and Nayarit
Oplonaeschna magna	Small streams transition zone between template and tropical forests states of Edo. de Mex.and Morelos
GOMPHIDAE	
Epigomphus crepidus	West coast of Nayarit
*E.donnellyi**	Tropical rain forest, Los Tuxtlas, state of Veracruz
E. flinti *	Tropical rain forest, state of Oaxaca
E. paulsoni *	Tropical rain forest, state of Chiapas
*E. sulcatystila**	Tropical rain forest, state of Veracruz
*Erpetogomphus agkistrodon**	Cloud forest, state of Veracruz
E. boa	Cloud forest, states of Puebla and Veracruz
E. cophias	Cloud forests, states of Gro, Mich. and Mor
*E. erici**	Cloud forest, state of Veracruz
*E. liopeltis**	Cloud forests, states of Hgo, Mich, NL and SLP
E. sipedon	Tropical and temperate areas of Dgo, Jal. and Mor.
E. viperinus *	Cloud forest, state of Veracruz
*Ophiogomphus purepecha**	Pine forests Los Azufres, state of Michoacan
Phyllogomphoides apiculatus	West coast states of Nayarit andJalisco
P. danieli	Tropical deciduous forest state of Morelos
*P. indicatrix**	Tropical rain forest, state of Chiapas
P. luisi	Tropical deciduous forest states of Mor. and Jal.
P. nayaritensis	Tropical deciduous forests states of Nay. and Sin.
*Progomphus amarillus**	Tropical forest state of SLP
P. belyshevi	Tropical deciduous forest states of Mor. and Qro
Stylurus falcatus	Tropical dry forest states of Jal. and Michoacan?
LIBELLULIDAE	
Macrothemis ultima	Tropical and template forests states of Jal. and Mich.
Paltothemis cyanosoma	Transitional Pine-oak forest state of Jalisco

Six new records for Mexico have been accounted during this period. The platystictids, *Palaemnema angelina* and *P. paulina* both distributed in Central America. The protoneurid *Neoneura aaroni* was previously recorded from southern USA alongwith one species of Aeshnidae, *Aeshna constricta* from southern USA. The gomphids *Epigomphus clavatus* had been recorded for Guatemala while *Progomphus anomalus* had been recorded for Costa Rica and Panama (Förster, 2001). On the other hand in our 1996 list we omitted involuntarily *E. compositus* for México (previously cited as *E. coluber* considered now as a synonym [Garrison, 1994b]).

The species cited as *Argia fissa* for many authors for Mexico (see González & Novelo, 1996) resulted to be a new species: *A. anceps* Garrison. According to Garrison (1996) the true *A. fissa* is restricted to southern Central America and South America. On the other hand, *O. discolor* was recognized as a good species and not a synonymy of the widely distributed *O. ferruginea* (Donnelly, 1995; Paulson, 1998).

From the 349 species listed here, the females of 329 (94%) are known either from a formal description, partial description or at least can be accurately recognized with the help of a key (see Appendix I).

Table 3. Percentage and number of larvae described in relation to the number of species recorded for Mexico.

Families	# of Species	# Spp with Known Larva	%
Polythoridae	1	1	100
Amphipterygidae	2	1	50
Calopterygidae	11	9	82
Lestidae	11	8	73
Perilestidae	1	1	100
Megapodagrionidae	7	3	43
Pseudostigmatidae	5	4	80
Platystictidae	9	4	44
Protoneuridae	9	3	33
Coenagrionidae	91	59	65
Aeshnidae	32	25	78
Gomphidae	61	41	67
Cordulegastridae	3	2	66
Corduliidae	4	4	100
Libellulidae	102	74	73
TOTAL	349	239	6

Larvae

In our 1996 paper, we recorded 327 species of Odonata occurring in Mexico, of which 181 had their larvae described or in process of publication. To date, we have recorded the larva of 239 species, corresponding to 68% of the immature stages of the species of Mexico (see Table 3).

As shown in Table 3, three groups can be distinguished according to the number of larvae described: (1) families with more than 75% of the larvae described (Polythoridae, Calopterygidae, Perilestidae, Pseudostigmatidae, Aeshnidae, Corduliidae); (2) families between 50-75% (Lestidae, Coenagrionidae, Gomphidae, Cordulegastridae, Libellulidae, Amphipterygidae) and (3) families with less than 50% of their larvae described (Megapodagrionidae, Platystictidae, Protoneuridae). However, this situation could be favorably reverted in the short term because in some families such as Megapodagrionidae, Protoneuridae, Coenagrionidae, Gomphidae and Cordulegastridae, the larvae have been reared and will be described.

Most of the the new larvae described were coenagrionids with 16 species, followed by gomphids (13), libellulids (10), aeshnids (4), protoneurids (3), lestids (2), pseudostigmatids (2), platystictids (2), calopterygids (1), and corduliids (1). Species and authors are provided in Table 4.

Table 4. Larvae described for Mexico during the 1994-2006 period

SPECIES	AUTHOR(S)	YEAR
AMPHIPTERYGIDAE		
Amphipteryx longicaudata González	Novelo-Gutiérrez, R.	1995
CALOPTERYGIDAE		
Hetaerina infecta Calvert	Novelo-Gutiérrez, R.	2000
LESTIDAE		
Lestes alfonsoi González & Novelo	Novelo-G.,R. & E. González-S.	2003
L. sigma Calvert*	Westfall, M.J. & M.L. May	1996
PSEUDOSTIGMATIDAE		
Mecistogaster ornata Rambur	Ramírez, A.	1995
Megaloprepus caerulatus Drury	Ramírez, A.	1997
PLATYSTICTIDAE		
Palaemnema domina Calvert	Novelo-Gutiérrez, R.	2003
P. paulicoba Calvert**	Westfall, M.J. & M.L. May	1996
PROTONEURIDAE		
Neoneura aaroni Calvert *	Westfall, M.J. & M.L. May	1996
Protoneura aurantiaca Selys	Novelo-Gutiérrez, R.	1994
P. cupida Calvert	Novelo-Gutiérrez, R.	1994
COENAGRIONIDAE		
Acanthagrion quadratum Selys*	Westfall, M.J. & M.L. May	1996

Apanisagrion lais Selys*	Westfall, M.J. & M.L. May	1996
Argia agrioides Calvert*	Westfall, M.J. & M.L. May	1996
A. cuprea Hagen*	Westfall, M.J. & M.L. May	1996
A. extranea Hagen*	Westfall, M.J. & M.L. May	1996
A. funcki Selys	Novelo-G.R. & J.A. Gómez-A.	2006
A. hinei Kennedy*	Westfall, M.J. & M.L. May	1996
A. immunda Hagen*	Westfall, M.J. & M.L. May	1996
A. pallens Calvert*	Westfall, M.J. & M.L. May	1996
A. pima Garrison	Hoekstra, J.D. & R.L. Smith	1999
A. sabino Garrison	Hoekstra, J.D. & R.L. Smith	1999
Enallagma novaehispaniae Calvert	Novelo-Gutiérrez, R.	2005
E. semicirculare Selys*	Westfall, M.J. & M.L. May	1996
Telebasis boomsmae Garrison	Garrison, R.W.	1994
T. digiticollis Calvert	Novelo-G.R. & J.A. Gómez-A.	2005
T. filiola Perty*	Westfall, M.J. & M.L. May	1996
AESHNIDAE		
Aeshna williamsoniana Calvert	Novelo-Gutiérrez, R.	Received
Oplonaeschna magna Glez. & Nvlo.	González-S., E. & R. Novelo-G.	1998
Remartinia secreta Calvert	Novelo-Gutiérrez, R.	1998
Triacanthagyna septima Selys	Calil, E.R. & A.L. Carvalho	1999
GOMPHIDAE		
Epigomphus subobtusus Selys	Ramírez, A.	1996
Erpetogomphus agkistrodon Garrison	Novelo-Gutiérrez, R.	2002a
E. boa Selys	Novelo-Gutiérrez, R.	2002b
E. bothrops Garrison	Novelo-Gutiérrez, R.	2005
E. constrictor Ris	Ramírez, A.	1996
E. cophias Selys	Novelo-Gutiérrez, R.	2002b
E. elaps Selys	Novelo-Gutiérrez, R.	2005
E. erici Novelo	Novelo-Gutiérrez, R.	2002a
E. eutainia Calvert	Novelo-Gutiérrez, R.	2005
E. liopeltis Garrison	Novelo-Gutiérrez, R.	2005
E. ophibolus Calvert	Novelo-Gutiérrez, R.	2002a
E. viperinus Selys	Novelo-Gutiérrez, R.	2005
Phyllogomphoides bifasciatus Hagen	Ramírez, A.	1996
CORDULIIDAE		
Neocordulia batesi longipollex Calvert	Novelo-G.R. & A. Ramírez	1995
LIBELLULIDAE		
Brechmorhoga pertinax Hagen	Novelo-Gutiérrez, R.	1995
B. praecox Hagen	Novelo-Gutiérrez, R.	1995
Dythemis maya Calvert	Novelo-G.R. & E. González-S.	2004
Macrothemis ultima González	Novelo-Gutiérrez, R.	2006

Micrathyria ocellata Martin	Assis, C.V. & J.M. Costa	1994
Elasmothemis aliciae Gzlez. & Nvlo.	González-S., E. & R. Novelo-G.	2006
Perithemis intensa Kirby	Novelo-Gutiérrez, R.	2002c
Tholymis citrina Hagen	Fleck, De Marmels & Grand	2004
Tramea binotata Rambur*	Needham, Westfall & May	2000
T. calverti Muttkowski	Souza, Costa & Santos	1999

* Keyed only, no formal description provided

** Drawings only, no formal description provided

BIOGEOGRAPHICAL NOTES

The Mexican territory is located at the confluence of two big biogeographic regions: the Neartic and the Neotropical. For this reason, some authors (Halffter, 1976) have considered Mexico as a transitional zone, an area that also includes the south and southwestern portions of the United States, and most part of Central America (until Nicaragua Lake). Morrone (2001) established a subdivision of the Mexican territory into 14 biotic Provinces (5 Neartic, 5 Transitional and 4 Neotropical); no other country in Latin America has such diversity of biotic zones. This privilege situation and the elevated number of endemic species makes Mexico one of the megadiverse countries in the world (Wilson, 1988; Ramamoorthy *et al.*, 1993). Although a lot of collecting work is needed, we can describe in a preliminary way the composition and affinities of the Mexican odonate fauna as follows.

The two suborders apparently follow similar distributional patterns with Zygoptera having more restricted areas of distribution than Anisoptera due to the barrier effect (Paulson, 1982). Of the 29 genera of Zygoptera recorded for Mexico, 16 (55%) (*Cora*, *Hetaerina*, *Archilestes*, *Perissolestes*, *Heteragrion*, *Mecistogaster*, *Palaemnema*, *Neoneura*, *Protoneura*, *Psaironeura*, *Acanthagrion*, *Anisagrion*, *Argia* (in part), *Chrysobasis*, *Leptobasis* and *Telebasis*) follow the Neotropical Pattern (*sensu* Halffter, 1976), with different grades of penetration into the Mexican Transition Zone (MTZ). Of these genera, *Argia* deserves a special analysis. This genus is particularly speciose (Westfall & May, 1996) with more than 150 names proposed, although Garrison (1994a) stated that about only 110 may be considered valid species. Nevertheless, there are many undescribed species in Mexico, and in South and Central America. Some authors have suggested that *Argia* is a phylogenetically young taxon (Kennedy, 1919; Kiauta & Kiauta, 1980, both cited by Pritchard, 1982). Likewise, Pritchard (1982) suggests that Central and South America America are important centers of speciation, with many unique species. Thus an open question persists as to where was the center of origin of this genus? We speculate that Mexico could be such center based upon two arguments: (1) 49 out of the 110 valid species occur in Mexico, and (2) the two unique representatives of the primitive subgenus *Hyponeura, A. (H.) funcki* and *A.(H.) lugens*, also occur in Mexico. Only five of the 49 species of *Argia* recorded for Mexico would strictly follow the Neotropical dispersal pattern (*chelata, indicatrix, oculata, pocomana,* and *variabilis*); these species would integrate a recent group of invaders from South America. The rest of Argias (44 spp) could have evolved at the MTZ probably by vicariance processes, with six endemic species : *A. carlcooki, A. deami, A. garrisoni, A. harknessi, A percellulata* and *A. westfalli.*

Other five genera of Zygoptera (17%) have septentrional affinities, following the Paleoamerican dispersal pattern (Halffter 1976): *Lestes, Enallagma, Ischnura, Nehalennia* and *Amphiagrion.* The remaining 28% is represented by genera endemic to MTZ: *Amphipteryx* (relictual, probably from the late Jurassic [*ca.* 150 million years] as stated by Dumont *et al.,* (2005) for other amphipterygids, and restricted to cloud mountainous forests, its closest relative is the Oriental genus *Devadatta* [Novelo-Gutiérrez, 1995c]); *Paraphlebia* (larvae semi-aquatic, inhabiting seepages); *Megaloprepus* and *Pseudostigma* (the giants of the extant Odonata, both genera with larvae inhabiting exclusively phytotelmata, belonging to the family Pseudostigmatidae which has been sustained as a Gondwanian relict [Rhen, 2003]), these last 3 genera related to South American elements; *Neoerythromma* (with only two species, one of them also occurring at the Antilles [*N. cultellatum*], and probably related to the Palearctic genus *Erythromma* [Westfall & May, 1996]); *Apanisagrion* (related to the Central American genus *Anisagrion* [Westfall & May, 1996]), and *Hesperagrion* (probably related to *Ischnura* [Paulson & Cannings, 1980], both *Apanisagrion* and *Hesperagrion* probably following the High Plateau dispersal pattern [Halffter *op. cit.*]).

On the other hand, 29 (54%) of the 54 genera of Anisoptera have neotropical affinities, thus following the Neotropical pattern (*Coryphaeschna, Remartinia, Rhionaeschna, Triacanthagyna, Agriogomphus, Aphylla, Archaeogomphus, Phyllocycla, Phyllogomphoides, Progomphus, Neocordulia, Anatya, Brachymesia, Brechmorhoga, Cannaphila, Dythemis, Elasmothemis, Erythemis, Erythrodiplax, Idiataphe, Macrothemis, Miathyria, Micrathyria, Nephepeltia, Orthemis, Perithemis, Planiplax, Tauriphila* and *Uracis*). Other five (9%) are exclusively North American genera (*Epiaeschna, Dromogomphus, Octogomphus, Celithemis, Pachydiplax*), five (9%) have a wide distribution throughout the northern hemisphere (*Aeshna, Gomphus, Cordulegaster, Libellula, Sympetrum*), five (9%) probably originated in Mexico (*Oplonaeschna, Epigomphus* (*cf.* Kennedy, 1936), *Erpetogomphus, Paltothemis, Pseudoleon*), four (7.5%) are mainly circumtropical (*Anax* [cf. Belyshev & Haritonov, 1978], *Gynacantha, Pantala, Tramea*), three (5.5%) probably have a holarctic distribution (*Ophiogomphus, Stylurus, Epitheca*), one (2%) has a worldwide distribution (*Macromia*), and two (4%) with only one species in America but with one or two relatives in the Old World tropics (*i.e. Macrodiplax* and *Tholymis*).

An interesting pattern that some species show is the following: some species that are common and abundant in one of the two Mexican coasts (*e.g.* the Pacific or Gulf slopes) are on the contrary rare, less abundant or gradually replaced by other species in the opposite coast. In every case these species are taxonomically related. For example *Argia translata* which is common and abundant along the Gulf stretch is on the other hand rare or less abundant in the Pacific slope, in which *A. tezpi* is dominant. Other example is *Neoerythromma cultellatum*, which is abundant in ponds and lakes in coastal plains in the Gulf slope, while *N. gladiolatum* occupies the same habitats but in the Pacific slope (both species coexists in some localities of the Pacific slope but *N. cultellatum* occurs at lower densities). Regarding the Anisopterans some examples are similar, *e.g. Erythrodiplax umbrata* is a common species in both ephemeral ponds and more permanent lentic habitats along the Gulf coast but on the other hand *E. funerea* is more common at the Pacific coast with some degree of overlapping of the two at some localities in Central Mexico but with the pattern described above, *i.e. umbrata* occurring at lower densities on the Pacific and *funerea* being rare at the Gulf coast. Another couplet within this genus occurs with *E. fusca* and *E. basifusca*, the former distributed more commonly along the Gulf and the latter in the Pacific coast. One last example is found in the couplet *Dythemis sterilis*/*D. nigrescens*, both species are present in both slopes but *D. sterilis* been more abundant along the Gulf and *D. nigrescens* been dominant on the Pacific one. Similar examples can be found for other genera. A complete explanation to this strange phenomenon is not known. However, this phenomenon precludes vicariance, with those species from the Gulf coast plain favored by a more wet climatic regime, and their counterparts on the Pacific plain adapted to a drier conditions and different patterns of rains. Even more, in the same species (*e.g. Argia oenea*) individuals of populations inhabiting humid regions differ notoriously in coloration respect to populations inhabiting drier zones, the former being blue and the later violet.

Endemic species

In 1996 we cited a total of 40 endemic species for Mexico (González & Novelo 1996). However, due to some new distributional findings and the discovery of several new taxa the original statement should be modified. Thus, at the present we consider a total of 46 species (13.2%) restricted to the limits of this country (Table 2). It is interesting to note that Gomphidae contribute with 45.6% of the endemic species (21 spp.) followed by Coenagrionidae with 21.7% (10 spp) Megapodagrionidae, Platystictidae and Aeshnidae with 6.5% each (three spp each famiy), Lestidae and Libellulidae with 4.3% (two spp. each), and Amphypterygidae and Protoneuridae with 2.2% each (one spp. each).

From these endemics the case of *O. purepecha* is remarkable because the genus has an holartic distribution and in America is distributed in Canada and USA. The presence of a relictual population in Central Mexico open the possibility that more species of this genus could be found at the limits of the country. The presence of several undescribed taxa specially within the genera *Amphipteryx*, *Paraphlebia*, *Palaemnema*, *Argia* and *Phyllogomphoides* should modify the endemic species pattern in the future.

Mexican diversity and Global Odonata Diversity

Mexico has been considered by several authors as a megadiverse country along with other countries of the world (Wilson, 1988). Although it is the fourteenth largest country in the world, it ranks third in biological diversity (Mittermeier, 1988). A common pattern that has been observed in several groups of plants and animals is that in this country is located around the 10% of the global diversity. Besides, Mexico comprises almost the 50% of the so-called Mesoamerican biodiversity 'hotspot', which is second in species diversity and endemism for several groups of plants and animals of the main 25 'hotspots' in the world (Mittermeier *et al.,* 1999). For the case of Odonata the 349 spp here presented represent the 6.3% of the total species recorded (5574 spp) for the world (Schorr *et al.,* 2006). Even though some species could be discovered in this order in the future we consider that the common pattern observed for other groups do not apply for this group. We speculate that even though some groups are rather speciose (e.g. *Argia*, *Palaemnema*, *Erpetogomphus, Phyllogomphoides,*) and in which it is expected that most of the new taxa could be found, it will never reach the critical number of 10% or closer of species of the world diversity that it would expected to be found for some other insects (*e.g.* Papilionidae – Llorente & Luis Martínez, 1993; Rophalocera – Llorente *et al.,* 1996).

Critical species

The increasing rate of environmental deterioration such as stream and river pollution by uncontrolled industries and domestic sewages, the alarming deforestation for grazing, agriculture and logging purposes, as well as wetlands drying, are causes that are already affecting the Mexican odonate diversity, and probably with unknown or catastrophic consequences in the middle and long term.

We consider that the most vulnerable groups could be those whose species have long life cycles as larvae (*e.g.,* Gomphidae, Cordulegastridae), low density populations (Amphipterygidae, some Gomphidae), remarkable seasonal reproductive periods (Polythoridae, Platystictidae, Gomphidae), occurring in streams or creeks in forested situations (Polythoridae, Amphipterygidae, *Palaemnema* [Platystictidae], *Heteragrion* [Megapodagrionidae], *Perissolestes* [Perilestidae], *Agriogomphus* and *Epigomphus* [Gomphidae]), and those that use specialized substrates for reproduction and/or for larval development: Pseudostigmatidae in phytotelmata, and *Paraphlebia* (Megapodagrionidae) and some *Argia* (Coenagrionidae) in shaded seepages. In all these cases, the primary factor in odonate population declines is the habitat destruction (Paulson, 2004).

Paulson (2004) highlights the difficulty to designate critical species in the neotropics basically for the lack of information on the real distribution and abundance of those species considered as endangered, threatened or vulnerable because of their rarity, monotypy or endemicity. However, it is interesting to stand out that 5 species out of the 11 Neotropical species red-listed by IUCN (2003) reported for Mexico (Paulson *op.cit*), inhabit a bioma considered as threatened in this country: the cloud mountainous forest. This ecosystem is highly fragmented in its natural distribution in Mexico occupying about the 0.8% of the total area of the country (Rzedowski, 1978), and many of the relatively recent new species described from Mexico were from this bioma: *Amphipteryx longicaudata, Lestes alfonsoi*, *Erpetogomphus agkistrodon*, *E. erici* and *Brechmorhoga latialata.* However, the anthropogenic activities (agriculture, induced fires, overexploitation of freshwater sources, and others) have reduced and eventually confined this bioma to abrupt ravines which have been serving as a refugees for the fauna associated with this ecosystem. In summary, we would consider the species of Odonata inhabiting the cloud mountainous forest as vulnerable in general, *Amphipteryx "agrioides"* and *A. logicaudata* are critically endangered because they are relicts and have disjunct and low populations. *Hetaerina rudis, Paraphlebia hyalina, Palaemnema paulicaxa, Erpetogomphus erici, Cordulegaster godmani* and *Brechmorhoga latialata* are endangered for their low populations and/or restricted distribution. Likewise, but in other ecosystem, *Heteragrion azulum* and *E. donnellyi* are also endangered species. They are restricted to the tropical rain forest of "Los Tuxtlas" region, which has been heavily disturbed in the last 3 decades.

Work in progress and perspectives

Based on the previous information the most obvious projects to do in the future are: (1) an intensive collecting work in those states that remain almost unexplored (*e.g.* Aguascalientes, Guanajuato, Tlaxcala and Zacatecas) or little explored (Coahuila, Queretaro) all of which have less than 15 species reported, to obtain at least a general overview on their odonate fauna and (2) the elaboration of a data base from all specific localities of Mexico (including geographic coordinates) is indispensable in order to obtain a better understanding on the geographic distribution of Mexican odonates. This point is critical to establish programs for conservation and biogeography. Regarding to the first point, one student is currently working with the Odonate fauna from the state of Aguascalientes and he has reported until now a total of 54 species (Escoto, 2006). Recent advances are now on the way in the study of the Odonata fauna of critical states with the help of several north American workers [(*e.g.* Bershstock *et al.* (unpublished) for the state of Chihuahua; Upson, *et al.* (unpublished) and González *et al.* (unpublished) for the state of Sonora]. For Sonora there is an indication that the odonate diversity for such state could increase to 116 species (Upson *et al.,* unpublished data). Additional work have been done in other critical states or areas [(*e.g.* Calakmul Reserve in the state of Campeche (González *et al.* unpublished data)], the areas of Huatulco and Cuicatlan in state of Oaxaca (González *et al.* unpublished data); coastal regions of the states of Jalisco, and Colima (González *et al.,* unpublished data); coastal and mountainous regions of Michoacan (Novelo *et al.,* unpublished); and the states of Queretaro and Guanajuato (Novelo, unpublished data).

On the other hand, we have been gathering information in order to publish the complete inventories of species collected in the states of Veracruz and Morelos, providing keys and illustrations for adults and larvae, which will be the baseline for the elaboration of dot maps. At the same time, we will continue with the improvement of the Odonata collections, to cover most of the representatives of the Mexican odonate fauna, in both adult and larval stages.

REFERENCES

Assis, C.V. & J.M. Costa, 1994. Seis novas larvas do gênero *Micrathyria* Kirby e notas sobre a distribuição no Brasil (Odonata: Libellulidae). Revista Brasileira de Zoología, Curitiba, 11(2): 195-209.

Belle, J. 1988. A synopsis of the species of *Phyllocycla* Calvert, with descriptions of four new taxa and a key to the genera of Neotropical Gomphidae. (Odonata : Gomphidae). Tijdschrift voor Entomologie 131:73-102

Belle, J. 1991. The ultimate instar larvae of the Central American species of *Progomphus* Selys, with a description of *P. belyshevi* spec. nov. from Mexico (Anisoptera: Gomphidae). Odonatologica 20(1): 9-27.

Belyshev, B.F. & A.Y. Haritonov. 1978. The distribution of the cosmopolitan genus *Anax* Leach and its probable origin (Anisoptera:Aeshnidae). Odonatologica 7(2)115-121.

Bick, G.H. and J.C. Bick. 1995. A review of the genus *Telebasis* with descriptions of eight new species (Zygoptera:Coenagrionidae). Odonatologica 24(1):11-44.

Calil, E.F. & A.L. Carvalho, 1999. Descriçôes da larva de ultimo estádio e do adulto de *Triacanthagyna septima* (Selys, 1857) (Odonata, Aeshnidae), com notas sobre a biologia da espécie. Revista Brasileira de Entomologia, Sâo Paulo, 43(1/2): 73-83.

Daigle, J.J. 1995. *Argia carlcooki* spec. nov. from Mexico (Zygoptera : Coenagrionidae). Odonatologica 24(4):467-471.

Donnelly, T.W. 1967. The discovery of *Chrysobasis* in Central America, with the description of a new species (Odonata:Coenagrionidae). The Florida Entomologist 50(1):47-52.

Donnelly, T.W. 1981. A new species of *Archilestes* from Mexico and Central America, with further notes on the status of *Cyptolestes* Williamson (Odonata:Lestidae). The Florida Entomologist 64(3):412-417.

Donnelly, T.W. 1995. *Orthemis ferruginea* – an adventure in Caribbean biogeography. Argia 7(4): 9-12.

Dumont, H.J., J.R. Vanfleteren, J.F. De Jonckheere & P.H.H. Weekers, 2005. Phylogenetic relationships, divergence time estimation, and global biogeographic patterns of Calopterygoid damselflies (Odonata, Zygoptera) inferred from ribosomal DNA sequences. Systematic Biology 54(3): 347-362.

Dunkle, S.W. 2000. Dragonflies through binoculars. A field guide to dragonflies of North America. Oxford University Press. 266 p.

Escoto M.J. 2006. Odonata de Aguascalientes: diversidad y distribución. Universidad Autónoma de Aguascalientes, Depto de Biología. (unpubl. Thesis).

Fleck, G., J. De Marmels & D. Grand, 2004. La larve de *Tholymis citrina* Hagen, 1867 (Odonata, Anisoptera, Libellulidae). Bulletin de la Société entomologique de France 109(5): 455-457.

Förster, S. 2001. 2nd. ed. The dragonflies of Central America exclusive of Mexico and the West Indies. Odonatological Monographs 2. Gunnart, Rehfeldt Braunschweig Ed.141.p

Garrison, R.W. 1990. A synopsis of the genus *Hetaerina* with descriptions of four new species (Odonata:Calopterygidae). Transactions of the American Entomological Society 116(1):175-259.

Garrison, R.W. 1994a. A synopsis of the genus *Argia* of the United States with keys and descriptions of new species, *Argia sabino*, *A. leonorae*, and *A. pima* (Odonata: Coenagrionidae). Transactions of the American Entomological Society 120(4): 287-368.

Garrison, R.W. 1994b. A revision of the new world genus *Erpetogomphus* Hagen in Selys (Odonata:Gomphidae). Tijdschrift voor Entomologie 137:31-269.

Garrison, R.W. 1994c. *Telebasis boomsmae* spec. nov., a new damselfly from Belice (Zygoptera: Coenagrionidae). *Odonatologica* 23(3): 277-282.

Garrison, R.W. 1996. A synopsis of the *Argia fissa* group, with descriptions of two new species, *A. anceps* sp. n. and *A. westfalli* sp. n. (Zygoptera: Coenagrionidae). Odonatologica 25(1): 31-47.

Garrison, R.W. 1999. The genus *Neoneura*, with keys and description of a new species, *Neoneura jurzitzai* spec. nov. (Zygoptera:Protoneuridae). Odonatologica 28(4):343-375.

González-Soriano, E. 1992. *Protoneura rojiza* spec. nov., a new damselfly from Mexico (Zygoptera:Protoneuridae). Odonatologica 21(4):489-493.

González-Soriano E. 1999. *Brechmorhoga latialata* spec. nov., from Mexico (Odonata:Libellulidae). Odonatologica 28(1):83-86.

González-Soriano E. 2002. *Leptobasis melinogaster* spec. nov., a new species from Mexico (Zygoptera:Coenagrionidae). Odonatologica 31(2):181-185.

González-Soriano, E. & R. Novelo-Gutiérrez, 1996. Odonata, pp. 147-167. In: Llorente-Bousquets, J., A.N. García-Aldrete & E. González-Soriano (Eds.). Biodiversidad, taxonomía y biogeografía de artrópodos de México: Hacia una síntesis de su conocimiento. UNAM, México.

González-Soriano, E. & R. Novelo-Gutiérrez, 1998. *Oplonaeschna magna* sp. nov. (Odonata: Aeshnidae), from Mexico with a description of its larva. Revista de Biología Tropical 46(3): 705-715.

González-Soriano, E. & R. Novelo-Gutierrez. 2001. *Lestes alfonsoi* spec. nov., a new damselfly from Mexico (Zygoptera:Lestidae). Odonatologica 30(4)441-444.

González-Soriano, E. & R. Novelo-Gutiérrez, 2006. *Elasmothemis aliciae* spec. nov., a new dragonfly from Mexico, Belice and Costa Rica with a description of its larva and a key to the known larvae of the genus (Anisoptera: Libellulidae). Odonatologica 35(3): 243-253.

Halffter, G. 1976. Distribución de los insectos en la Zona de Transición Mexicana. Relaciones con la entomofauna de Norteamérica. Folia Entomológica Mexicana 35: 5-64.

Hoekstra, J.D. & R.W. Garrison. 1999. Range extensions of *Palaemnema domina* Calvert (Odonata:Platystictidae) to southeastern Arizona, USA: a new odonate family for the United States. Proceedings of the Entomological Society of Washington. 101(4):756-759.

Hoekstra, J.D. & R.L. Smith, 1999. Descriptions of the final instar larvae of *Argia sabino* Garrison and *Argia pima* Garrison (Odonata: Coenagrionidae). Proceedings of the Entomological Society of Washington 101(4): 887-896.

Hedström, I. & G. Sahlén, 2003. An extended description of the larva of *Megaloprepus caerulatus* from Costa Rica (Odonata: Pseudostigmatidae). International Journal of Odonatology 6(1): 23-31.

Kennedy, C.H. 1936. *Epigomphus crepidus* a new dragonfly (Odonata: Gomphidae) from Nayarit, Mexico with notes on the genus. Annals Entomological Society 29:126-135.

Llorente, B.J. & A. Luis Martínez. 1993. Conservation-oriented analysis of Mexican butterflies:Papilionidae (Lepidoptera:Papilionoidea) pp. 147-177. In: Ramammorthy, T.P., R. Bye, A. Lot & J. Fa (eds.). Biological diversity of Mexico. Oxford University Press, New York.

Llorente, B.J., E. González-S., A.N. García A. & C. Cordero. 1996. Breve Panorama de la Taxonomía en México. In: Llorente-Bousquets, J., A.N. García-Aldrete & E. González-Soriano (Eds.). Biodiversidad, taxonomía y biogeografía de artrópodos de México: Hacia una síntesis de su conocimiento. UNAM, México.

Llorente, B.J., A. Luis Martínez, I. Vargas Fernández & J. Soberón Mainero. 1996. Papilonoidea (Lepidoptera) pp. 531-548. In: Llorente-Bousquets, J., A.N. García-Aldrete & E. González-Soriano (Eds.). Biodiversidad, taxonomía y biogeografía de artrópodos de México: Hacia una síntesis de su conocimiento. UNAM, México.

May, M.L. 1992. *Telebasis aurea* (Odonata:Zygoptera:Coenagrinidae), a new species of damselfly from Costa Rica. Entomological News 103(5):161-168

May, M.L. 1993. *Lestes secula*, a new species of damselfly (Odonata: Zygoptera: Lestidae) from Panama. Journal of the New York Entomological Society 101(3): 410-416.

Mittermeier, R.A. 1988. Primate diversity and the tropical forest: case studies from Brazil and Madagascar and the importance of the megadiversity countries. In: E. Wilson (ed.) Biodiversity. Washington D.C.: National Academic Press.

Mittermeier, R. A., N. Myers & C. G. Goettsch Mittermeier. 1999. Hotspots: earth's biologically richest and most endangered terrestrial ecoregions. CEMEX-Conservation International. 430 p.

Morrone, J.J. 2001. Biogeografía de América Latina y el Caribe. M&T-manuales & Tesis SEA, vol. 3 Zaragoza. 148 pp.

Needham, J.G., M.J. Westfall & M.L. May, 2000. Dragonflies of North America. Scientific Publishers, Gainesville. 939 pp.

Novelo-Gutiérrez, R. 1994. Las náyades de *Protoneura aurantiaca* Selys y *P. cupida* Calvert (Odonata: Zygoptera: Protoneuridae). Folia Entomológica. Mexicana 90:25-31.

Novelo-Gutiérrez, R. 1995a. La náyade de *Brechmorhoga pertinax* Hagen, 1861 (Odonata: Libellulidae). Anales del Instituto de Biología Universidad Nacional Autónoma de México, Serie Zoología 66(2):181-187.

Novelo-Gutiérrez, R. 1995b. La náyade de *Brechmorhoga praecox* (Hagen, 1861), y notas sobre las náyades de *B. rapax* Calvert, 1898, *B. vivax* Calvert, 1906 y *B. mendax* (Hagen, 1861)(Odonata: Libellulidae). Folia Entomológica Mexicana 94:33-40.

Novelo-Gutiérrez,.R. 1995c. The larva of *Amphipteryx* (Zygoptera: Amphipterygidae) and a reclassification of Amphipterygidae *sensu lato* based upon the larvae. Odonatologica 24(1):73-87.

Novelo-Gutiérrez, R. 1998. Description of the larva of *Remartinia secreta* and notes on the larva of *Remartina luteipennis florida* (Odonata: Aeshnidae). The Canadian Entomologist 130:893-897.

Novelo-Gutiérrez R. 1999. *Erpetogomphus erici* spec. nov. from Mexico, and a description of the male of *E. agkistrodon* Garrison (Anisoptera: Gomphidae). Odonatologica 28(2):171-179.

Novelo-Gutiérrez, R. 2000. Description of the larva of *Hetaerina infecta* Calvert (Odonata: Calopterygidae). Proceedings of the Entomological Society of Washington 102(1):99-104.

Novelo-Gutiérrez, R. 2002a Larvae of the *ophibolus*-species group of *Erpetogomphus* Hagen in Selys from Mexico and Central America (Anisoptera: Gomphidae). Odonatologica 31(1):35-46.

Novelo-Gutiérrez, R. 2002b. Two new Mexican larvae of the genus *Erpetogomphus* Hagen in Selys (Odonata: Gomphidae). Journal of the New York Entomological Society 110(3-4): 370-375.

Novelo-Gutiérrez, R. 2002c. Descripción de las larvas de *Perithemis intensa* Kirby, 1889 y *P. domitia* (Drury, 1773), con notas sobre otras larvas del género en México (Odonata: Anisoptera: Libellulidae). Folia Entomológica Mexicana 41(3): 321-327.

Novelo-Gutiérrez, R. 2003. The larva of *Palaemnema domina* Calvert, 1903 (Odonata: Platystictidae). Transactions of the American Entomological Society 129(1): 71-75.

Novelo-Gutiérrez, R. 2005a. Five new *Erpetogomphus* Hagen in Selys larvae from Mexico, with a key to the known species (Anisoptera: Gomphidae). Odonatologica 34(3): 243-257.

Novelo-Gutiérrez, R. 2005b. La larva de *Enallagma novaehispaniae* Calvert, 1902 (Odonata: Zygoptera: Coenagrionidae). Folia Entomológica Mexicana 44(2): 219-224.

Novelo-Gutiérrez, R. 2006. The larva of *Macrothemis ultima* González-Soriano, 1992 (Odonata: Anisoptera: Libellulidae). Transactions of the American Entomological Society 132 (1+2): 151-156.

Novelo-Gutiérrez, R. The larva of *Aeshna williamsoniana* (Odonata: Anisoptera: Aeshnidae). The Canadian Entomologist. Received.

Novelo-Gutiérrez, R. & J.A. Gómez-Anaya, 2005. Description of the larva of *Telebasis digiticollis* (Odonata: Zygoptera: Coenagrionidae). The Canadian Entomologist 137: 61-66.

Novelo-Gutiérrez, R. & J.A. Gómez-Anaya, 2006. A description of the larva of *Argia funcki* (Selys, 1854) (Odonata: Zygoptera: Coenagrionidae). Proceedings of the Entomological Society of Washington 108(2): 261-266.

Novelo-Gutiérrez, R. & E. González-Soriano, 2003. The larva of *Lestes alfonsoi* González & Novelo (Zygoptera: Lestidae). Odonatologica 32(3): 289-294.

Novelo-Gutiérrez, R. & E. González-Soriano, 2004. The larva of *Dythemis maya* Calvert, 1906 and a redescription of the larva of *D. sterilis* Hagen, 1861 with a key to the larvae of the genus (Anisoptera: Libellulidae). Odonatologica 33(3): 279-289.

Novelo-Gutiérrez, R. & A. Ramírez. 1995. The larva of *Neocordulia batesi longipollex* Calvert, 1909 (Odonata: Corduliidae). Journal of the New York Entomological Society 103(2):195-199.

Novelo-Gutiérrez, R. & A. Ramírez, 1998. The larva of *Macrothemis inacuta* (Odonata: Libellulidae). Entomological News 109(5):301-306.

Paulson, D.R. 1982. Odonata, pp. 249-277. In: Hurlbert, S.H. & A. Villalobos-Figueroa (eds.). Aquatic Biota of Mexico, Central America and the West Indies. San Diego State University, San Diego, California.

Paulson, D.R. 1998. The distribution and relative abundance of the sibling species *Orthemis ferruginea* (Fabricius, 1775) and *O. discolor* (Burmeister, 1839) in North and Middle America (Anisoptera: Libellulidae). International Journal of Odonatology 1:89-93.

Paulson, D.R. 2003. Comments on the *Erythrodiplax connata* (Burmeister, 1839) group, with the elevation of *E. fusca* (Rambur, 1842), *E. minuscula* (Rambur, 1842), and *E. basifusca* (Calvert, 1895) to full species. Bulletin of American Odonatology 6(4):101-110.

Paulson, D.R. 2004. Critical species of Odonata in the Neotropics. International Journal of Odonatology 7(2): 163-188.

Paulson, D.R. & R.A. Cannings, 1980. Distribution, natural history and relationships of *Ischnura erratica* Calvert (Zygoptera: Coenagrionidae). Odonatologica 9: 147-153.

Paulson, D.R. & E. González-Soriano. 1994 (last revised March 2006). Mexican Odonata: Odonata of Mexico by state. http://www.ups.edu/x 6530.xml

Pritchard, G. 1982. Life-history strategies in dragonflies and the colonization of North America by the genus *Argia* (Odonata: Coenagrionidae). Advances in Odonatology 1: 227-241.

Ramammorthy,T.P., R. Bye, A. Lot & J. Fa. 2003 . Introduction pp. xxix-xxxix. In: Ramammorthy, T.P., R. Bye, A. Lot & J. Fa (eds.). Biological diversity of Mexico. Oxford University Press, New York. 812 p.

Ramírez, A. 1995. Descripción e historia natural de las larvas de odonatos de Costa Rica. IV. *Mecistogaster ornata* (Rambur, 1842) (Zygoptera, Pseudostigmatidae). Bulletin of American Odonatology 3:43-47.

Ramírez, A. 1996. Six new dragonfly larvae of the family Gomphidae in Costa Rica, with a key to the Central American genera (Anisoptera). Odonatologica 25(2): 143-156.

Ramírez, A. 1997. Description and natural history of the Costa Rican Odonata larvae. 5. *Megaloprepus caerulatus* (Drury, 1782) (Zygoptera: Pseudostigmatidae). Odonatologica 26(1): 75-81.

Ramírez, A. & R. Novelo-Gutiérrez, 1999. The neotropical dragonfly genus *Macrothemis*: new larval descriptions and an evaluation of its generic status based on larval stages (Odonata: Libellulidae). Journal of the North American Benthological Society 18(1):66-72.

Rhen, A.W. 2003. Phylogenetic analysis of higher-level relationships of Odonata. Systematic Entomology 28: 181-239.

Rzedowski, J. 1978. Vegetación de México. Limusa, México. 432 p.

Schorr, M., M. Lindeboom & D.R. Pauslon. 2006. World Odonata List. http://www.ups.edu/x 6140.xml

Souza, L.O.I., J.M. Costa & T.C. Santos, 1999. Redescrição da larva de *Tramea calverti* Muttkowski, 1910, com chave para identifição das larvas conhecidas do gênero (Odonata, Libellulidae). Boletim do Museu nacional, Nova Série, Zoologia, Rio de Janeiro 409: 1-7.

von Ellenrieder, N. 2003. A synopsis of the Neotropical species of '*Aeshna*' Fabricius: The genus *Rhionaeschna* Förster (Odonata:Aeshnidae). Tijdschrift voor Entomologie 146:67-207.

von Ellenrieder, N. & R.W. Garrison. 2003. A synopsis of the genus *Triacanthagyna* (Odonata:Aeshnidae). International Journal of Odonatology 6(2):147-184.

Westfall, M.J. & M.L. May, 1996. Damselflies of North America. Scientific Publishers, Gainesville. 649 p.

Wilson, E.O. (ed.). Biodiversity. Washington, DC. National Academic Press.

Williamson, E.B. 1919. Results of the University of Michigan-Williamson expedition to Colombia 1916-1917. Occasional Papers of the Museum of Zoology, University of Michigan 68:1-89.

Williamson, E.B. 1915. Notes on Neotropical dragonflies or Odonata. Proceedings of the United States National Museum 48:601-638.

APPENDIX - 1. Data on species in various Mexican states

SPECIES/STATE	1	2	3	4	5	6	7	8	9	10	11	12	13	14	15	16	17	18	19	20	21	22	23	24	25	26	27	28	29	30	31	32	33
POLYTHORIDAE																																	
Cora marina					X							X								X										X			♀
AMPHIPTERYGIDAE																																	
Amphipteryx "agrioides"					X																												♀
A. longicaudata																				X										X			♀
CALOPTERYGIDAE																																	
Hetaerina americana		X	X		X	X	X	X		X	X	X	X	X	X	X	X	X	X	X	X	X		X	X	X	X	X		X			♀
H. caja												¿?																					♀
H. capitalis					X							X	X					X		X	X			X						X			♀
H. cruentata					X				X			X	X	X		X	X	X	X	X	X	X		X			X			X			♀
H. infecta					X							X	X					X		X		X		X						X			♀
H. occisa					X							X	X	X				X		X	X			X			X	X		X			♀
H. pilula					X																						X			X			♀
H. rudis					X																												♀
H. sempronia					X															X							X			X			♀
H. titia					X			X				X	X	X				X	X				X	X			X	X		X			♀
H. vulnerata					X	X			X	X		X	X			X	X	X	X	X	X	X			X	X				X			♀
LESTIDAE																																	
Archilestes californicus			X			X																											♀
A. grandis			X		X	X			X	X	X	X	X	X			X	X	X	X	X	X		X	X		X	X		X			♀
A. latialatus					X															X										X			♀
A. regalis													X											X						X			♀
Lestes alacer					X		X		X	X			X	X	X	X	X		X	X	X							X		X			♀
L. alfonsoi																														X			♀
L. forficula				X																X			X	X			X	X		X	X		♀
L. sigma					X							X		X			X	X		X				X	X	X		X		X			♀
L. simplex												X					X																♀
L. tenuatus					X							X		X			X	X		X			X	X						X	X		♀

SPECIES/STATE	1	2	3	4	5	6	7	8	9	10	11	12	13	14	15	16	17	18	19	20	21	22	23	24	25	26	27	28	29	30	31	32	33
L. tikalus					X			X						X						X			X							X	X		♀
PERILESTIDAE																																	
Perissolestes magdalenae					X															X										X			♀
MEGAPODAGRIONIDAE																																	
Heteragrion albifrons					X															X										X			♀
H. alienum					X															X			X				X			X			♀
H. azulum																														X			
H. tricellulare					X							X								X							X			X			♀
Paraphlebia hyalina																														X			♀
P. quinta																				X										X			♀
P. zoe													X								X			X						X			♀
PSEUDOSTIGMATIDAE																																	
Mecistogaster modesta					X								X								X			X				X		X			♀
M. ornata					X					X		X		X			X	X		X				X	X		X	X		X			♀
Megaloprepus caerulatus					X															X										X			♀
Pseudostigma aberrans					X														X	X				X			X	X		X			♀
P. accedens																							X				X			X			♀
PLATYSTICTIDAE																																	
Palaemnema angelina					X																												♀
Palaemnema desiderata					X															X							X			X			
P. domina					X	X		X									X	X		X													♀
P. nathalia					X															X													♀
P. paulicaxa																					X									X			
P. paulicoba																			X	X	X			X						X			♀
P. paulitaba					X																						X			X			♀
P. paulitoyaca					X								X						X	X				X						X			♀
P. paulina																																	♀
PROTONEURIDAE																																	
Neoneura aaroni																			X														♀
Neoneura amelia				X	X			X										X		X			X	X		X	X			X			♀
N. paya					X															X							X			X			♀

SPECIES/STATE	1	2	3	4	5	6	7	8	9	10	11	12	13	14	15	16	17	18	19	20	21	22	23	24	25	26	27	28	29	30	31	32	33
Protoneura aurantiaca					X																			X			X			X			♀
P. cara								X				X		X			X	X		X				X		X		X		X			♀
P. corculum				X																			X							X	X		♀
P. cupida					X								X											X						X			♀
P. rojiza												X								X													♀
Psaironeura remissa																				X							X			X			♀
COENAGRIONIDAE																																	
Acanthagrion inexpectum				X																										X	X		
A quadratum					X								X						X	X	X		X	X			X	X		X	X		♀
Amphiagrion abbreviatum		X																															♀
Anisagrion allopterum					X							X																					♀
A. truncatipenne					X																												♀
Apanisagrion lais					X						X	X	X			X	X	X	X	X	X	X		X	X	X		X		X			♀
Argia agrioides		X	X																														♀
A. anceps					X							X	X	X		X	X	X	X	X	X	X			X	X		X		X			♀
A. apicalis																			X														♀
A. barretti																			X		X			X				X					♀
A. calida													X								X			X				X		X			
A. carlcooki								X						X			X																♀
A chelata					X															X										X			
A. cuprea					X								X					X	X			X		X			X			X			♀
A. deami													X		X																		♀
A. extranea					¿?			X				X	X	X		X	X	X		X	X	X		X	X	X		X		X			♀
A. frequentula					X							X					X			X	X		X	X			X			X			♀
A. fumipennis										X			X						X					X									♀
A. funcki					X	X			X			X	X	X		X	X	X		X						X				X			♀
A. funebris					X																									X			
A. garrisoni																			X					X				X					♀
A. gaumeri																							X							X	X		♀
A. harknessi												X					X	X		X										X			♀
A. herberti												X					X			X													

SPECIES/STATE	1	2	3	4	5	6	7	8	9	10	11	12	13	14	15	16	17	18	19	20	21	22	23	24	25	26	27	28	29	30	31	32	33
A. hinei		X					X																		X								♀
A. immunda					X	X		¿?				¿?	X	¿?			¿?	¿?	X	¿?	X	X		X	X			X		X			♀
A. indicatrix					X																						X			X			♀
A. lacrimans						X				X		X	X	X	X		X		X						X					X			♀
A. leonorae																			X														♀
A. lugens		X			X	X				X			X					X		X		X				X				X			♀
A. medullaris													X																				
A. moesta		X				X	X							X		X	X		X									X					♀
A. munda						X				X														X									♀
A. nahuana	X	X				X			X	X			X	X	X	X	X	X	X														♀
A. oculata					X							X	X	X			X	X	X	X		X		X			X	X		X			♀
A. oenea			X		X	X	X					X	X	X			X	X	X	X	X	X		X	X			X		X			♀
A. pallens												X	X	X			X	X		X	X	X											♀
A. percellulata					X																X									X			♀
A. pima																										X							♀
A. pipila					X																						X			X			♀
A. plana					¿?	X	X		X	X		X	X		X	X	X		X					X	X								♀
A. nr. pocomana																	X																♀
A. popoluca																											X			X			♀
A. pulla					X			X				X	X	X		X	X	X		X	X			X	X	X	X			X			♀
A. rhoadsi													X						X		X			X				X		X			♀
A. sabino														X																			♀
A. sedula						X	X			X			X						X		X	X		X		X		X		X			♀
A. tarascana										X	X	X	X	X		X	X			X	X											X	♀
A. tezpi			X		X							X	X	X		X	X	X		X	X				X	X				X			♀
Argia tonto						X				X									X							X							♀
A. translata					X	X				X			X						X	X	X	X	X	X	X		X	X		X	X		♀
A. ulmeca					X							X	X				X	X	X		X	X		X				X		X			♀
A. variabilis					X							X	X					X		X										X			♀
A. vivida		X	X				X																										♀
A. westfalli					X														X			X		X				X		X			♀

SPECIES/STATE	1	2	3	4	5	6	7	8	9	10	11	12	13	14	15	16	17	18	19	20	21	22	23	24	25	26	27	28	29	30	31	32	33
Chrysobasis lucifer																														X			♀
Enacantha caribbea				X	X								X							X			X	X							X		♀
Enallagma basidens						X	X												X									X					♀
E. boreale										X																							♀
E. carunculatum		X																															♀
E. civile	X	X	X		X	X	X		X	X			X	X	X	X	X	X	X	X	X	X		X	X	X		X	X	X			♀
E. cyathigerum		X	X																														♀
E. durum																												X					♀
E. eiseni			X																														♀
E. exsulans													X						X									X					♀
E. novaehispaniae			X		X			X				X	X	X			X	X	X	X	X	X	X	X	X	X	X	X		X			♀
E. praevarum	X	X				X	X		X	X		X	X	X	X	X	X	X	X	X	X	X		X	X	X				X			♀
E. rua					X																												♀
E. semicirculare					X		X					X	X	X			X	X	X	X	X	X		X	X	X		X		X			♀
H. heterodoxum						X	X			X			X			X	X		X	X	X	X		X						X			♀
Ischnura capreolus				X	X			X						X				X					X	X			X	X		X			♀
I. cervula		X	X																							X							♀
I. demorsa	X					X	X		X	X			X	X	X	X				X		X		X	X	X							♀
I. denticollis		X	X		X	X			X	X			X	X	X	X		X		X	X	X				X			X	X			♀
I. hastata			X		X			X				X		X		X	X	X	X	X	X		X	X	X	X	X	X		X	X		♀
I. posita													X										X	X									♀
I. ramburii			X	X	X	X		X	X			X	X	X	X	X	X	X	X	X			X	X	X	X	X	X		X	X		♀
Leptobasis candelaria				X																										X			♀
L. melinogaster														X						X													
L. vacillans					X			X									X	X			X		X	X	X		X			X	X		♀
Nehalennia minuta				X	X													X		X			X	X						X	X		♀
Neoerythromma cultellatum				X	X			X												X			X	X				X		X	X		♀
N. gladiolatum								X				X		X				X							X								♀
Telebasis boomsmae																											X						♀
T. collopistes																											X			X			♀
T. digiticollis					X			X												X			X				X			X			♀

SPECIES/STATE	1	2	3	4	5	6	7	8	9	10	11	12	13	14	15	16	17	18	19	20	21	22	23	24	25	26	27	28	29	30	31	32	33
T. filiola					X			X						X				X						X				X		X	X		♀
T. griffinii														X				X		X				X			X			X			♀
T. incolumis			X																														♀
T. isthmica					X													X															♀
T. salva			X	X	X	X		X	X	X		X	X	X		X	X	X	X	X	X	X	X	X	X	X	X	X		X	X		♀
AESHNIDAE																																	
Aeshna constricta		X																															♀
A. persephone						X												X															♀
A. walkeri		X																															♀
A. williamsoniana																X	X													X			♀
Anax amazili				X	X			X					X	X	X			X						X	X			X		X			♀
A. concolor												X								X			X							X	X		♀
A. junius		X	X			X	X		X	X			X	X		X	X	X		X			X		X	X	X	X		X	X		♀
A. walsinghami		X	X							X		X	X				X	X	X	X	X	X											♀
Coryphaeschna adnexa					X			X						X			X	X					X	X	X		X	X		X	X		♀
C. apeora																														X			♀
C. diapyra								X						X			X	X					X							X			♀
C. viriditas					X			X						X									X				X	X		X	X		♀
Epiaeschna heros																																	♀
Gynacantha helenga								X						X			X	X					X	X			X			X			♀
G. laticeps																														X			♀
G. mexicana																							X	X						X	X		♀
G. nervosa				X	X							X		X			X			X			X	X	X					X	X		♀
Oplonaeschna armata					X				X	X	X		X		X	X	X								X					X			♀
O. magna												X	X		X		X																♀
Remartinia luteipennis			X		X					X			X	X	X	X	X	X	X	X										X			♀
R. secreta														X										X						X			
Rhionaeschna cornigera					X															X													♀
R. dugesi			X							X	X		X		X					X	X					X				X			♀
R. jalapensis					X					X		X	X			X	X				X									X			♀
R. manni			X																														♀

SPECIES/STATE	1	2	3	4	5	6	7	8	9	10	11	12	13	14	15	16	17	18	19	20	21	22	23	24	25	26	27	28	29	30	31	32	33
R. multicolor	X	X	X						X	X			X	X	X	X	X					X				X			X	X			♀
R. psilus	X	X	X		X				X				X	X			X	X	X		X		X	X	X			X		X	X		♀
R. vazquezae												X						X															♀
Triacanthagyna caribbea				X	X									X				X					X				X	X		X	X		♀
T. dentata																														X			♀
T. ditzleri																											X			X			♀
T. septima				X	X				X									X					X				X			X			♀
GOMPHIDAE																																	
Agriogomphus tumens					X																									X			♀
Aphylla angustifolia														X					X	X			X	X			X	X		X			♀
A. sp. aff caraiba (*)																							X										♀
A. protracta						X	X	X						X			X	X	X					X	X	X	X	X		X			♀
A. sp. aff williamsoni (*)																							X										♀
Archaegomphus furcatus					X																							X		X			♀
Dromogomphus spoliatus																			X									X					♀
Epigomphus clavatus					X																												
Epigomphus crepidus																		X		X													♀
E. donnellyi																														X			♀
E. flinti																				X													
E. paulsoni					X																												
E. subobtusus					X																												♀
E. sulcatystila																														X			
Erpetogomphus agkistrodon																														X			♀
E. boa																					X									X			♀
E. bothrops					X			X				X		X		X	X	X		X				X	X	X		X		X			♀
E. compositus		X	X																							X							♀
E. constrictor																			X					X				X		X			♀
E. cophias											X	X				X	X																♀
E. crotalinus						X			X	X		X	X	X	X	X	X	X			X	X											♀
E. designatus						X	X			X									X									X					♀
E. elaps					X							X	X	X		X	X	X	X	X		X		X	X			X		X			♀

SPECIES/STATE	1	2	3	4	5	6	7	8	9	10	11	12	13	14	15	16	17	18	19	20	21	22	23	24	25	26	27	28	29	30	31	32	33
E. erici													X																	X			♀
E. eutainia					X							X				X	X			X	X			X				X		X			♀
E. heterodon						X																											♀
E. lampropeltis			X			X				X																X							♀
E. liopelltis													X			X			X					X									♀
E. ophibolus					X															X										X			♀
E. sipedon										X				X	X		X	X			X												♀
E. viperinus																														X			♀
G. (Gomphurus) gonzalezi																								X									♀
Gomphus (G.) militaris																			X														♀
Octogomphus specularis			X																														♀
Ophiogomphus purepecha																X																	♀
Phyllocycla breviphylla																								X				X		X			♀
P. elongata							X					X		X			X	X		X				X	X					X			♀
P. volsella					X															X							X			X			♀
Phyllogomphoides albrighti																			X					X				X		X			♀
P. apiculatus					X							X		X		X		X		X					X								♀
P. bifasciatus					X									X						X													♀
P. danieli																	X			X													
P. duodentatus					X																X			X						X			♀
P. indicatrix					X																												
P. luisi																	X																♀
P. nayaritensis																		X															
P. pacificus					X			X				X		X		X	X	X		X													♀
P. pugnifer					X																									X			♀
P. stigmatus																			X														♀
P. suasus					X							X	X					X	X	X	X			X				X		X			♀
Progomphus amarillus																								X									♀
P. anomalus					X																												♀
P. belyshevi																	X					X											♀
P. borealis			X			X				X			X			X										X							♀

SPECIES/STATE	1	2	3	4	5	6	7	8	9	10	11	12	13	14	15	16	17	18	19	20	21	22	23	24	25	26	27	28	29	30	31	32	33
P. clendoni					X			X				X		X			X	X		X		X		X				X		X			♀
P. longistigma					X																												♀
P. mexicanus														X						X					X					X			♀
P. pygmaeus					X															X										X			♀
P. risi					X																												♀
P. zonatus					X																				X								♀
Stylurus falcatus														X		X																	
S. plagiatus																			X														♀
CORDULEGASTRIDAE																																	
Cordulegaster diadema						X			X	X					X	X	X		X	X						X				X			♀
C. dorsalis		X																															♀
C. godmani												X	X																	X			♀
CORDULIIDAE																																	
Epitheca cynosura										X																							♀
Macromia annulata																			X					X									♀
M . magnifica						X				X			X																				♀
Neocordulia batesi																				X													♀
LIBELLULIDAE																																	
Anatya guttata					X									X				X					X					¿?		X	X		♀
Brachymesia furcata			X	X									X	X			X	X					X	X	X		X	X		X	X		♀
B. herbida				X								X						X						X	X		X	X		X	X		♀
B. latialata													X																	X			♀
Brechmorhoga mendax			X																X	X				X		X		X					♀
B. nubecula					X								X											X			X			X			♀
B. pertinax					X	X											X	X		X						X				X			♀
B. praecox					X							X	X	X			X	X		X		X		X		X	X	X		X			♀
B. rapax					X															X										X			♀
B. tepeaca									¿?								¿?											X		X			♀
B. vivax					X								X					X			X			X				X		X			♀
Cannaphila insularis					X			X						X				X					X	X			X	X		X	X		♀
C. vibex					X								X							X										X			♀

SPECIES/STATE	1	2	3	4	5	6	7	8	9	10	11	12	13	14	15	16	17	18	19	20	21	22	23	24	25	26	27	28	29	30	31	32	33
Celithemis eponina																			X														♀
Dythemis fugax																												X					♀
D. maya												X	X	X			X	X	X	X					X			X		X			♀
D. multipunctata					X							X	X					X	X		X	X		X				X		X			♀
D. nigrescens			X					X				X	X	X			X	X	X	X	X	X		X	X	X		X		X			♀
D. sterilis			X		X			X					X	X			X	X	X	X	X		X	X	X		X	X		X	X		♀
D. velox							X			X									X									X					♀
Elasmothemis aliciae																								X						X			♀
E. cannacrioides					X																									X			♀
Erythemis attala					X			X						X			X	X		X				X			X			X	X		♀
E. collocata			X			X	X		X	X				X		X													X				♀
E. haematogastra														¿?													X	X		X			♀
E. mithroides				X	X									X				X		X				X				X		X	X		♀
E. peruviana					X									X										X			X	X		X			♀
E. plebeja			X	X	X			X	X			X	X	X			X	X		X	X	X		X	X		X	X		X	X		♀
E. simplicicollis				X				X	X					X			X						X	X			X	X		X	X		♀
E. vesiculosa				X	X			X	X	X		X		X	X		X	X		X				X	X	X	X	X		X	X		♀
Erythrodiplax abjecta					X												¿?																♀
E. andagoya																														X			♀
E. basifusca		X	X			X				X		X		X		X	X	X		X	X		.		X	X				X			♀
E. berenice		X	X	X										X						X			X		X					X	X		♀
E. fervida				X	X			X	X			X		X				X		X			X	X			X	X		X	X		♀
E. funerea					X		X	X	X			X	X	X		X	X	X		X	X				X	X				X			♀
E. fusca				X	X			X					X				X			X	X		X	X			X	X		X	X		♀
E. umbrata				X	X							X	X	X			X	X	X	X	X		X	X			X	X		X	X		♀
Idiataphe cubensis				X	X	X																	X					X		X	X		♀
Libellula auripennis																												X					♀
L. comanche						X																				X							♀
L. croceipennis			X		X			X				X	X	X			X	X	X	X	X	X		X	X	X		X		X			♀
L. foliata					X																												♀
L. gaigei																							X		X					X	X		♀

SPECIES/STATE	1	2	3	4	5	6	7	8	9	10	11	12	13	14	15	16	17	18	19	20	21	22	23	24	25	26	27	28	29	30	31	32	33
L. herculea				X	X								X							X	X			X						X	X		♀
L. luctuosa						X				X																							♀
L. lydia																			X														♀
L. needhami														X									X				X	X					♀
L. nodisticta									X	X				X	X	X													X				♀
L. saturata			X			X	X		X	X	X		X	X		X	X	X	X	X	X	X				X				X			♀
L. subornata						X								X												X							♀
Macrodiplax balteata		X																					X		X					X	X		♀
Macrothemis delia					X																												♀
M. extensa					X																												♀
M. hemichlora					X			X										X		X			X				X	X		X			♀
M. imitans					X																			X				X		X			♀
M. inacuta					X			X				X		X			X	X	X	X				X	X		X	X		X			♀
M. inequiunguis			X		X							X	X	X			X	X	X					X			X	X		X			♀
M. musiva					X																						X			X			♀
M. pseudimitans			X		X				X		X	X	X	X			X	X	X	X	X	X		X	X		X	X		X			♀
M. ultima														X		X																	
Miathyria marcella					X			X	X				X	X			X	X		X		X		X	X	X	X	X		X	X		♀
M. simplex					X							X		X				X		X			X	X			X	X		X			♀
Micrathyria aequalis			X	X	X			X					X	X			X	X		X	X		X	X	X		X	X		X	X		♀
M. atra					X													X												X			♀
M. debilis				X	X															X			X	X			X	X		X	X		♀
M. dictynna					X																									X			♀
M. didyma			X	X	X								X	X			X	X		X			X	X	X		X	X		X	X		♀
M. dissocians					X			X												X				X			X			X			♀
M. hagenii			X	X	X							X	X	X				X	X	X	X		X	X	X			X		X	X		♀
M. ocellata					X			X				X		X				X		X				X			X			X			♀
M. schumanni					X			X				X		X				X		X													♀
Nephepeltia chalconota																				X										X			♀
Orthemis discolor			X		X							X	X	X		X	X	X	X	X				X	X	X	X	X		X	X		♀
Orthemis ferruginea	X		X	X	X	X	X	X	X	X		X	X	X		X	X	X	X	X	X	X	X	X	X	X	X	X		X	X		♀

SPECIES/STATE	1	2	3	4	5	6	7	8	9	10	11	12	13	14	15	16	17	18	19	20	21	22	23	24	25	26	27	28	29	30	31	32	33
O. levis					X									X				X		X					X		X			X	X		♀
Pachydiplax longipennis			X				X		X	X				X		X		X							X	X		X		X			♀
Paltothemis cyanosoma												X		X																			♀
P. lineatipes		X	X		X	X		X		X	X	X	X	X	X		X		X	X		X		X	X	X				X			♀
Pantala flavescens			X	X	X	X	X	X	X	X		X	X	X	X		X	X	X	X	X	X	X	X	X	X	X	X		X	X		♀
P. hymenaea		X	X	X		X	X		X	X		X		X			X	X	X		X		X		X	X	X	X		X	X		♀
Perithemis domitia					X			X				X	X	X			X	X	X	X	X	X	X	X	X	X	X			X	X		♀
P. intensa			X				X	X				X	X	X		X	X	X			X				X	X							♀
P. mooma				X	X							X	X	X		X	X	X		X	X		X	X	X	X	X	X		X	X		♀
P. tenera							X			X																		X					♀
Planiplax sanguiniventris				X																			X	X				X		X	X		♀
Pseudoleon superbus		X	X		X				X	X	X	X	X	X			X	X	X	X	X	X		X	X	X		X		X			♀
Sympetrum corruptum	X	X	X			X	X		X	X	X	X	X	X	X	X	X	X	X	X				X		X		X		X			♀
S. illotum			X		X				X	X		X	X	X	X	X	X	X	X	X	X	X							X	X			♀
S. signiferum										X								X															♀
Tauriphila argo								X															X	X									♀
T. australis					X									X				X						X			X			X			♀
T. azteca					X			X						X				X					X	X	X		X			X			♀
Tholymis citrina				X	X				X			X		X				X					X	X			X			X			♀
Tramea abdominalis				X	X									X		X	X			X			X	X			X	X		X	X		♀
T. binotata					X																		X				X			X			♀
T. calverti			X	X	X				X	X				X				X					X	X	X		X			X	X		♀
T insularis				X																													♀
T. lacerata			X																				X	X				X		X	X		♀
T. onusta		X	X		X			X	X			X	X	X			X	X	X	X	X		X	X	X	X	X	X		X	X		♀
Uracis fastigiata																																	♀
U. imbuta					X															X			X										♀
TOTALS BY STATE	**8**	**29**	**56**	**40**	**170**	**48**	**29**	**53**	**40**	**55**	**12**	**91**	**104**	**118**	**26**	**54**	**102**	**119**	**84**	**141**	**72**	**46**	**71**	**131**	**72**	**58**	**89**	**108**	**6**	**220**	**59**	**1**	

1. AGS; 2. BC; 3. BCS; 4. CAM; 5. CHIS; 6. CHIH; 7. COAH; 8. COL; 9. DF; 10. DGO; 11. GTO; 12. GRO; 13. HGO; 14. JAL; 15. MEX; 16. MICH; 17. MOR; 18. NAY; 19. NL; 20. OAX; 21. PUE; 22. QRO; 23. QR; 24. SLP; 25. SIN; 26. SON; 27. TAB; 28. TAMP; 29. TLAX; 30. VER; 31. YUC; 32. ZAC; 33. ♀

Odonata : Biology of Dragonflies
Ed.: B.K. Tyagi
Scientific Publishers (India), 2007, pp. 137-151

10

ODONATA INVENTORIES IN BRITISH COLUMBIA, CANADA: DETERMINING THE CONSERVATION STATUS OF ODONATA SPECIES

ROBERT A. CANNINGS[1], LEAH R. RAMSAY[2] and SYDNEY G. CANNINGS[3]

[1] *Royal British Columbia Museum, 675 Belleville Street, Victoria, British Columbia Canada V8W 9W2*
e.mail: RCannings@royalbcmuseum.bc.ca
[2] *British Columbia Conservation Data Centre, British Columbia Ministry of Environment, 395 Waterfront Crescent, Victoria, British Columbia, Canada V8W 9M2*
[3] *Nature Serve Yukon, Yukon Territorial Government, Box 2703, Whitehorse, Yukon, Canada Y1A 2C6*

ABSTRACT

The Order Odonata in British Columbia, Canada, is reasonably well known but, before 1996, systematic inventories were confined to small areas or even single localities. From 1996 to 2006, concentrated surveys were conducted annually throughout much of the province. The main goals of these surveys were to determine the status and habitat needs of Odonata, with an emphasis on the species considered to be at risk; to increase public awareness of dragonflies, their ecology and conservation; and to encourage interest in dragonfly monitoring and research in the various regions. Each year, public lectures about dragonflies were given in local communities around the province and volunteer collectors from these places participated in the project.

To demonstrate how such inventory provides information for assigning conservation status ranks to dragonfly species, the changes in these ranks over a nine-year period were examined. Preliminary conservation status ranks were assigned to British Columbia's Odonata species in 1993. Subsequently, inventory efforts focused on those species considered rare or at risk in order to determine more accurately their status and habitat requirements. During these surveys, the geographical distributions of many species were expanded, our knowledge of habitat preferences increased and five species were added to the provincial list. Many of the targeted species were found to be more abundant than previously thought and their conservation ranks were changed accordingly. Others were recorded only rarely or not at all. Accurately ranking poorly known species is challenging, particularly if samples are small or if much of their range is inaccessible. By increasing our knowledge of these dragonflies and their habitats, we can assign species ranks more with more confidence, thus ensuring that conservation efforts will target the species and habitats that truly require them.

INTRODUCTION

In a multi-year project in British Columbia, Canada (BC), the Royal British Columbia Museum (RBCM) and the British Columbia Conservation Data Centre (CDC) of the provincial Ministry of Environment joined forces to study the Odonata of the province in a more systematic manner than had occurred in the past. Except for 2004, field work was undertaken every year from 1996 to 2006. The project will end in 2007.

This project addresses the problem of monitoring aquatic ecosystem integrity and diversity in BC through an extensive and, at certain sites, intensive survey of the Odonata. The broad goals of the overall project were to: (i) determine the status and habitat management needs of the 82 species then known in the province, with an emphasis on the species considered to be at risk; (ii) increase public awareness of dragonflies, their ecology and conservation; and (iii) foster an ongoing, local interest in dragonfly monitoring and research.

Linking the programs of the RBCM and CDC in a joint project is logical. The RBCM is responsible for collecting and documenting the natural world of BC and disseminating information about it to the public and the scientific community. Thus, the collection of Odonata specimens and ecological data, the creation of specimen databases, distribution maps and publications and the involvement of the public in these activities are good examples of its mission. The CDC gathers and manages information on rare, threatened and endangered species and natural plant communities in BC. One of its major goals is to portray, as accurately as possible, the risk of extinction or vulnerability of species and habitats through status ranks.

The first stage in this process is to draw up lists of all species in the province and give each species a numerical value indicating its conservation status rank. The ranks are based on weighted criteria as determined by NatureServe, an organization that represents an international network of agencies (known as Natural Heritage Programs or Conservation Data Centres) operating in the United States, Canada and Latin America that document biological diversity. These agencies are a source of information about rare and endangered species and threatened natural plant communities and they promote the use of standardized procedures for determining the status of these species and natural plant communities. The criteria used by Nature Serve are the number of populations, the number of individuals, population trends (long and short term), threats (broken down into scope, severity, and immediacy) and distribution (separated into total range and area of occupancy). In BC, environmental specificity and intrinsic vulnerability estimates are also included. Each species is assigned an 'S' (sub-national or provincial) rank ranging from 1 (most at risk) to 5 (secure) (Nature Serve 2006). British Columbia's Red and Blue Lists are based on S ranks -- S1, S1S2, S1S3, and S2 are Red List ranks; S2S3, S3, and S3S4 are Blue List ranks. All other ranks place a species on the Yellow List. These terms are more fully defined in the Methods section.

The Odonata is a relatively well-known order of insects whose members breed in a wide variety of aquatic habitats. Some species are specialists that use discrete habitats and others are generalists that are able to survive in a wide variety of environments. In BC, wetland habitats have been, and continue to be, altered, lost, or destroyed because of urban development, agriculture and resource extraction (Stevens *et al.,* 1995). Estimates indicate that 65 to 80 per cent of the wetlands in southern BC have been altered or destroyed, depending on the region. Odonata species are considered a priority for inventory for several reasons (Scudder, 1996). Unlike most invertebrates, they can be relatively easily identified, even in the field in many cases. They are upper-level predators in the invertebrate food chain and have often been identified as indicators of ecosystem health (Walker &

Corbet, 1975; Carle, 1979; Takamura *et al.,* 1991; Clarke & Samways 1996; Trevino, 1997; Corbet, 1999). Many species are habitat-specific, and their presence can be used to characterize healthy wetlands of all sorts. Within the constraints of weather, Odonata surveys are well suited for long-term monitoring programs. Finally, because they are large, colourful, diurnal creatures with interesting behaviours, Odonata are excellent subjects for nature interpretation programs and public education programs about aquatic ecosystems in general.

In 1993, the CDC ranked BC's odonates (82 species at that time) with respect to their known conservation status. However, although dragonflies are among the best known insects in BC, detailed knowledge of species distributions, relative abundance and habitat preferences or requirements was very limited, particularly for the less common species. The main sources of information were Cannings & Stuart (1977), records gathered by collectors over the years (specimens housed in the RBCM and the University of BC), Whitehouse's (1941) notes, and habitat descriptions from more general guides or references (*e.g.,* Walker, 1953). Before 1996, systematic inventories were confined to small areas or even single localities (e.g., Cannings & Cannings, 1983, 1987). To improve scientific knowledge about the Odonata, to gather information for use in wetland management and conservation planning and to enable more accurate species ranking, we decided to inventory the province as systematically and as quickly as possible, given its huge area of almost 1,000,000 square kilometres.

METHODS

Between 1996 and 2005, Odonata surveys were conducted throughout much of BC. Staff from the CDC and the RBCM organized these surveys; government staff, contractors and volunteer naturalists participated in a number of teams each year. The sites visited were chosen based on historical records, recommendations by local biologists and naturalists, examination of topographical maps and air photographs, and by scouting the region by vehicle and, in a few cases, by helicopter.

Study Areas and Habitats

The annual survey regions are mapped in Fig. 1.

1996: Lower Mainland and southern Vancouver Island (Hutchings 1997, Kenner 1997).

1997: Okanagan (Cannings 1998) and northeastern BC (Peace River-Fort Nelson regions).

1998: East Kootenay region (Cannings *et al.*, 2000).

1999: West Kootenay region (Cannings *et al.*, 2000).

2000: East and south of Prince George including the western slopes of the Rocky Mountains from Tête Jaune Cache north to Pine Pass; the northeastern Chilcotin Plateau (Nazko area) and the northern Cariboo Mountains (Likely, Quesnel Lake) (Cannings *et al.*, 2004).

2001: Vanderhoof-Omineca-Williston regions (Cannings *et al.*, 2004).

2002: North Tweedsmuir-Babine-Bulkley-Skeena regions (Cannings *et al.*, 2004).

2003: BC Northwest, including the Highway 37 corridor, the Atlin area, Skagway Highway and the Haines Road (Cannings *et al.*, 2004).

2004: No inventory

2005: North coastal region from the Nass River south through Prince Rupert and Terrace to Kitimat.

2006: Queen Charlotte Islands (minimal); northern Vancouver Island; Bute Inlet on central mainland coast.

2007: This final year of the surveys will concentrate on the southern Cariboo and Shuswap regions in the southern BC Interior.

Dragonfly habitats in BC are varied - they range from warm, eutrophic *Typha/Scirpus* marshes at the lowest elevations of the southern valleys (Fig. 2) to cold *Carex* marshes (Fig. 3), rain-drenched coastal bogs and a variety of bog and fen types (Fig. 4) from low to high elevations throughout the province. Grasslands in the semi-arid southern Interior support saline lakes and ponds; these are often ephemeral (Fig. 5). Lakes of various types with diverse ecosystems along their shorelines abound. Warm and cool springs and their outflows produce a few special species. Odonata species typical of running waters (*e.g.*, Gomphidae) are not diverse; because mountain physiography dominates the province, warm streams are uncommon.

METHODOLOGY

To maximize the chances of recording all species present in an area, we surveyed throughout the odonate flying season and critical sites were visited several times. Some early- or late-flying species have been under-represented in earlier collections because entomologists collect primarily in mid-summer. At each site, we collected voucher specimens of each species observed. We collected adults using aerial nets and estimated relative abundance for each species at a site. Adult voucher specimens were killed with ethyl acetate and fixed in acetone for 8 to 24 hours; they were then dried and stored in glassine envelopes until accessioned by the RBCM. We collected larvae in aquatic dip nets and preserved them in 70% ethanol; exuviae were also collected into ethanol. We identified specimens using Walker (1953, 1958), Walker & Corbet (1975), Cannings & Stuart (1977), Westfall & May (1996), Cannings (1996, 2002a) & Paulson (1998), and by comparing the specimens to material in the RBCM and the Spencer Entomological Museum at the University of BC. We described the habitat for each collection site by recording wetland type, dominant vegetation, water and air temperature, pH and amount of disturbance. Wetland site associations (MacKenzie & Moran 2004) were assigned to each dragonfly species depending on its apparent habitat requirements.

In 1995 the criteria used for conservation rank assessments of dragonfly species were the number of element occurrences, population size, trend (increase or decrease in population size, geographical range or number of occurrences), threats and the amount of protection of habitats. The 2004 assessments further separated threats into severity, scope and immediacy; the trends were divided into long and short term, and environmental specificity and intrinsic vulnerability estimates were added. The criteria were only used when relevant information was known.

Definitions of Provincial (sub-national) Conservation Ranks

S1: Critically Imperiled -- extremely rare (often 5 or fewer occurrences) or has some factor(s) (*e.g.,* steep declines) making it especially vulnerable to extirpation.

S2: Imperiled -- rare due to restricted range, few populations (often 20 or fewer), steep declines or other factors making it very vulnerable to extirpation.

S3: Of Special Concern -- restricted range, relatively few populations (often 80 or fewer), recent and widespread declines or other factors making it vulnerable to extirpation.

S4: Apparently Secure -- uncommon but not rare; some cause for long-term concern due to declines or other factors such as a small range.

S5: Secure -- common, widespread, and abundant in the province.

S#S#: Range Rank -- A numeric range rank (*e.g.,* S2S3) is used to indicate any range of uncertainty about the status of the species.

Definitions of Colour Lists

Red List: Species are extirpated, endangered or threatened. *Endangered* species are those facing imminent extirpation or extinction. *Threatened* species are those likely to become endangered if limiting factors are not reversed). Provincial ranks S1, S1S2, S1S3 S2 and SH.

Blue List: Species *of special concern.* Particularly sensitive to human activities or natural events but not endangered or threatened. Provincial ranks S2S3, S3 and S3S4.

Yellow List: Species that are *not at risk* in British Columbia. Includes all provincial ranks not listed in red and blue above.

RESULTS

We collected 83 of the province's 87 known species from over 1580 sites; however, we recorded 85 species during the past eight years, leaving only two (*Pantala hymenaea,* an accidental species observed only once, and *Enallagma civile*, collected only once before) unrecorded during this period. We added five new species to the provincial species list (Cannings *et al.*, 2005) and collected approximately 21,000 specimens for the RBCM collections. We also changed the conservation status ranks of 23 species; when the odonata were last ranked (in 2004) the Red and Blue Lists for BC contained 21 species at risk (Appendix 1) and the provincial list stood at 87 species.

New Species

These records are considered in more detail by Cannings *et al.* (2005). *Somatochlora kennedyi* Walker (Fig. 6) was known from Loon Lake in the Yukon Territory, 3 km north of the BC border (Cannings *et al.,* 1991) and from the southern Northwest Territories (Walker & Corbet, 1975); therefore, it was a target species during the 1997 survey in the Northeast. We found and collected only one that year, a female flying near a sedge fen near Andy Bailey Provincial Park southeast of Fort Nelson (Kenner, 2000). Since then, this emerald has been collected in a variety of habitats and several other localities across the northern half of the province.

Edmund Walker collected *Somatochlora forcipata* (Scudder) in the 1920s at Boom Creek, Alberta, about 3 km from the BC-Alberta border in Banff National Park (Walker & Corbet, 1975). For decades this was the only record west of Manitoba. In 1998, by focusing on sites that matched Walker's descriptions of this species' habitat (small, spring-fed boggy streams "following a devious course" in "glades in spruce forests" (Walker & Corbet, 1975)), we discovered three breeding localities in BC -- two in Yoho National Park (Fig. 7) and one in Kootenay National Park. Subsequently, we found the species in three more locations in central BC.

In 1997, *Lestes forcipatus* Rambur (Fig. 8) was discovered in Washington State, the first record west of Montana (Paulson, 1997). In Canada, it was not known west of Saskatchewan (Walker, 1953). In 1998, we collected it in the Bluewater Creek area in the Rocky Mountain Trench north of Golden. *Lestes forcipatus* is very similar to the widespread and abundant *L. disjunctus* Selys, and although females are readily separated, the males of the two species are almost indistinguishable (Simaika & Cannings, 2004). Examination of specimens in the RBCM revealed that *L. forcipatus* had been collected previously in BC but, because often only the males are collected, specimens had been misidentified as *L. disjunctus. Lestes forcipatus,* it turns out, is widespread in southern and central BC, although it is not often as abundant as other *Lestes* species at any site (Cannings & Simaika, 2005)

We discovered *Calopteryx aequabilis* Say at Christina Creek in the Boundary region in 1998. Previously, the nearest locality was in Stevens County, about 100 km to the south in Washington State (Paulson, 1997). It is a striking species that requires relatively warm streams, a rather uncommon habitat in BC. It is still known only from Christina Creek in BC and it is unlikely that it occurs in many other locations in the southern parts of the province.

Somatochlora brevicincta Robert was one of the most surprising discoveries; previously, it was known from only a few locations in Quebec, the Atlantic provinces and the state of Maine. After its initial discovery on Bell Mountain near McBride in 2000, we collected it at another nine locations, all along the western margin of the Rocky Mountains.

List Changes

We changed the rank of 23 species (26.4%) -- 20 are now less at risk and 3 are considered more at risk (Table 1). These numbers include changes made to conservation status ranks of species that were unknown in the province before the surveys. These were initially listed in higher risk categories than they are now after further inventory. This pattern is consistent with other taxonomic groups both in BC and globally (Butchart *et al.*, 2005; Quayle and Ramsay, 2005).

Table 1. Changes in the conservation status of British Columbia Odonata species between 1995 and 2004

Scientific name	English name	1995	2000	2004	Rank change
Lestes forcipatus Rambur	Sweetflag Spreadwing	unknown	S3	S4	+2.0
Coenagrion angulatum Walker	Prairie Bluet	S1S2	S4	S3S4	+2.0
Coenagrion interrogatum (Hagen)	Subarctic Bluet	S4	S4	S4S5	+0.5
Ischnura erratica Calvert	Swift Forktail	S3?	S4	S4	+0.5
Nehalennia irene (Hagen)	Sedge Sprite	S4	S5	S5	+1.0
Aeshna constricta Say	Lance-tipped Darner	S2S3	S2S3	S2	-0.5
Aeshna septentrionalis Burmeister	Azure Darner	S4S5	S3S4	S4	+0.5
Aeshna tuberculifera Walker	Black-tipped Darner	S2S3	S3	S4	+1.5
Gomphus graslinellus Walsh	Pronghorn Clubtail	S2	S2	S2S3	+0.5
Ophiogomphus colubrinus Selys	Boreal Snaketail	S3?	S3?	S4	+0.5
Stylurus olivaceus (Selys)	Olive Clubtail	S2	S2	S1S2	-0.5
Epitheca canis MacLachlan	Beaverpond Baskettail	S2S3	S2S3	S3	+0.5
Somatochlora brevicincta Robert	Quebec Emerald	unknown	S1	S3	+2.0
Somatochlora cingulata (Selys)	Lake Emerald	S2S3	S3	S4	+1.5
Somatochlora forcipata (Scudder)	Forcipate emerald	unknown	S1S2	S2S3	+1.0
Somatochlora franklini (Selys)	Delicate Emerald	S4S5	S4S5	S5	+0.5
Somatochlora hudsonica (Selys)	Hudsonian Emerald	S5	S4S5	S4S5	-0.5
Somatochlora kennedyi Walker	Kennedy's Emerald	unknown	S1S2	S3S4	+2.0
Somatochlora septentrionalis (Hagen)	Muskeg Emerald	S3S4	S3?	S4	+0.5

Somatochlora whitehousei Walker	Whitehouse's Emerald	S4	S4	S5	+1.0
Erythemis collocata (Hagen)	Western Pondhawk	S2S3	S2	S3	+0.5
Leucorrhinia patricia Walker	Canada Whiteface	S3S4	S3?	S4	+0.5
Pachydiplax longipennis (Burmeister)	Blue Dasher	S2S3	S2S3	S3S4	+1.0

Other Knowledge

In addition to increasing our knowledge of the habitats and distributions of the species of Odonata that were considered to be at risk, we learned much more about the habitat needs, status and behaviour of the more common species. For example, *Somatochlora walshii* (Scudder) was originally known from only half a dozen locations scattered throughout the southern half of the province. It is now clear that this species is widespread across the southern two-thirds of the province and that it inhabits many spring-fed wetlands that contain slow-moving water. *Nehalennia irene* (Hagen), another example of an early-flying species missed by entomologists active in mid-summer, is now known as far north as the Yukon and southcentral Alaska (about 63°N) in the Northwest, as well as on Vancouver Island and the Columbia Basin in the Southeast. When *The Dragonflies of British Columbia* (Cannings & Stuart, 1977) was published in 1977, this tiny iridescent damselfly had only been found in the interior of the province and no further north than about 54°N. Our data also indicate that *N. irene* has a longer flying season than previously known. Additionally, we had always assumed that *Aeshna subarctica* (Walker) was more common than the scanty records indicated; our surveys confirmed this (Fig. 9). We recorded it at many peatlands in the Columbia-Kootenay region and northern BC. Even though *Somatochlora minor* Calvert was known from the southern Yukon (Cannings *et al.,* 1991), it had been found only in BC south of 52° N; we now have several records from as far north as Fort Nelson. Some other species that were found to be more common than earlier evidence suggested are *Aeshna septentrionalis* Burmeister, *A. tuberculifera* Walker, *Ophiogomphus colubrinus* Selys (Fig. 10), *Somatochlora cingulata* (Selys), *S. septentrionalis* (Hagen) and *Leucorrhinia patricia* Walker (Fig. 11). However, some rare species remain truly rare, including *Aeshna constricta* Say, *Octogomphus specularis* (Hagen), *Stylurus olivaceus* (Selys) and *Macromia magnifica* MacLachlan.

DISCUSSION

We performed BC Odonata surveys on a relatively small scale. Usually three to five individuals or teams examined an area for a limited amount of time over a single flight season. However, even with this limited amount of effort (there are many areas left unsurveyed, localities to revisit and species to look for), the surveys have produced a much more defensible set of conservation status ranks than previously existed.

The revised S ranks, in addition to their provincial applications, were used to inform the General Status ranks nationally. The first national ranking of dragonflies was completed in 2003 (Canadian Endangered Species Conservation Council, 2006). These ranks are then used to help set priorities for the Committee on the Status of Endangered Wildlife in Canada (COSEWIC), which has recently added the Odonata to the taxa that it assesses for conservation action across Canada.

We were able to visit some critical sites (with unusual habitats or especially diverse dragonfly faunas) repeatedly and obtained a good estimate of the relative abundance of each species as well as accurate species lists for those sites. These baseline data can be used to monitor populations at these important localities. Our data are also useful in monitoring individual species over a number of sites.

When the assemblages of Odonata and their habitats are associated, one can begin to predict species occurrence in particular habitats within a certain geographical range. Such predictions can be applied to rare as well as common species. The reverse is also true - certain dragonfly species can be used to predict wetland types. For example, in BC the presence of *Argia vivida* Hagen is a sure sign of spring-fed waters. Dragonfly populations used in conjunction with vegetation type can predict and define the presence of rare ecosystems.

The fostering of interest about the Odonata amongst amateurs is a less tangible but beneficial result of our surveys. We held dragonfly collection workshops and field trips for local naturalists. These, along with slide shows, lectures, and informal information sessions encouraged local participation. The results of the surveys were assembled to help produce a new pictorial field guide, *Introducing the Dragonflies of British Columbia and the Yukon* (Cannings, 2002a). A 'Species at Risk' brochure was also produced which highlighted the Odonata that were red-listed in BC at the time of printing (Cannings, 2002b). The project was instrumental in the inclusion of up-to-date BC information in the recent checklist of the Odonata of Canada (Catling *et al.*, 2005). New distribution maps for species were developed and included a series of maps showing records collected both before and during the surveys (Fig. 7). These publications make it easier and more rewarding for naturalists to identify dragonflies and damselflies in the field; they also encourage and allow naturalists to provide reliable information about the group throughout their region. We need more students of Odonata: because of BC's large size and the strong seasonality of the dragonfly species, the recruitment of enthusiastic, well-informed amateurs is essential to furthering our understanding of dragonflies and damselflies in the province.

Our results highlight the value of intensive, targeted surveys and illustrate the dynamic nature of the conservation status rank lists. A common question posed by wildlife or land managers and policy makers is, "How can we reduce the number of species on tracked or 'at risk' lists?" Undertaking inventories that are focused on particular species or group of species is the most straightforward way to answer the question. Intensive collecting often reveals that many species appear rare only because they are poorly sampled. In the course of establishing species' distributions and numbers during inventories, other criteria that are used to establish conservation status ranks can be determined, including habitat or threats. The resulting lists can then be used to focus efforts and resources on those species or habitats that are truly at risk, either by addressing the threats or considering recovery planning. This is, of course, true for not only the Odonata but for all taxa. Inventories not only fill key gaps in our knowledge, they also focus future studies on species and regions for which there is still a lack of information.

ACKNOWLEDGEMENTS

This work is dedicated to Bastiaan Kiauta on his 70th birthday in appreciation for his important contributions to odonatology around the world. Bastiaan has long had a special affection for British Columbia, especially its diverse landscapes and wildlife.

All of these surveys were cooperative efforts funded by a variety of sources and involving several agencies and many volunteers. Financial support and aid in kind came from the Habitat Conservation Trust Fund, the Royal BC Museum, the BC Ministries of Sustainable Resource Management and Water, Land and Air Protection (formerly Ministry of Environment, Lands and Parks and now the Ministry of Environment), Parks Canada, the Columbia Basin Trust through the *Living Landscapes* project of the Royal BC Museum, and Forest

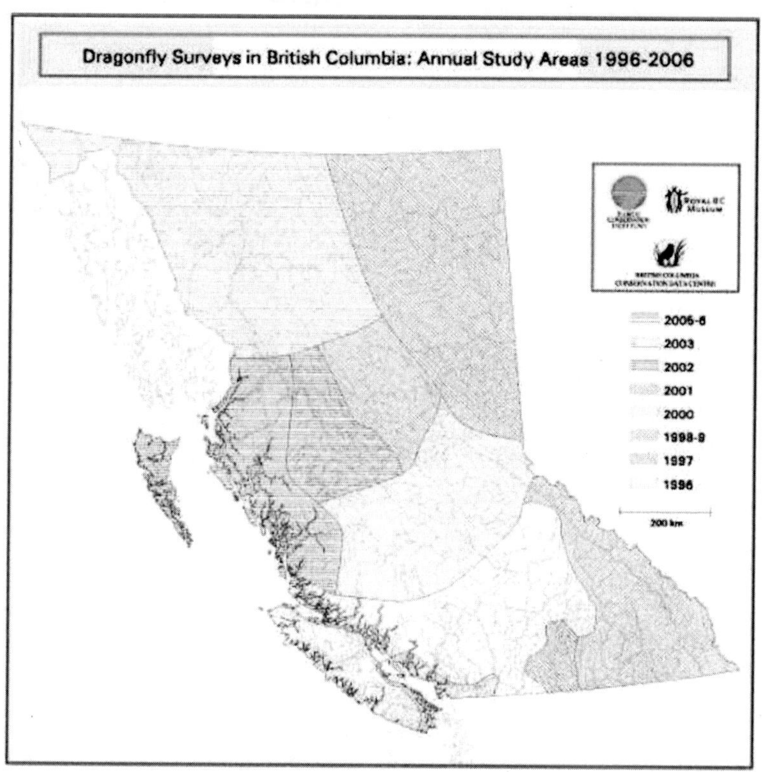

Figure 1. Odonata survey areas in British Columbia, 1996-2006.

Figure 2. Eutrophic marsh dominated by *Carex*, *Scirpus* and *Typha*, Columbia River Valley near Invermere, British Columbia. Photo: Robert A. Cannings.

Figure 3. *Carex* marsh at Horsethief Creek near Invermere, British Columbia. Photo: Robert A. Cannings.

Figure 4. Fen near Takla Landing, British Columbia. Photo: Robert A. Cannings.

Figure 5. Saline lake in grassland near Kamloops, British Columbia. Photo: Richard J. Cannings.

Figure 6. *Somatochlora kennedyi*, male -- a species found in British Columbia for the first time during the surveys of 1996-2006. Photo: Old Crow, Yukon; Robert A. Cannings.

Figure 7. Rob Cannings collecting some of the first *Somatochlora forcipata* specimens in British Columbia, 1998. Photo: Kicking Horse Pass, Yoho National Park, British Columbia; Sydney G. Cannings.

Figure 8. *Lestes forcipatus*, female. First collected in British Columbia in 1998 after being confused with *L. disjunctus* for decades, this damselfly is now known to be widespread in the province. Photo: Prince George, British Columbia; Robert A. Cannings.

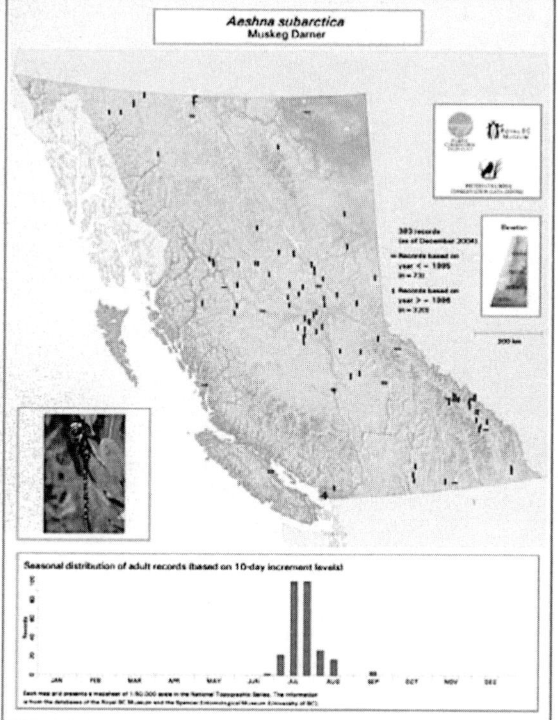

Figure 9. Distribution map of *Aeshna subarctica* collection localities in British Columbia. Horizontal red bars indicate records from or before 1995; vertical blue bars are records from the surveys of 1996 to 2004. The histogram represents records of adult dragonflies, indicating the general flight period across British Columbia.

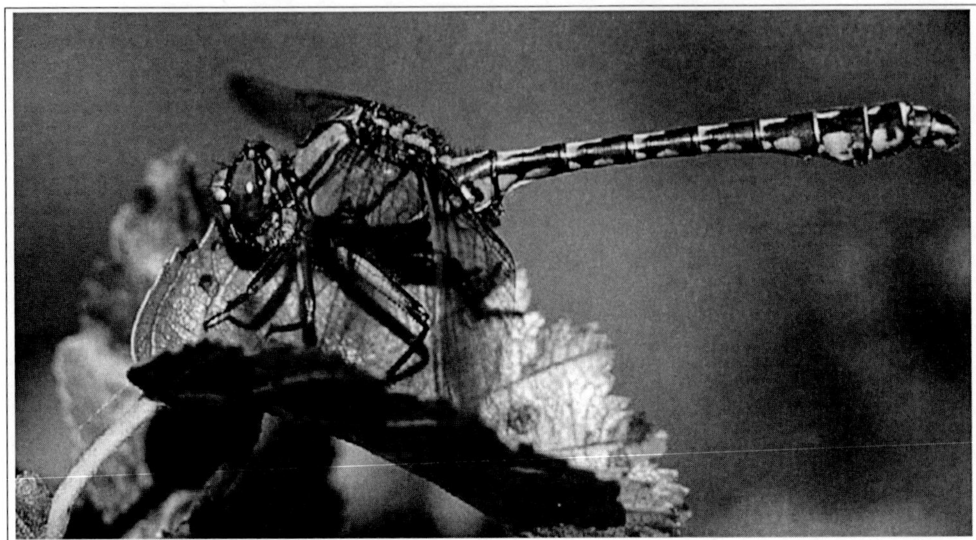

Figure 10. *Ophiogomphus colubrinus*, male a boreal species rarely collected before intensive surveys in 1996-2006 showed it to be common in suitable habitat across northern and central British Columbia. Photo: Tezzeron Creek, Fort St. James, British Columbia; Robert A. Cannings.

Figure 11. *Leucorrhinia patricia*, male. In British Columbia this boreal Nearctic dragonfly is known to range south to only about 52.5°N in the Coast Mountains. Inventories have shown it to be widespread in peatlands in the northern two-thirds of the province. Photo: Heckman Pass, British Columbia; Netta Smith

Renewal BC. Leah Ramsay, Robert Cannings, Sydney Cannings and Gord Hutchings were the principal investigators. Additional collectors are too numerous to list, but several people participated year after year, including Mike Badry, David Fraser, Andrew Harcombe, Ian Hatter and Leah Westereng. BC Parks and Parks Canada provided collecting permits and Dennis Paulson helped with advice and identifications. David Blades, Claudia Copley and Moretta Frederick (Royal BC Museum) managed the database. Many others enthusiastically swung nets, curated specimens, pointed out wonderful wetlands and provided access to collecting locations.

REFERENCES

Butchart, S.H.M., A.J. Stattersfield, L.A. Bennun, S.M. Shutes, H.R. Akçakaya, J.E.M. Baillie, S.N. Stuart, C. Hilton-Taylor and G.M. Mace. 2005. Measuring global trends in the status of biodiversity: Red List indices for birds. Public Library of Science, Biology 2: 2294-2304.

Canadian Endangered Species Conservation Council (CESCC). 2006. Wild Species 2005: The General Status of Species in Canada. On the internet at www.wildspecies.ca/wildspecies2005/index.cfm?lang=e. Accessed 15 October 2006.

Cannings, R.A. 1996. The blue darners: dragonflies of the genus *Aeshna* in British Columbia. Cordillera 3(1): 28-38.

Cannings, R.A. 2002a. Introducing the dragonflies of British Columbia and the Yukon. Royal British Columbia Museum, Victoria, BC. 96 pp.

Cannings, R.A. 2002b. Rare dragonflies of British Columbia. British Columbia Ministry of Environment, Lands and Parks, and Ministry of Sustainable Resource Management, Victoria, BC. 6 pp.

Cannings, R.A. and S.G. Cannings. 1983. The Odonata of the Brooks Peninsula, Vancouver Island, British Columbia. Journal of the Entomological Society of British Columbia 80: 46-51.

Cannings, R.A. and S.G. Cannings. 1987. The Odonata of some saline lakes in British Columbia, Canada: ecological distribution and zoogeography. Advances in Odonatology 3: 7-21.

Cannings, R.A., S.G. Cannings, and L.R. Ramsay. 2000. The dragonflies (Insecta: Odonata) of the Columbia Basin, British Columbia: field surveys, collections development and public education. Royal British Columbia Museum, Victoria, British Columbia. 287 pp. On the internet at http://livinglandscapes.bc.ca/cbasin/www_dragon/pdf/dragonflies4.pdf. Accessed 15 October 2006.

Cannings, R.A., S.G. Cannings, L.R. Ramsay and G.E. Hutchings. 2005. Four species of Odonata new to British Columbia, Canada. Notulae odonatologicae 6(5): 45-49.

Cannings, R.A., L.R. Ramsay and S.G. Cannings. 2005. The dragonflies (Insecta: Odonata) of Northern British Columbia: field surveys, collections development and public education. Royal British Columbia Museum and BC Ministry of Sustainable Resource Management, Victoria, BC. Unpublished report to the Habitat Conservation Trust Fund.

Cannings, R.A. and J.P. Simaika. 2005. *Lestes disjunctus* Selys and *L. forcipatus* Rambur (Odonata: Lestidae): An evaluation of status and distribution in British Columbia. Journal of the Entomological Society of British Columbia 102: 57-63.

Cannings, R.A. and K.M. Stuart. 1977. The dragonflies of British Columbia. Handbook No. 35. British Columbia Provincial Museum, Victoria, BC. 254 pp.

Cannings, R.J. 1998. A survey of rare dragonflies and damselflies (Odonata) in the Okanagan and Similkameen Valleys. Unpublished report submitted to the British Columbia Ministry of Environment, Lands and Parks, Victoria, BC. 19 pp.

Cannings, S.G., R.A. Cannings, and R.J. Cannings. 1991. Distribution of the dragonflies (Insecta: Odonata) of the Yukon Territory, Canada with notes on ecology and behaviour. Contributions to Natural Science, Number 13. Royal British Columbia Museum, Victoria, BC. 27 pp.

Carle, F.L. 1979. Environmental monitoring potential of the Odonata, with a list of rare and endangered Anisoptera of Virginia, United States. Odonatologica 8: 319-323.

Catling, P.M., R.A. Cannings and P.M. Brunelle. 2005. An annotated check-list of the Odonata of Canada. Bull. of American Odonatology 9(1):1-20

Clark, T.E. and M.J. Samways. 1996. Dragonflies (Odonata) as indicators of biotope quality in the Krüger National Park, South Africa. Journal of Applied Ecology 33: 1001-1012.

Corbet, P.S. 1999. Dragonflies: behavior and ecology of Odonata. Cornell University Press, Ithaca, New York. 829 pp.

Dragonfly Society of the Americas. 2006. The Odonata of North America. On the internet at http://www.ups.edu/x7015.xml. Accessed 20 October 2006.

Hutchings, G. 1997. Status of dragonflies of conservation concern in the Georgia Depression Ecoregion. Unpublished report submitted to the British Columbia Ministry of Environment, Lands & Parks, Victoria, BC. 22 pp.

Kenner, R.D. 1997. Database for a survey of the dragonflies of conservation concern in the Georgia Depression Ecoregion (Fraser Valley component). Unpublished database submitted to the British Columbia Ministry of Environment, Lands and Parks, Victoria, BC.

Kenner, R.D. 2000. *Somatochlora kennedyi* (Odonata: Corduliidae): a new species for British Columbia, with notes on geographical variation in size and wing venation. Journal of the Entomological Society of British Columbia 97: 47-49.

MacKenzie, W.H. and J.R. Moran. 2004. Wetlands of British Columbia: a guide to identification. Land Management Handbook No. 52. Research Branch, BC Ministry of Forests, Victoria, BC. 287 pp.

NatureServe. 2006. NatureServe Explorer: an online encyclopedia of life [web application]. Version 4.2. NatureServe, Arlington, Virginia. On the internet at http://www.natureserve.org/explorer/. Accessed 30 October 2006.

Paulson, D. 1997. The dragonflies of Washington. Bulletin of American Odonatology 4(4): 75-90.

Paulson, D. 1998. Field key to adult Washington dragonflies (Odonata). On the internet at http://www.ups.edu/x6518.xml. Accessed 20 October 2006.

Quayle, J.F. and L.R. Ramsay. 2005. Conservation status as a biodiversity trend indicator: recommendations from a decade of listing species at risk in British Columbia. Conservation Biology 19: 1306-1311.

Scudder, G.G.E. 1996. Terrestrial and freshwater invertebrates of British Columbia: priorities for inventory and descriptive research. Working Paper 09/1996. Research Branch, BC Ministry of Environment, Lands and Parks. Victoria, BC. 206 pp.

Simaika, J.P., and R.A. Cannings. 2004. *Lestes disjunctus* Selys and *L. forcipatus* Rambur (Odonata: Lestidae): some solutions for identification. Journal of the Entomological Society of British Columbia 101: 101-109.

Stevens, V.F., F. Backhouse, and A. Eriksson. 1995. Riparian management in British Columbia: an important step towards maintaining biodiversity. Working Paper 13/1995. Research Branch, BC Ministry of Forests, and Habitat Protection Branch, BC Ministry of Environment, Lands and Parks, Victoria, BC.

Takamura K., S. Hatakeyama and H. Shiraishi. 1991. Odonate larvae as an indicator of pesticide contamination. Applied Entomology and Zoology 26: 321-326.

Trevino, J. 1997. Dragonfly naiads as an indicator of water quality. Technical Note No. 99. Watershed Protection Techniques 2: 533-535.

Walker, E.M. 1953. The Odonata of Canada and Alaska. Vol. 1. University of Toronto Press, Toronto, ON. 292 pp.

Walker, E.M. 1958. The Odonata of Canada and Alaska. Vol. 2. University of Toronto Press, Toronto, ON. 318 pp.

Walker, E.M., and P.S. Corbet. 1975. The Odonata of Canada and Alaska. Vol. 3. University of Toronto Press, Toronto, ON. 307 pp.

Westfall, M.J., and M.L. May. 1996. Damselflies of North America. Scientific Publishers, Gainesville, FL. 650 pp.

Whitehouse, F.C. 1941. British Columbia dragonflies (Odonata), with notes on distribution and habits. American Midland Naturalist 26: 488-557.

Odonata : Biology of Dragonflies
Ed.: B.K. Tyagi
Scientific Publishers (India), 2007, pp. 153-173

11

CYTOGENETICS OF AMERICAN ODONATA

LILIANA M. MOLA

Laboratorio de Citogenética y Evolución; Depto Ecología, Genética y Evolución; Facultad de Ciencias Exactas y Naturales; Universidad de Buenos Aires. Int. Güiraldes 2620, Ciudad Universitaria, Pabellón II, 4° Piso. (C1428EHA) Ciudad Autónoma de Buenos Aires.
e.mail: limola@ege.fcen.uba.ar

ABSTRACT

The current knowledge of the cytogenetics of American odonates is reviewed. Chromosome studies have been performed in nearly 330 species and subspecies of 9 Zygoptera and 6 Anisoptera families. The species analysed were collected in 15 countries: Canada, the US and Mexico from North America; Guatemala and Costa Rica from Central America; Dominica and Jamaica from the Caribbean; and Venezuela, Surinam, Brazil, Peru, Bolivia, Chile, Argentina and Uruguay from South America. The proportion of species analysed differs from one country to another. Although the most frequent haploid numbers are 12, 13 and 14, there is a wide variation from n=3 in *Macrothemis hemichlora* from Bolivia and Brazil to n=21 in *Orthemis nodiplaga* from Argentina. Two distinctive characters of the order are the presence of holokinetic chromosomes and the post-reductional type of meiosis. The most frequent sex chromosome mechanism is XX/X0 (female/male), the derived neo-XY system occurs in approximately 5.5% of the species, and the $X_1X_1X_2X_2/X_1X_2Y$ multiple system is only present in *Micrathyria ungulata*. The presence of a small pair of chromosomes, the m-chromosomes, is found in nearly 80% of the species. Studies of the distribution and composition of the heterochromatin are scarce. C-banding showed that autosomes usually have heterochromatic blocks in the telomeric regions, and that, in general, the sex chromosome in males is completely C-positive, or shows an intermediate staining. In the few species analysed with fluorochromes, the heterochromatin seems to have a heterogeneous molecular composition. Odonata shows a high degree of karyotypic constancy at both intra- and inter-specific levels. However, polytypisms for the number of autosomes, the sex chromosome mechanism, and the size of the *m*-chromosomes were described in some species. Inter-specific variation in the chromosome number and/or in the sex chromosome mechanism were seen in some genera. Fusions and, less frequently, fragmentations are the main chromosome rearrangements involved in karyotype evolution.

INTRODUCTION

Cytogenetics is a field of investigation which developed from genetic and cytology sciences. It comprises the behaviour of chromosomes during mitosis and meiosis, their origin and their relation to the transmission and recombination of genes (Rieger *et al.*, 1991). The analysis of the morphological features of chromosomes aiming to define the karyotype, the determination of DNA content, and the analysis of the meiotic behaviour of species are useful from both cytotaxonomic and phylogenetic points of view because they allow us to infer the evolution of karyotypes and, hence, to establish relationships among different taxa.

The number, size and morphology of chromosomes as well as the position of the constrictions are some of the features to be considered for a more complete analysis of the karyotype. In addition, there are a variety of techniques currently employed to reveal specific chromosome regions, such as C-banding, base-specific fluorochrome banding and NOR-banding. These techniques, which can also be performed in meiotic cells, have not only contributed enormously to our understanding of chromosome structure but also to a better knowledge of the role played by chromosomes in evolution and speciation. Although these techniques showed to be useful for characterizing the karyotype of insects, they have rarely been used in odonates.

The meiotic analysis can be used to diagnose karyotype abnormalities and deviations. The occurrence of unusual meiotic configurations (I, II heteromorphic and multivalent) suggests structural changes and is the result of chromosome pairing and presence of chiasmata during meiosis.

The order Odonata comprises more than 5300 species; on the basis of the number of species described in the last six decades, Tennessen (1997) assumes that the number of known species is, at least, half of the world's species, which are estimated to be around 10000. This order had historically been divided into the suborders Zygoptera, Anisozygoptera and Anisoptera. However, the suborders currently recognised are Zygoptera and Epiprocta, with the latter including the infra-orders Anisoptera and Epiophlebioptera. Anisoptera is the most representative infraorder within Epiprocta, while Epiophlebioptera is only represented by two species (*Epiophlebia superstes* and *E. laidlawi*, in Asia) (Fraser, 1957; Lohmann, 1996; Muzón & von Ellenrieder, 1998).

CHROMOSOME NUMBER

The earliest cytological studies in odonates were performed by Carnoy at the end of the nineteenth century (1885), in a work dealing with the cytology of arthropods. In the Americas, chromosome studies started at the beginning of the twentieth century, with the analysis of *Anax junius, Libellula luctuosa (sub L. basalis)* and *Sympetrum semicinctum* from the US made by Mc Gill (1904), Lefevre & Mc Gill (1908) and Smith (1916). From these pioneer studies to the present, the chromosome number has been determined in almost 700 species and subspecies belonging to 20 families (13 of Zygoptera, 1 of Epioplhebioptera and 6 of Anisoptera). Dr. B. Kiauta, who published numerous works dealing with the cytotaxonomy and evolution of odonates for over 30 years (1965-1995), stands out among the many authors that have contributed to the knowledge of the cytogenetics of this order worldwide.

Although the most frequent haploid numbers are 12, 13 and 14, being found in almost 93% of the analysed species, there is a wide variation from n=3 in *Macrothemis hemichlora* from Bolivia and Brazil to n=21 in *Orthemis nodiplaga* from Argentina (Fig. 1A, B, C, F) (Cumming, 1964; Kiauta, 1972a; Souza Bueno, 1982; Tyagi, 1982, 1985; Agopian & Mola, 1984a; Mola, 1992).

To establish the modal number for each suborden is a difficult task. Of the 20 families of suborder Zygoptera so far described (Davies & Tobin, 1984), only 13 have been studied at the chromosome level. In this suborder, half of the cytogenetically analysed species belong to the family Coenagrionidae, which shows the haploid modal number

14. Instead, the haploid modal number of the families Calopterygidae, Lestidae, Platycnemididae is 13. It is also the most frequent number in the analysed species of the Euphaeidae, Megapodagrionidae, Platystictidae, Protoneuridae and Synlestidae. In Polythoridae and Chlorocyphidae, the most frequent haploid number is 12, but the modal number cannot be ascertained because of the few species that have been studied. The haploid chromosome complement of the single species of the Pseudolestidae analysed is 9, and one of the two species of the Pseudoestigmatidae is n=6 and the other n=15 (Kiauta, 1972a; Tyagi, 1982; Mola, 1992; Sandhu & Walia, 1993, 1994a, 1996, 1997; Sandhu *et al.,* 1994, Durani & Sharma, 1995).

In the suborder Epiprocta, within the infraorder Epiophlebioptera, the chromosome number has been ascertained only for *Epiophlebia superstes* (n=13). In the infraorder Anisoptera, chromosome studies have been performed in six out of the eight families taxonomically described (Davies & Tobin, 1985). The family Aeshnidae shows the haploid modal number 14, the family Gomphidae, 12, and in the families Cordulegasteridae, Cordullidae and Libellulidae it is 13. The modal number in the family Petaluridae could not be so far established because, among the four studied species, two have n=9, one has n=10 and the remaining n=13 (Kiauta, 1972a; Tyagi, 1982, 1985; Mola, 1992; Sandhu & Gulati, 1994; Sandhu & Malhotra, 1994; Sandhu & Walia, 1994b, 1994c; Kiauta & Kiauta, 1995; Sharma & Durani, 1996; Perepelov *et al.,* 1998; Perepelov & Bugrov, 2001, Perepelov *et al.,* 2001).

In the Americas, chromosome studies have been performed in nearly 330 species and subspecies (ca. 47%) of 15 families (9 of Zygoptera and 6 of Anisoptera) the chromosome number being determined in about 85 species of Zygoptera and 240 of Anisoptera. The most analysed families, both in the Americas and worldwide, are Coenagrionidae of Zygoptera and Libellulidae of Anisoptera, together accounting for 56% of the species. The species analysed were collected from 15 countries: Canada, the US and Mexico from North America; Guatemala and Costa Rica from Central America; Dominica and Jamaica from the Caribbean; and Venezuela, Surinam, Brazil, Peru, Bolivia, Chile, Argentina and Uruguay from South America. The proportion of species analysed differs from one country to another. In North America, most of the available data pertain to US species (155 spp.), while in Canada the number of species analysed is 17 and in Mexico 6. In Central America and the Caribbean, only a limited number of species were studied: 12 from Jamaica, 3 from Guatemala, and 2 from Costa Rica and Dominica. In South America, the most

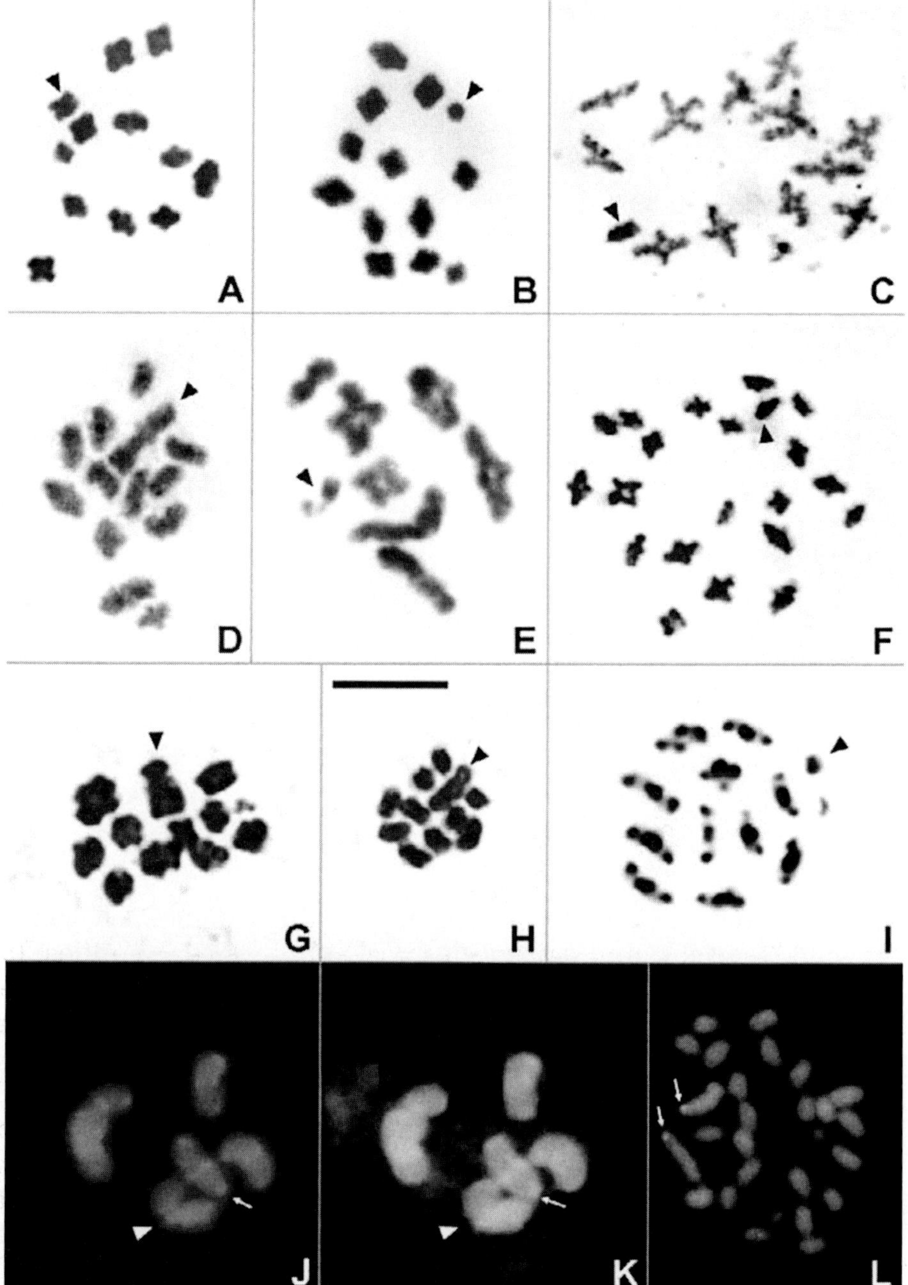

Figure 1: Meiosis in: A) *Aphylla* sp. Diakinesis (n=11+X0); B) *Perithemis icteroptera* Diakinesis (n=12+X0); C) *Rhionaeshna confusa* Diplotene (n=13+X0); D) *Rhionaeshna bonariensis* Diakinesis (n=12+neo-XY); E) *Rhionaeshna planaltica* Diakinesis (n=7+neo-XY); F) *Orthemis nodiplaga* Diakinesis (n=20+X0); G) *Micrathyria longifasciata* Diakinesis (n=11+neo-XY); H) *Micrathyria ungulata* Prometaphase I (n=10+X_1X_2Y); I) *Tauriphila risi* Diakinesis (n=12+X0), C-banding; J-K) *Orthemis ambinigra* Prometaphase I (n=5+neo-XY), J- DAPI banding, K- CMA banding; L) *Coryphaeschna perrensi* Spermatogonial Prometaphase (2n=27), CMA banding. Arrowheads point to sex chromosomes. Arrows in J-L show DAPI or CMA bright bands. Bar=10 øm

representative studies were performed in species collected from Brazil (91 spp.), Bolivia (78 spp.) and Argentina (38 spp.) (Table 1), followed by Surinam (16 spp.), Uruguay (8 spp.), Peru (3 spp.), and Chile and Venezuela (1 sp. in each country) (Mc Gill, 1904; Lefevre & Mc Gill, 1908; Smith, 1916; Cumming,1964; Cruden, 1968; Kiauta, 1970, 1971a; Hung, 1971; Kiauta & Boyes, 1972; Kiauta, 1972b, c, 1973a, 1978; Kiauta & van Brink, 1978; Kiauta, 1979; Ferreira *et al.,* 1979; Kiauta & Kiauta, 1979, 1980a; Gama *et al.,* 1981; Goñi & Abenante,1982; Souza Bueno, 1982; Agopian & Mola, 1984a,b; Mola & Agopian, 1985; Agopian & Mola, 1988; Rodríguez Capítulo *et al.,* 1991; Mola, 1992; Mola & Papeschi, 1994; Mola, 1995, 1996; Mola *et al.,* 1999; Papeschi *et al.,* 2001; de Gennaro, 2004).

Table I: Chromosomal data and provenience of Odonata species from Argentina

Taxa	2n	n (male)	*m*	Provenience	References
Aeshnidae					
Anax amazili	27	13+X0	+	Argentina (ER)	Mola *et al.,*1999
Coryphaeschna perrensi	27	13+X0	+	Argentina (M)	Mola *et al.,*1999 De Gennaro, 2004
Rhionaeshna bonariensis	26	12+neo-XY	+	Argentina (BA) Uruguay	Mola & Papeschi, 1994 Mola, 1995 *(1)*
R. confusa	27	13+X0	+	Argentina (BA)	Mola & Papeschi, 1994 Mola, 1995*(1)*
R. planaltica	16	7+neo-XY	-	Argentina (M)	Mola & Papeschi, 1994 Mola, 1995 *(2)*
Corduliidae					
Rialla villosa		12+X0	-	Argentina (N,Ch)	De Gennaro, 2004
Gomphidae					
Aphylla sp.	23	11+X0	+	Argentina (BA)	unpublished
Phyllocycla propinqua		10+X0	-	Argentina (M)	De Gennaro, 2004
Phyllocycla sp. 1		11+X0	+	Argentina (BA)	unpublished
Phyllocycla sp. 2	23	11+X0	-	Argentina (C,M)	unpublished
Libellulidae					
Brachymesia furcata	25	12+X0	-	Argentina (C) Brazil	Agopian & Mola, 1988 Ferreira *et al.,* 1979 Souza Bueno, 1982
Erythemis attala	25	12+X0 12+X0	+ -	Argentina (BA,M) Bolivia	Agopian & Mola, 1988 Cumming, 1964
Erythrodiplax	25	12+X0	+	Argentina (Cb)	Mola, 1996

atroterminata	26h		-	Uruguay	Goñi & Abenante,1982
E. corallina	25	12+X0	+	Argentina (BA)	Mola, 1996
E. fusca	25	12+X0	+	Argentina (BA,M)	Mola, 1996
			-	Bolivia	Cumming, 1964
			-	Guatemala	Cruden, 1968
			+	Surinam	Kiauta, 1979
			+	Brazil	Ferreira *et al.,* 1979
					Souza Bueno, 1982
E. lygaea	25	12+X0	+	Argentina (ER)	Mola, 1996
E. media	22	10+neoXY	+	Argentina (ER,C,M)	Mola, 1996
		10+X0	+	Bolivia	Cumming, 1964
	22h		+	Brazil	Kiauta & Boyes, 1972
		10+X0	+	Brazil	Ferreira *et al.,* 1979
					Souza Bueno, 1982
E. melanorubra	25	12+X0	+	Argentina (M)	Mola, 1996
				Bolivia	Cumming, 1964
				Venezuela	Kiauta & Boyes, 1972
E. nigricans	25	12+X0	-	Argentina (BA)	De Gennaro, 2004
	25	12+X0	+	Argentina (BA,ER)	Mola, 1996
				Uruguay	Goñi & Abenante,1982
E. ochraceae	25	12+X0	+	Argentina (M)	Mola, 1996
E. umbrata	25	12+X0	+	Argentina (M)	Mola, 1996
				Bolivia	Cumming, 1964
				Dominica	Cruden, 1968
				Surinam	Kiauta, 1979
				Brazil	Ferreira *et al.,* 1979
Macrothemis sp.	25	12+X0	+	Argentina (Cb)	unpublished
Miathyria marcella	25	12+X0	+	Argentina (BA)	Mola & Agopian, 1985
				Bolivia	Cumming, 1964
				Brazil	Ferreira *et al.,* 1979
Micrathyria catenata		12+X0	+	Argentina (C)	unpublished
				Brazil	Souza Bueno, 1982
M. hesperis		12+X0	+	Argentina (M)	Mola *et al.,* 1999
				Brazil	Ferreira *et al.,* 1979
					Souza Bueno, 1982
M. hypodidyma	25	12+X0	+	Argentina (BA,ER)	Agopian & Mola, 1988
	23	11+X0	+	Brazil	Souza Bueno, 1982
M. longifasciata	24	11+neo-XY	+	Argentina (ER)	Agopian & Mola, 1988
M. spuria		12+X0	+	Argentina (ER,C)	Mola *et al.,* 1999
				Bolivia	Cumming, 1964
M. ungulata	23	$10+X_1X_2Y$	-	Argentina (ER)	Mola *et al.,* 1999
Oligoclada laetitia	23	11+X0	+	Argentina (BA)	Mola *et al.,* 1999
		10+X0	+	Brazil	Souza Bueno, 1982
Orthemis ambinigra	12	5+neo-XY	-	Argentina (BA,M)	Agopian & Mola,

					1984b
O. discolor	23	11+X0	+	Argentina (M)	unpublished
O. nodiplaga	41	20+X0	-	Argentina (BA)	Agopian & Mola, 1984a
Pantala flavescens	25	12+X0	+	Argentina (BA,ER,C,M)	Agopian & Mola, 1988
				Uruguay	Cumming, 1964
				Bolivia	Ferreira *et al.,* 1979
				Brazil	Souza Bueno, 1982
Perithemis icteroptera	25	12+X0	-	Argentina (BA)	Mola & Agopian, 1985
P. mooma	25	12+X0	+	Argentina (BA)	Mola & Agopian, 1985
				Uruguay	Goñi & Abenante,1982
				Bolivia	Cumming, 1964
				Brazil	Ferreira *et al.,* 1979
					Souza Bueno, 1982
Planiplax erythropyga	25	12+X0	+	Argentina (ER)	Mola *et al.,* 1999
					De Gennaro, 2004
Tauriphila risi	25	12+X0	+	Argentina (BA)	Mola & Agopian, 1985
				Uruguay	

(1) sub Aeshna (von Ellenrieder, 2003)

(2) sub Aeshna cornigera planaltica (von Ellenrieder, 2003)

BA: Buenos Aires Province
ER: Entre Rios Province
C: Corrientes Province
M: Misiones Province
Cb: Córdoba Province
N: Neuquén Province
Ch: Chubut Province

CHROMOSOME MORPHOLOGY

A distinctive character of order Odonata is the presence of holokinetic chromosomes, that is to say, without a localized centromere. In contrast, monocentric chromosomes are found in the majority of the organisms.

Different authors, based on light microscopy studies, have proposed that chromosomes might be monocentric (metacentric and/or acrocentric), dicentric or holokinetic (Oksala, 1943; Seshachar & Bagga, 1962; Cumming, 1964; Kiauta, 1975; Frankovic and Jurecic, 1987). The single study on chromosome ultrastructure was performed in *Enallagma cheliferum* from Brazil; although mitotic chromosomes were not analysed, the particular disposition of the kinetochore plates on each of the four chromatids of the meiotic bivalents, prompted Gama et al. (1981) to propose that chromosomes were polycentric. This disposition of kinetochore plates at meiosis was also found in *Luzula*, a rush of the family Juncaceae, in which the presence of holokinetic chromosomes has been verified at mitosis (Braselton, 1971; Bokhari &

Godward, 1980; Braselton, 1981; Madej, 1998). Therefore, the holokinetic nature of the chromosomes can be assumed taking into account the results of the ultrastructural study, and the morphology and behaviour of mitotic chromosomes (Fig. 1L).

SEX CHROMOSOME DETERMINING SYSTEMS

The most frequent sex chromosome mechanism is XX/X0; male being the heterogametic sex (Fig. 1 A,B,C,F,I,J). The X chromosome is one of the smallest chromosomes of the complement in most of the species studied, except for some species of the family Gomphidae whose sex chromosome is one of the largest of the complement (Kiauta, 1968b, 1969a; Oksala, 1943; Suzuki & Saitoh, 1988, 1990; Perepelov *et al.,* 2001). The neo-XY sex chromosome system is derived from the X0 system by the fusion of the atavic X chromosome with an autosome. This system was first described by Oksala (1943), who provided a detailed analysis of the features and behaviour of this bivalent. This system is infrequent, as it has only been reported in approximately 5.5% of the species, and shows a heterogeneous distribution. It has been observed in about 3% of the species of Zygoptera (families Coenagrionidae, Lestidae, Megapodagri-onidae and Pseudostigmatidae), and in about 7% of the species of Anisoptera (families Aeshnidae, Gomphidae and Libellulidae) (Kiauta, 1968b, 1969a; Tyagi, 1985; Agopian & Mola, 1988; Mola, 1995, 1996, Mola *et al.,* 1999; Perepelov et al. 1998; Perepelov & Bugrov, 2002). Taking into account the small size of the X chromosome, the actual recognition of a heteromorphic sex bivalent at different meiotic stages depends on both the size of the autosome with which the X chromosome became fused and the degree of bivalent contraction. The male sex bivalent is homomorphic in approximately half of the species with this neo-system, as, for example, in *Orthemis ambinigra* (Fig. 1J,K). Its presence is inferred from the even number of chromosomes in spermatogonial cells and/or the absence of a univalent chromosome during the first meiotic division, and from the absence of a chromosome placed outside of the metaphase II plate. The latter behaviour is typical of the free-X chromosome in most of the species studied. In contrast, the sex bivalent is heteromorphic at both the first and second meiotic divisions in *Gynacantha interioris*, *Aeshna grandis*, *A. coerulea*, *Rhionaeshna planaltica* and *Micrathyria longifasciata* (Fig. 1E,G)). In some species, as for example *Erythrodiplax media*, *Crocothemis servilia mariannae*, *Hemianax ephippiger*, *Rhionaeshna bonariensis* and *Aeshna juncea*, the heteromorphism of the sex bivalent can only be recognized at diplotene and diakinesis, because at metaphase I it is masked by the high chromosome contraction (Fig. 1D) (Oksala, 1943; Seshachar & Bagga, 1962; Kiauta, 1969a; Kiauta, 1971b; Ferreira *et al.,* 1979; Kiauta, 1983; Agopian & Mola, 1884b, 1988; Mola, 1995, 1996). Some genera comprise a large proportion of species exhibiting this neo-system. This is the case, for example, in *Aeshna* (Aeshnidae), *Onychogomphus* (Gomphidae), and the American genus *Orthemis* (Libellulidae). Kiauta (1972a) considers that the presence of the neo-XY system is not related to any taxonomic group, but the examples mentioned above may indicate that some genera show a greater tendency for this type of system than others.

Another derived sex chromosome system is the $X_1X_1X_2X_2/X_1X_2Y$, present in *Micrathyria ungulata* (Fig 1H). This system may have derived from a neo-XY system, by a further fusion between the neo-Y and an autosome, named Y. The original neo-X would be the X_1 , and the free homologous autosome, which co-segregates with the X_1, would be the X_2 (Mola *et al.,* 1999).

The presence of multiple sex chromosome determining systems derived from the fragmentation of sex chromosomes has been documented in other insect groups with holokinetic chromosomes such as Heteroptera and Lepidoptera, but none has been reported for Odonata. (White, 1973; Ueshima, 1979; Mola & Papeschi, 2006).

m-CHROMOSOMES

In a study involving many heteropteran species of the family Coreidae, Wilson described in 1905 the presence of a small pair of chromosomes with a particular meiotic behaviour since they were unpaired at early prophase and associated non-chiasmatically at M I, and coined the term "*m*-chromosomes" for these elements of the complement. Lefevre and Mc Gill, in 1908, while studying the chromosome complement of the heteropteran *Anasa tristis* and the dragonfly *Anax junius,* also observed a small pair of chromosomes in the latter species and referred to them as *m*, in analogy with those of Heteroptera.

Makalovskaja (1940) first, and Srivastava and Das (1953) and Das (1956) later, argued that it was meaningless to call the smallest pair of chromosomes in odonates "*m*", since they form a chiasmatic bivalent and their meiotic behaviour is similar to that of the rest of the autosomes. Despite the validity of this statement, species continued to be cytogenetically characterized according to the presence or absence of the *m*-chromosomes.

The regular meiotic behaviour of the *m* pair was subsequently verified in all odonate species so far analysed, and is solely characterized by its size and in some cases by its negative heteropycnosis. The size of the smaller bivalent is highly variable within different species, and therefore its identification as *m*-chromosomes depended upon the investigator. In order to unify criteria, it has been proposed to denominate as *m* only those chromosomes smaller than or equal to the half of the immediately larger chromosome (Mola, 1992).

The origin and role of these chromosomes in the genetic systems of the species are still unknown and many theories have been proposed to answer these questions. Oguma (1930) hypothesized that the evolution of the chromosome numbers was accomplished by a gradual and continuous reduction of the chromatin in an autosomal pair, which first was transformed into *m* chromosomes and then disappeared ("*m*-chromosome theory"). Kiauta (1968a) instead proposed that *m*-chromosomes represent fragments of normal autosomes and that the variation in size observed among the different species or populations within a same species is only accidental since the fragmentation can take place at any time, at any place in the chromosome and in any geographic population. Consequently, this author concluded that although the *m*-chromosomes are found in nearly 80% of the species, their presence is not associated with any particular genus. Souza Bueno (1982) also postulated that the *m*-chromosomes were originated by fragmentations, but she considered that these occurred in ancestral individuals in response to an intrinsic and/or extrinsic variation; they then persisted over generations through the action of natural selection, and may explain their wide presence in odonate species. Although Souza Bueno (1982) acknowledges that her hypothesis on the origin of the *m*-chromosomes is in disagreement with that of Kiauta (1968a), she does not rule out the occurrence of potential fragmentations in currently living species, as well as the fact that the *m*-chromosomes can –or cannot– persist in the populations by means of natural selection.

MEIOTIC BEHAVIOUR

There are two types of meiosis in the groups with holokinetic chromosomes, the pre- and post-reductional meiosis. From a cytological viewpoint, White (1973) proposed that bivalents with a terminal or sub-terminal chiasma orientate axially or equatorially at metaphase I. In the pre-reductional meiosis, bivalents orientate axial to the spindle, *i.e.* with their long axis perpendicular to the equatorial plane; while in the post-reductional meiosis, bivalents orientate equatorial to the spindle, *i.e.* with their long axis parallel to the equatorial plane. From a genetic viewpoint, gametes produced by both types of meiosis are indistinguishable from each other. Although in theory the two possible bivalent orientations are easy to distinguish from each other, they are difficult to assess at metaphase I because of the small size and the high degree of bivalent condensation in most of odonate species.

Oksala (1943) was the first to postulate that meiosis is post-reductional based on the orientation of the bivalents on the equatorial plane, and on the equational division of the heteromorphic sex bivalents in the Aeshnidae. Later, Cumming (1964), provided an important evidence for this type of division describing the behaviour of a fusion autosomal trivalent in *Orthemis levis*. Such trivalent is orientated equatorially at metaphase I, and its three chromatids can be distinguished at all metaphases II as the result of equational division. Some authors further considered the possibility that the meiotic division of the autosomes and/or sex chromosomes in odonates was pre-reductional (Handa & Kochhar, 1985; Handa *et al.,* 1984; Mittal & Gandhi, 1982, 1984; Nokkala *et al.,* 2002). Kiauta (1975) and Souza Bueno (1982) instead believed that the division of the X chromosome in males was post-reductional, but were uncertain about the type of division of the autosomes.

A detailed analysis of the meiotic behaviour has allowed to determine the equatorial orientation of the bivalents at metaphase I in species with few chromosomes of large size; the equational division of sex bivalents, and autosomal and sex univalents (as, for example, in species of the American genera *Rhionaeshna* and *Erythrodiplax*), and the equational division of the sex trivalent in *Micrathiria ungulata* and the heteromorphic autosomal bivalent in *Brachymesia furcata* (Agopian & Mola, 1988; Mola, 1995, 1996; Mola *et al.,* 1999). These results, along with the descriptions provided by Oksala (1943) and Cumming (1964), lead to the conclusion that odonates possess a pre-reductional type of meiosis.

The typical meiotic development in males exhibits the following features: the X chromosome is usually positively heteropycnotic during early prophase I, bivalents regularly show only one chiasma from diplotene onwards, the bivalents and the sex chromosome do not present a particular disposition on the equatorial plane at metaphase I, and chromosomes migrate synchronously and almost parallel to the equatorial plane at anaphase I. At prometaphase II, the autosomal chromatids are joined by one of their telomeric regions adopting a characteristic "ε" shape, and the X chromosome is present in all the cells as a consequence of the post-reductional division, at metaphase II the sex chromosome is usually outside the metaphasic plane, and at anaphase II it migrates asynchronously with the autosomes, generally lying ahead.

HETEROCHROMATIN DISTRIBUTION AND COMPOSITION

Heterochromatin is an important component of the genome, since differences in its composition and distribution are likely to be related with the karyotypic evolution and genetic differentiation among related species.

C-banding is one of the most used techniques for detecting heterochromatin because it stains almost all constitutive heterochromatin segments. Heterochromatin can be either AT- or GC-rich, and its composition may vary among species, among different chromosomes within a same species and even in different regions of the same chromosome. The molecular composition can be determined by using base-specific fluorochromes. Among these, the most commonly employed are CMA (chromomycin A3) showing a preference for GC-rich DNA regions, and DAPI (4'-6-Diamidino-2-phenylindole) showing a preference for AT-rich DNA regions (Appels *et al.,* 1998; Sumner, 2003).

Studies of the distribution and composition of the heterochromatin in odonates are scarce (Agopian & Mola, 1984b; Meeldijk, 1984; Thomas & Prasad 1984; Prasad & Thomas, 1992; Suzuki & Saitoh, 1988; Franković & Jurečić, 1989; Perepelov *et al.,* 1998, 2001; Papeschi *et al.,* 2001; Perepelov & Burgrov, 2001, 2002; Nokkala *et al.,* 2002; De Gennaro, 2004).

In different species, C-banding showed that autosomes usually have heterochromatic blocks in both telomeric regions (Fig. 1I). Perepelov *et al.* (1998) classified the distribution pattern of the terminal heterochromatin into three categories: (a) presence of small blocks at pachytene, which are no longer observed towards metaphase I, (b) large asymmetric blocks and (c) large symmetric blocks. The sex chromosome in males appears to be completely C-positive, and may show an intermediate staining or may only exhibit C-positive terminal or interstitial bands. In the neo-X, the differential staining of the X chromosome still persists even when it is fused with an autosome (Agopian & Mola, 1984b; Suzuki & Saitoh, 1988; Prasad & Thomas, 1992; Perepelov *et al.,* 1998; Perepelov & Bugrov, 2001).

In the few species analysed using fluorochromes, the telomeric heterochromatin was found to be AT-rich (DAPI-positive bands), as in

Orthemis nodiplaga, Erythrodiplax nigricans and *Rhionaeshna planaltica*, or to have interspersed AT and GC repeats (equally localized positive DAPI and CMA bands), as in *Orthemis ambinigra* (Fig. 1J,K). Besides, a pair of positive telomeric CMA bands were observed in an autosomal pair of *Erythrodiplax nigricans* and *Coryphaeschna perrensi*, and in the neo-X of *Rhionaeshna planaltica* (Fig. 1L) (Papeschi *et al.,* 2001; De Gennaro, 2004). Despite the limited number of species analysed, these results may indicate that the heterochromatin of odonates has a heterogeneous molecular composition.

KARYOTYPE EVOLUTION

Fusions and fragmentations are the main chromosome rearrangements involved in the karyotype evolution of the holokinetic systems. Although it may be assumed that both types of rearrangements are equally probable due to the kinetic characteristics of these chromosomes, fusions are more frequent than fragmentations in many groups. This may be explained by the fact that even though chromosomes formed by fragmentation can persist through successive cell divisions, there is a limit imposed by the need to generate new telomeres, which are essential for chromosome structure (Mola & Papeschi, 2006).

Odonata seems to show a high degree of karyotypic constancy at both intra- and inter-specific levels, but population studies have revealed in some species the presence of cytotypes that differ in the number of chromosomes (e.g. *Oligoclada laetitia* and *Micrathyria hypodidyma*), or in the sex chromosome determining system (e.g. *Erythrodiplax media*) (Table 1). In some genera like *Orthemis* and *Rhionaeshna*, there is great inter-specific variation in the chromosome number and/or in the sex chromosome determining system due to fusions and, to a lesser extent, to autosomal fragmentations (Cumming, 1964 Cruden, 1968; Kiauta, 1969b; Ferreira *et al.,* 1979, Kiauta & Boyes, 1972; Kiauta, 1979; Souza Bueno, 1982; Agopian & Mola, 1988, Mola & Papeschi, 1994, Mola, 1995, 1996, Mola *et al.,* 1999).

Other karyotypic variations often observed in populations of the same species are related to the size of the *m*-bivalent, which in some cases was considered to be absent (Cruden, 1968; Ferreira *et al.,* 1979; Kiauta, 1968c, 1972a, 1973b, 1975, 1979; Kiauta & Kiauta, 1980b, 1983; Souza Bueno, 1982; Yadav, 1979). In the Argentinean species analysed, the presence of polytypism for the size of the *m*-chromosomes has been described in *Pantala flavescens, Miathyria marcella, Erythemis attala, Erythrodiplax atroterminata, E. fusca,* E. nigricans and *E. umbrata* (Mola & Agopian, 1985; Agopian & Mola; 1988; Mola,

1996; Mola *et al.,* 1999; De Gennaro, 2004). Such size variations may be the result of unequal reciprocal translocations between an *m*-chromosome and a member of any pair of autosomes. This type of rearrangement may lead to a change in the relative size of the involved chromosomes, but given the karyotypic characteristics of odonates, the difference in size would be easily detected only in the *m*-chromosomes. Likewise, variations in size might be due to changes in the heterochromatin content affecting exclusively the m-chromosomes, as for example in *E. nigricans* (de Gennaro, 2004). These changes could be associated with the speciation process, as observed in *Calopteryx virgo*, in which the different subspecies –and even some populations of *C. virgo virgo*–, exhibit a great variation in the size of the *m*- chromosomes (Kiauta, 1968c).

Great progress has been made in cytogenetics of odonates although much is still left to be done. There are some families in which cytological studies have only been conducted in a few species and others that have never been explored. Novel molecular cytogenetic techniques, such as fluorescent *in situ* hybridisation, will enhance and complement previous results obtained with classic cytogenetic methods. Indeed, the use of fluorescent labeled DNA probes (FISH) will allow to localise specific DNA sequences on the chromosomes (e.g. ribosomal DNA and highly repetitive DNA), and the use of total genomic DNA (GISH) will be useful to compare the genomic variations among related taxa. The information gathered by these techniques will contribute to a better understanding of the biosystematics and evolution of the order.

ACKNOWLEDGMENTS

The present work was supported by grants from CONICET (PIP 5927) and UBA (Ex 317) to Dr. L. Poggio and Dr. L. Mola. I wish to thank to Lic. S. Agopian, Lic. E. Taboada Carrillo, Dr. A. Papeschi, Lic. D. De Gennaro, Dr. J. Muzón, Dr. A. Rodríguez Capítulo and Dr. G. Jurzitza for their collaboration in different aspects of Odonata research. I also wish to thank Dr. L. Poggio, Lic. P. Rebagliati, Dr. A. Wulff and Dr. S. Rodríguez Gil for their continuous encouragement and friendship.

REFERENCES

Agopian, S.S. & L.M. Mola. 1984a. An exceptionally high chromosome number in *Orthemis nodiplaga* Karsch (Anisoptera, Libellulidae). Notulae Odonatologica 2(3): 45.

Agopian, S.S. & L.M. Mola. 1984b. Bandeo C en Odonata. Res. XV Cong. Soc. Arg. de Genética p. 17.

Agopian, S.S. & L.M. Mola. 1988. Intra and interspecific karyotype variability in five species of Libelluidae (Anisoptera, Odonata). Caryologia 41(1): 69-78.

Appels, R.; R. Morris; B.S. Gill & C.E. May. 1998. Chromosome Biology. Kluwer Academic Publishers, Boston. xv+401 pp.

Bokhari, F.S., M.B.E. Godward. 1980. The ultrastructure of the diffuse kinetochore in *Luzula nivea*. Chromosoma 79, 125-136.

Braselton, J.P. 1971. The ultrastructure of the non-localized kinetochores of *Luzula* and *Cyperus*. Chromosoma 36, 89-99.

Braselton, J.P. 1981. The ultrastructure of meiotic kinetochores of *Luzula*. Chromosoma 82, 143-151.

Carnoy, B.Y. 1885. La cytodiérèse chez les arthropoda.(4) Pseudo-Nèuroptères. Cellule 1: 279-282.

Cruden, R.W. 1968. Chromosome numbers of some North American dragonflies (Odonata). Can. J. Genet. Cytol. 10(1): 200-214.

Cumming, R.B. 1964. Cytogenetic studies in the order Odonata. Ph. D. Thesis, Univ. Texas, Austin. v+93 pp.

Das, C. 1956. Studies on the association between non-homologous chromosomes during meiosis in four species of the Indian dragonflies (Odonata). J. Zool. Soc. India 8(2): 19-132.

Davies, D.A.L. & P. Tobin. 1984. The dragonflies of the world: A systematic list of the extant species of Odonata. Vol. 1. Zygoptera, Anisozygoptera. Societas Internationalis Odonatologica Rapid Commun-ications (Suppl.) 3: 1-127.

Davies, D.A.L. & P. Tobin. 1985. The dragonflies of the world: A systematic list of the extant species of Odonata. Vol. 2. Anisoptera. Societas Internationalis Odonatologica Rapid Communications (Suppl.) 5: 1-151.

De Gennaro, D. 2004. Análisis meiótico y caracterización de la heterocromatina en especies argentinas de Anisoptera (Odonata). Dir. Dra. L. M. Mola. Tesis de Licenciatura. Fac. Cs. Exactas y Naturales, U.B.A, Buenos Aires. ii+66 pp.

Durani S. & O.P. Sharma 1995. On the chromosomes of four species of dragonfly from Shillong, Noth East India. Chromosome Information Service 59: 4-5.

Ferreira, A., B. Kiauta, & A. Zaha. 1979. Male germ cell chromosomes of thrirty-two Brazilian dragonflies. Odonatologica 8(1): 5-22.

Franković, M. & R. Jurečić. 1987. Comparative cytogenetic analysis of karyotype morphology and organization in males of species *Libellula depressa* L. and *L. fulva* Müll. (Insecta, Odonata). Proc. Abstr. Pap. 3rd Congress of Croatian Biologists. pp 292-293.

Franković, M. & R. Jurečić. 1989. Comparative cytogenetic analysis of karyotypes morphology and organization in males of species *Libellula depressa* L. and *Libellula fulva* Müller (Insecta, Odonata). Period. Biol. 91(1): 32-33.

Fraser, F.C. 1957. A reclassification of the order Odonata. Royal Zoological Society, New South Wales. Sydney. 133pp.

Gama, V., C. Da Cruz Landim & A. Ferreira. 1981. Estudos morfológicos da estrutura dos cinetocores de *Enallagma cheliferum* (Odonata). Naturalia (Sao Paulo) 6: 75-87.

Goñi, B. & Y.P. De ABENaNTE. 1982. Cytological notes on five dragonfly species from Uruguay. Odonatologica 11(4): 323-329.

Handa, S.M. & N. Kochhar. 1985. Chromosomal architecture in two species of damselflies from Chandigarh and its surroundig areas. Nat. Seminar Current trends Chrom. Dynamics, Chandigarh. p. 34. En: Odonatological Abstracts No 6195, Odonatologica 17(3), septiembre 1988. p 291.

Handa, S.M.; O.P. Mittal & H.N. Batra. 1984. Chromosomes in ten species of dragonflies (Anisoptera: Odonata). Research Bulletin (Science) of the Panjab University 35 (III-IV): 65-73.

Hung, A.C.F. 1971. Cytological studies of five species of dragonflies (Odonata, Anisoptera). Ent. News 82: 103-106.

Kiauta, B. 1965. The chromosome behaviour in spermatocyte meiosis of *Anax imperator* Leach (Aesshnidae). Tombo VII (3/4): 18-20.

Kiauta, B.. 1968a. Variation in size of the dragonfly m- chromosome, with considerations on its significance for the chorogeography and taxonomy of the order Odonata, and notes on the validity of the rule of Reinig. Genetica 39: 64-74.

Kiauta, B. 1968b. Morphology and kinetic behaviour of the odonate sex chromosomes, with a review of the distribution of the sex determining mechanisms in the order. Genen en Phaenen 12(2): 21-24.

Kiauta, B. 1968c. Variation in size of the m-chromosome of the dragonfly *Calopteryx virgo* (L) and its significance for the chorogeografy and taxonomy of the *Calopteryx virgo* superspecies. Genen en Phaenen 12(1): 11-16.

Kiauta, B. 1969a. Sex chromosomes and sex determining mechanisms in Odonata, with a review of the cytological conditions in the family Gomphidae, and references to the karyotypic evolution in the order. Genetica 40(2): 127-157.

Kiauta, B. 1969b. Autosomal fragmentations and fusions in Odonata and their evolutionary implications. Genetica 40: 158-180.

Kiauta, B. 1970. The chromosome numbers of four neotropic dragonflies from Mexico (Odonata). Chromosome Information Service 11: 8-9.

Kiauta, B. 1971a. Cytotaxonomic peculiarities in the neotropical odonate genera *Leptagrion* Selys, *Orthemis* Hagen and *Macrothemis* Hagen. Abstr. Pap. 1st European Symp. Odonatol. Gent. p 27.

Kiauta, B. 1971b. Studies on the germ cell chromosome cytology of some cytotaxonomically interesting or hithertò not studied Odonata from the autonomus region Friuli-Venezia Giulia (Northern Italy). Atti. Mus. Civ. Stor. Nat. Trieste 27(2): 65-127.

Kiauta, B. 1972a. Synopsis of the main cytotaxonomic data in order Odonata. Odonatologica 1(2): 73-102.

Kiauta, B. 1972b. The karyotype of the damselfly *Leptagrion macrurum* (Burmeister, 1839) and its possible origin, with a note on the cytotaxonomic affinities of the genus (Zygoptera, Coenagrionidae). Odonatologica 1(1): 31-35.

Kiauta, B. 1972c. Notes on new or little known dragonfly karyotypes. I. The germ cell chromosomes of three Latin American species *Argia funebris* (Hagen), *Megapodagrion contortum* (Selys) (Zygoptera: *Coenagrionidae, Megapodagrionidae*) and *Castoraeschna castor* (Brauer) (Anisoptera: *Aeshnidae*). Genen Phaenen 15(1): 23-26.

Kiauta, B. 1973a. Notes on new or little known dragonfly karyotypes. lll. Spermatocyte chromosomes of four Neartic anisopterans, *Aeshna californica* Calvert (Aeshnidae), *Cordulia shutlefly* Scuder (Corduliidae), *Sympetrum internum* Montgomery, and *S. madidum* (Hagen) (Libellulidae). Genen Phaenen 16(1): 7-12.

Kiauta, B. 1973b. Notes on new or little known dragonfly karyotypes IV. Spermatocyte chromosomes of *Calopteryx splendens splendens* (Harris) (Zygoptera: *calopterygidae*), *Gomphus pulchellus selys*, and *Libellula depressa* Linnaeus (Anisoptera: *Gomphidae, Libellulidae*) from Northern France. Genen Phaenen 16(2): 55-60.

Kiauta, B. 1975. Cytotaxonomy of dragonflies, with special reference to the Nepalese fauna. Nepal Research Center, Katmandu. x+78 pp.

Kiauta, B. 1978. Two cytotaxonomically interesting cases of irreversible autosome fusion in dragonflies *Argia modesta* (Hagen) (Zygoptera: Coenagrionidae) and *Anaciaeschna isosceles* (Müller) (Anisoptera: Aeschnidae). Notulae Odonatologica 1(1): 7-9.

Kiauta, B. 1979. The karyotypes of some Anisoptera from Surinam. Odonatologica 8(4): 267-283.

Kiauta, B. 1983. The status of the japanese *Crocothemis servilia* Drury as revealed by karyotypic morphology (Anisoptera, Libellulidae). Odonatologica 12(4): 381-388.

Kiauta, B. & J.W. Boyes 1972. Cytology of ten South American Libellulidae, with cytophylogenetic considerations on the genera *Orthemis* Hagen and *Erythropidlax* Brauer (Odonata, Anisoptera). Genetica 43: 407-421.

Kiauta, B. & J.M. Van Brink 1978. Male chromosome complements of some Florida dragonflies, United States. Odonatologica 7(1): 15-25.

Kiauta, B. & M. A.J.E. Kiauta 1979. Cytotaxonomy of the genus *Argia* Rambur (Zygoptera: Coenagrionidae). Abstr. Pap. 5th Int. Symp. Odonatol. Montreal. pp 17-19.

Kiauta, B. & M.A.J.E. Kiauta 1980a. Introduction to the cytotaxonomy of the odonate genus *Argia* Rambur (Zygoptera: Coenagrionidae). Odonatologica 9(1): 35-56.

Kiauta, B. & M.A.J.E. Kiauta 1980b. On a small collection of dragonfly karyotypes from the Philippines. Odonatologia 9(3): 237-245.

Kiauta, B. & M.A.J.E. Kiauta 1980b. The karyotypes of *Aeshna subartica elizabeth* Djack. and *Somatochlora alpestris* (Sel.) from Switzerland (Anisoptera: Aeshnidae, Corduliidae). Notulae Odonatologica 1(6): 104-105.

Kiauta, B. & M. Kiauta 1983. The chromosome numbers of some Odonata from Thailand. Notulae Odonatologica 2(2): 27-28.

Kiauta, B. & M. Kiauta 1995. The karyotypes of *Somatochlora meridionalis* Nielsen from Slovenia, and *S. metallica* (Vander L.) from Switzerland, with a tentative note on the origin of the central european *S. metallica* (Odonata: Corduliidae). Opusc. zool. flumin. 137: 1-5.

Lefevre, G. & C.C. Mc Gill 1908. The chromosomes of *Anasa tristis* and *Anax junius*. Amer. Jour. Anat. 7(4): 467-487.

Lohmann, H. 1996. Das phylogenetische System der Anisoptera (Odonata). Entomologische Zeitschrift. 106(6): 209-252.

Madej, A. 1998. Spindle microtubules and chromosome behavior in mitosis of *Luzula luzuloides*, a species with holokinetic chromosomes. Acta Biologica Cracoviensia, Series Botanica 40, 61-67.

Makalovskaja, W.N. 1940. Comparative karyological studies of dragon-flies (Odonata). Arch. russ. d'anat., d'histol. et d'embryol, Leningrad 25(1): 24-39, 120-121.

Mc Gill, C.C. 1904. The spermatogenesis of *Anax junius*. Univ. Missouri Stud. 2(5): 236-250.

Meeldijk, M. 1984. C-banding in Syrphidae and Odonata. M. Sc. Thesis, Univ. Utrecht. iv+34pp.

Mittal, O.P. & V. Gandhi 1982. Meiotic chromosomes of two species of dragonflies (Odonata: Anisoptera). Chrom. Inf. Serv., Tokyo, 32: 15-17.

Mittal, O.P. & V. Gandhi 1984. Chromosome make-up in two species of damselflies (Odonata: Zygoptera). Abstr. Pap. 1st Indian Symp. Odonatol., Madurai. p 6.

Mola, L.M. 1992. Estudios cromosómicos en libélulas (Orden Odonata). Tesis Doctoral. Fac. Cs. Exactas y Naturales, U.B.A, Buenos Aires. iii+187 pp.

Mola, L.M. 1995. Post-reductional meiosis in *Aeshna* (Aeschnidae, Odonata). Hereditas 122: 47-55.

Mola, L.M. 1996. Meiotic studies in nine species of *Erythrodiplax* (Libellulidae, Odonata). Neo-XY sex chromosome system in *Erythrodiplax media*. Cytologia 61: 349-357.

Mola, L.M. & S.S. Agopian 1985. Observations on the chromosomes of four South American Libellulidae (Anisoptera). Odonatologica 14(2): 115-125.

Mola, L. M. & A.G. Papeschi 1994. Karyotype evolution of *Aeshna* (Aeshnidae, Odonata). Hereditas 121: 185-189.

Mola L. M. & A.G. Papeschi. 2006. Holokinetic chromosomes at a glance. Journal of Basic and Applied Genetics 17 (1): 17-33.

Mola, L. M., A.G. Papeschi, & E. Taboada Carrillo 1999. Cytogenetic of seven species of dragonflies. A novel sex chromosome determining system in *Micrathyria ungulata*. Hereditas 131: 147-153.

Muzón, J. & N. Von Ellenrieder 1998. Odonata. En: Biodiversidad de Artrópodos Argentinos. Una perspectiva biotaxonómica. Eds: Morrone, J. J. & Coscarón, S. Ediciones Sur. Capítulo 2, pp 14-25.

Nokkala, S., A. Laukkanev & C. Nokkala 2002. Mitotic and meiotic chromosomes in *Somatoclora metallica* (Corduliidae, Odonata). The absence of localized centromeres and inverted meiosis. Hereditas 133: 7-12.

Oguma, K. 1930. A comparative study of the spermatocyte chromosome in allied species of the dragonfly. Journ. Fac. Sci. Hokkaido Imp. Univ. Ser VI. 1: 1-32.

Oksala, T. 1943. Zytologische Studien an Odonaten I. Chromosomen verhältnisse bei der gattung *Aeschna* mit besonderer Berücksichtigung der postreduktionellen teilung der Bivalente. Ann. Acad. Sci. Fenn (Ser. A IV) 4: 1-64.

Papeschi A., L. Mola, P. Rebagliati, S. Rodríguez Gil & M. Bressa 2001. Heterochromatin characterization in the holokinetic chromosomes of some Heteroptera, Odonata and Araneae with DAPI-CMA. Chromosome Research 9 (Suplemento 1): 75 (P30).

Perepelov, E. & A.G. Bugrov 2001. C-heterochromatin in chromosomes of *Ophiogomphus cecilia cecilia* (Four.) (Anisoptera: Gomphidae) with notes on the sex chromosome origin in the species. Caryologia 54(2): 169-172.

Perepelov, E. & A.G. Bugrov 2002. Constitutive heterochromatin in chromosomes of some aeshnidae, with notes on the formation of the neo-XY/neo-XX mode of sex determination in Aeshna (Anisoptera). Odonatologica 31(1): 77-83.

Perepelov, E., A.G. Bugrov & E. Warchalowska – Sliwa 1998. C - banded karyotypes of some dragonfly species from Russia. Folia biológica 46(3/4): 137-142.

Perepelov, E., A.G. Bugrov & E. Warchalowska – Sliwa 2001. C - banded karyotypes of some dragonfly species from Russia II. The families

Cordulegasteridae, Corduliidae and Gomphidae. Folia biológica 49(3/4): 175-178.

Prasad, K & K.I. Thomas 1992. C- band pattern homogeneity in dragonflies (Odonata). Caryologia 45(1): 57-68.

Rieger, R., A. Michaelis & M.M. Green 1991. Glossary of Genetics: Classical and Molecular. Springer-Verlag 553pp.

Rodrigues Capítulo, A., L.M. Mola & S.S. Agopian 1991. Species catalogue and chromosome data of Odonata from Argentina. Revta Soc. ent. argent. 49(1/4): 59-72.

Sandhu, R. & I. Malhotra 1994. Karyological studies of four aeschnid dragonflies from states of Jammu & Kashmir, and Himachal Pradesh (India). Advances in Oriental Odonatology (Ed V. K. Srivastava) : 111-115.

Sandhu, R. & S. Gulati 1994. Karyological studies of some Libellulids from Himachal Pradesh State (India) (Anisoptera: Odonata). Bioved 5(1): 21-32.

Sandhu, R. & G.K. Walia 1993. Chromosome complements in six species of damsel flies. Chromosome Information Service 54: 22-23

Sandhu, R. & G.K. Walia 1994a. Karyological studies of four species of damselflies (Odonata: Zygoptera). Advances in Oriental Odonatology (Ed. V.K. Srivastava): 101-109.

Sandhu, R. & G.K. Walia 1994b. Karyological studies on four female Libellulids from Assam (India). Fraseria 1(1): 11-14.

Sandhu, R. & G.K. Walia 1994c. Chromosomal studies of three species of Libellulids (Anisoptera: Odonata). La Kromosomo II: 75-76.

Sandhu, R. & G.K. Walia 1996. Karyological studies of genus *Rhinocypha* (Odonata: Zygoptera: Chlorocyphidae). Proc. 83rd Indian Sci. Congr. 3: 3

Sandhu, R. & G.K. Walia 1997. Chromosome analysis of *Ischnura inarmata* (Coenagrionidae: Zygoptera: Odonata) Cromosome Sci. 1: 115-116

Sandhu, R., G. Walia & S. Gulati 1994. Chromosomal studies of three abundantly occurring damselflies from Himachal Pradesh (India). Advances in Oriental Odonatology (Ed. V.K. Srivastava): 93-100.

Seshachar, B.R. & Bagga, S. 1962. Chromosome number and sex-determining mechanism in the dragonfly *Hemianax ephippiger* Burminstein. Cytologia 27: 443-449.

Sharma, O.P. & S. Durani 1996. On the chromosomes of four species of genus *Orthetrum* (Libellulidae: Odonata). Proc. 83rd Indian Sci. Congr. 3: 2

Smith, E.A. 1916. Spermatogenesis of the dragon-fly *Sympetum semicinctum* (Say) with remarks upon *Libellula basalis*. Biol. Bull. 31: 269-302.

Souza Bueno, A.M. 1982. Estudos cromossômicos na ordem Odonata. M. Sc. Thesis, Univ. Estad. Paulista, Rio Claro. vi+140 pp.

Srivastava, M.D.L. & C.C. Das. 1953. Heteropycnosis in the autosomes segments of *Ceriagrion coromandelianum* (Odonata). Nature 172: 765.

Sumner, A.T. 2003. Chromosomes: organization and function. Blackwell Publishing, Oxford, U.K. vii+287.

Suzuki, K.J. & K. Saitoh 1988. Germ-line chromosome of two species of *Davidius*, with special references to the sex chromosomes (Anisoptera: Gomphidae). Odonatologica 17(3): 275-280.

Suzuki, K.J. & K. Saitoh 1990. A revised chromosome study of japanese odonates (II). Chromosomes of 13 species belonging to two families. Science Reports of the Hirosaki University 37(2): 111-123.

Tennessen, K.J. 1997. The rate of the species description in Odonata. Ent. News 108 (2): 122-126.

Thomas, K.I. & R. Prasad 1984. Chromosomes and the chromosomal basis of sex determination in a few species of Indian dragonflies (Odonata). Abstr. Pap. 1st Indian Symp. Odonatol., Madurai. p 10.

Tyagi, B.K. 1982. Cytotaxonomy of the Indian dragonflies. Indian Rev. Life Sci. 2: 149-161.

Tyagi, B.K. 1985. Cytotaxonomy of the genus *Onychogomphus* Selys (Odonata: Anisoptera, Gomphidae), with a Special Reference to the Evolution of the Sex-Determining Mechanism and Reduced Chromosome Number in the family Gomphidae. Proc. First Indian Symp. Odonatol. (Madurai) : 217-226.

Ueshima, N. 1979. Hemiptera II: Heteroptera. En: Animal Cytogenetics, Ed. B. John, Vol. 3, Insecta 6. Gebrüder Bornträger, Berlín, Stuttgart. v+117 pp.

Von Ellenrieder, N. 2003. A synopsis of the neotropical species of *Aeshna* Fabricius: the genus *Rhionaeshcna* Föster (Odonata: Aeshnidae). Tijdschrift voor Entomologie 146: 67-207.

White, M.J.D. 1973: Animal Cytology and Evolution. 3ª Ed. Cambridge Univ. Press, Cambridge. viii + 961pp.

Wilson, E.B. 1905. Studies on chromosomes. II. The paired microchromosomes, idiochromosomes and heterotropic chromosomes in Hemiptera. J. Exp. Zool. 2: 507-545.

Yadav. J.S. 1979. A note on the karyotypic variability in *Crocothemis erythraea* (Brullé) and *C. servilia* (Drury) (Anisoptera, Libellulidae). Notul. Odonatol. 1(4): 77-79.

Odonata : Biology of Dragonflies
Ed.: B.K. Tyagi
Scientific Publishers (India), 2007, pp. 175-182

12

ARE THE OBSERVED DISPERSAL CAPACITIES IN DAMSELFLY SPECIES SUFFICIENT TO COPE WITH THE ONGOING RAPID SHIFT OF CLIMATE ZONES?

JAN J. BEUKEMA

Royal Netherlands Institute for Sea Research, P.O. Box 59, 1790AB Den Burg, Texel, The Netherlands
e.mail: janb@nioz.nl

INTRODUCTION

To keep up with present polewards shifts in climate zones, species have to expand their distribution areas in polewards direction with an average speed of several kilometers per year. Therefore, their dispersal capacities and the use they make of these deserve special interest. Both results of mark-recapture studies as well as records of the time course of colonization of new water bodies would point to remarkably small proportions of populations of Zygopteran species actually moving over such distances within a flying season. Reasons are given why these observations may underestimate true dispersal capacities. Records of range shifts reveal that most Odonate species appear to be able to the present speed of polewards climate shift.

SHIFTS IN CLIMATE ZONES AND DISTRIBUTION AREAS

In the past century, both Europe and North America experienced a warming of the climate by almost 10°C. The accompanying northward shift in climatic isotherms amounted to about 100 km (Parmesan *et al.,* 1999). For the century to come, the Intergovernmental Panel on Climate Change expects similar or even larger shifts, up to 60°C (Houghton *et al.,* 2001), with a most likely warming of about 0.20 to 0.30°C per decade (Wigley & Raper, 2001). This would mean that populations of plant and animal species have to extend their ranges in a polewards direction at an average speed of minimally ~1 and most likely 2 or 3 km per year to keep up with the expected shifts of climate zones and to prevent range retraction or even extinction. More sophisticated models on the necessary migration rate to cope with climate change resulting from Greenhouse-induced warming yield rates of >1 km yr^{-1} in many species, particularly in temperate climate zones (Malcolm *et al.,* 2002).

Numerous examples in a wide variety of plant and animal species are available on polewards extension of distribution (Hickling *et al.,* 2006), including several species of butterflies (Parmesan *et al.,* 1999) and Odonata (Hickling *et al.,* 2005). Most of the 52 species of European non-migratory butterflies studied by Parmesan *et al.* (1999) recently extended their range in a northwards direction (and very few in a southward one), by distances between 35 and 240 km. Hickling *et al.* (2006) found that southern butterflies in Britain expanded their distribution area at an average rate of 1.7 km yr^{-1}. In British species of Odonata, Hickling *et al.* (2005) found northward shifts in range margins in >90% of the 37 species studied, with an average distance of 74 km (standard error 12 km) over ~25 years (i.e. between 1960/70 and 1985/95). More than half of the species showed a range margin shift exceeding 50 km, which will roughly have been sufficient to keep up with the climate shift. There was little difference between Zygoptera and Anisoptera in the extent of the range shifts, with mean values of 64 and 81 km, respectively. The mean speed of northward extension of distribution areas thus amounted to ~3 km yr^{-1} and was close to the northward shift of the climate zone in this part of the world. Parmesan *et al.* (1999) arrived at a similar conclusion in butterflies, both in Europe and in North America.

It should be added, that such observed range-extension rates may rest on an overestimate of actual dispersal in the recent past, due to more intensive searching for locally rare species by increasing numbers of "species hunters". Apart from this restriction, dispersal capacities in such groups as butterflies and odonates appear to have been sufficient to cope with the speed of the present polewards climate shift. This would imply that mean annual net movements of some km per year into the right direction did exist in a certain proportion of their populations.

DISPERSAL RATES IN ZYGOPTERA

In view of their excellent flying capacities, Anisoptera species are not expected to meet great difficulties in extending their range. Indeed, they have been encountered at islands or continental coasts hundreds or even thousands of km away from their probable area of origin (examples in Corbet, 1999: 649). Zygoptera species are generally weaker flyers, although examples are known of occasional long-distance dispersal, *e.g.* an observation by Kiauta (1963) of a specimen of *Calopteryx virgo* at an isolated lake and a recovery of a marked *Sympecma paedisca* at 20 km from the marking site (Manger & Beukema, 2006).

Some weak flyers appear to promote dispersion by rapidly flying upward at times of strong upward thermal currents, as observed by Fox (1989) in *Ischnura pumilio* and may thus be able to colonizes newly-formed biotopes despite their restricted flying capacities.

As Zygoptera comprise a more interesting group than Anisoptera with respect to possible problems in adapting their ranges to climate shifts, the following discussion will largely be restricted to this group. It will deal first with results of marking-recapture studies and further with recolonization studies.

Several recent studies are available on the movements of marked and released individuals of species of Zygoptera., *e.g.* Utzeri *et al.* (1984) in *Lestes barbarus*, Stettmer (1996) on *Calopteryx splendens* and *C. virgo*, Schutte *et al.* (1997) on *Calopteryx splendens*, Conrad *et al.* (1999) on *Ischnura elegans, Coenagrion puella, C. pulchellum, Lestes sponsa, Enallagma cyathigerum,* and *Pyrrhosoma nymphula*, Beukema (2002b) on *Calopteryx haemorrhoidalis*, Angelibert & Giani (2003) on *Coenagrion puella* and *C. scitulum*, and Purse *et al.* (2003) and Watts *et al.* (2004) on *Coenagrion mercuriale.*

Nearly all of these studies reveal that only minor fractions of the recovered part of the marked population were encountered at distances of more than a few hundred of meters away from the marking site. Distances of over 0.5 or 1 km had been covered by only 1 or 2% of the recovered individuals, with little difference between the species studied. Only in the populations of the two *Calopteryx* species studied by Stettmer (1996), ~10% moved over 1 km and the low proportion of ~1% was reached only at 2 km. Such slightly longer distances in Calopterygidae were confirmed by Schutte *et al.* (1997) who found that 2.2 % of the recovered individuals (and 3% of the males among them) had moved >1 km.

For the following reasons, the above studies may have underestimated the capacity to move over long distances:

(1) mark-recapture studies are characterized by an inherent limitation: the size of the study area usually does not exceed 1 x 1 km and individuals with net movements of > 1 km thus escape from the area that is searched intensively by the investigator. As a consequence, the recorded mean distances and particularly the maximal distances will generally be underestimates of the true distances over which the animals had actually moved. In all above studies, only about half (or even less) of the marked numbers were recovered. Most "missing" individuals will have died within the observation area, but there is no reason to believe that all of them did. A certain (unknown) proportion may have escaped beyond the area of observation. In this context, it may be meaningful that the two studies that yielded the longest mean distances and the highest proportions of long-distance movers (Stettmer, 1996; Schutte *et al.,* 1997) were executed in the largest areas (with diameters of 2 and 14 km, respectively).

(2) most individuals were marked when they were mature, several days after emergence, and at that time the stage of maximal movements may already have passed in some species, which again would result in an underestimate of the distances actually moved. For instance, Beukema (2002b) observed that in *Calopteryx haemorrhoidalis* maturing individuals moved on average about 10 times longer distances than either younger immature (with their initially low flying capacities) or older mature individuals (among which the males can occupy exactly the same territory site for many days in succession: Beukema, 2002a). In the six Zygoptera species studied by Conrad *et al.* (1999) and the two by Angelibert & Giani (2003) tendencies to disperse did not significantly differ between teneral and sexually mature individuals.

Another way to study dispersal capacities is by monitoring the arrival in the course of the flying season of adults at waters that were devoid of larvae but otherwise suitable habitat for Odonata. Lempert (cited from Corbet 1999: 422) monitored dragonfly arrivals at Mellum Island where no larvae were found. Though this German island is situated at ~7 km from the nearest land (and certainly even further from suitable breeding areas for most species), he observed 28 species (including 13 Zygoptera) to arrive in the course of the season. A detailed monitoring study was done by Beukema & Manger (in press) in a dune area in The Netherlands, where part of the ponds became completely dry for some months in the hot summer of 2003. At one of these ponds where all larvae had disappeared, the time pattern of appearance of adults of several species in 2004 was compared with the time patterns observed at the same pond in earlier years (without a preceding complete draining) and with the (almost identical) pattern in a nearby (slightly deeper) permanent pond. At the pond that had dried up in 2003 (but contained sufficient water again in 2004 and was well-vegetated: a significant feature for rapid repopulating, compare Kadoya *et al.,* 2004), adults of all species were found in 2004 only in small numbers as compared to earlier years and, more strikingly, they appeared only after substantial time lags as compared to the normal seasonal pattern. These time lags amounted to almost two months (mid-July versus mid-May) in *Ischnura elegans* as well as in *Enallagma cyathigerum*, about one month in *Lestes barbarus* and one week in *Lestes sponsa.* The intermediate distances (minimally ~300 m to a small and ~500 m to a large permanent pond) were thus not easily crossed by substantial numbers of these species, but finally (nearly) all species of the neighbouring permanent ponds did arrive at the non-permanent pond. Again, these observations may have yielded an underestimate of normal dispersal capacities, because the abundance of Odonata was lower in 2004 than in other years: nearly all species had probably suffered from the drought in 2003 (Manger & Beukema 2006). Thus, all reservoirs of odonates were small in 2004 and this may have slowed down repopulating rates of nearby temporary ponds.

DISCUSSION

A frequently drawn conclusion in the above mark-recapture studies is that the studied Zygoptera species would be poor dispersers, because large proportions of the marked individuals stayed within short distances (at most a few hundred of meters) of the first site where they were encountered (and marked). This conclusion may need some refining. Fully effective dispersion to remote unoccupied habitats can take place if only a certain (not necessarily large) fraction of the populations moves over longer distances. In all studies reviewed, a few percent of the individuals indeed were found to make use of their moving capacities and covered distances of in the order of 103 m. Exactly these individuals play a decisive part in the increase of a species' range.

As a consequence of the limited size of the study areas, the share in the total population of the move-prone individuals may consistently have been underestimated. Nevertheless, if the model by Conrad *et al.* (1999) of an exponentially declining relationship between dispersal and distance is generally valid, such long-distance movers will have been few in all studies. There is clearly a need for further testing this model for distances exceeding about 1 km. Only then, more exact estimates (rather than merely extrapolated values) can be obtained of the real magnitude of the dispersing parts of populations. In most of the present studies, a substantial part of dispersal-prone individuals may have escaped unnoticed from restricted areas of observation.

So far, only scarce evidence is available that certain individuals are more dispersal-prone than other more philopatric ones. Conrad *et al.* (2002) found in *Ischnura elegans* that dispersing males had significantly longer forewings than resident males. However, it appears questionable whether such size differences are general, as Thompson (1991) could not corroborate (in a study of *Coenagrion puella*) an observation by Anholt (1990) that dispersing adults of *Enallagma boreale* were heavier than residents.

CONCLUSION

A certain low proportion of the populations of even weak-flyer species appears to be able to regularly disperse for at least some kilometers per year and to bridge occasionally distances of even tens of kilometers, thus crossing uninhabitable areas. Such capacities would be sufficient to keep up with the present climate zone shifts. The observed correspondence between the recent polewards extension of distribution areas in many Odonate species (Hickling *et al.,* 2005) and the direction and speed of climate zone shift will not be coincidental. The similarity found in range extension between strong flying Anisoptera and weak flying Zygoptera might indicate that dispersal capacities in even the weaker flyers were not a limiting factor in range adjustment of Odonata to present climate change.

REFERENCES

Angelibert S, Giani N (2003) Dispersal characteristics of three odonate species in a patchy habitat. Ecography 26: 13-20.

Anholt BR (1990) Size-biased dispersal prior to breeding in a damselfly. Oecologia 83: 385-387.

Beukema JJ (2002a) Survival rates, site fidelity and homing ability in territorial *Calopteryx haemorrhoidalis* (Vanderlinden) (Zygoptera: Calopterygidae). Odonatologica 31: 9-22.

Beukema JJ (2002b) Changing distribution patterns along a stream in adults of *Calopteryx haemorrhoidalis* (Odonata: Calopterygidae):a case of larval-drift compensation? Int J Odonatol 5: 1-14.

Beukema JJ, Manger R. De trage terugkeer van libellen naar tijdelijk opdrogende duinplassen. Brachytron (in press).

Conrad KF, Wilson KH, Harvey IF, Thomas CJ, Sherratt TN. (1999) Dispersal characteristics of seven odonate species in an agricultural landscape. Ecography 22: 524-531.

Conrad KF, Wilson KH, Whitfield K, Harvey IF, Thomas CJ, Sherratt TN (2002) Characteristics of dispersing *Ischnura elegans* and *Coenagrion puella* (Odonata): age, size, morph and ectoparasitism. Ecography 25: 439-445.

Corbet PS (1999) Dragonflies: behaviour and ecology of Odonata. Harley Books, Great Harkesley/Colchester.

Fox AD (1989) *Ischnura pumilio* (Charpentier) (Odonata: Coenagriidae): a wandering opportunist? Ent Rec J Var 101: 25-26.

Hickling R, Roy DB, Hill JK, Thomas CD (2005) A northward shift of range margins in British Odonata. Global Change Biol 11: 502-506.

Hickling R, Roy DB, Hill JK, Fox R, Thomas CD (2006) The distributions of a wide range of taxonomic groups are expanding polewards. Global Change Biol 12: 450-455.

Houghton JT *et al.* (2001) Climate Change 2001: the scientific basis. Cambridge University Press, Cambridge.

Kadoya T, Suda S, Washitani I (2004) Dragonfly species richness on man-made ponds: effects of pond size and pond age on newly established assemblages. Ecol Res 19: 461-467.

Kiauta B (1963) A note on an unusual habitat of *Calopteryx virgo* (L.) (Calopterygidae). Tombo, Acta Odonatol. 6 (3-4): 25-26.

Malcolm JR, Markham A, Neilson RP, Garaci M (2002) Estimated migration rates under scenarios of global climate change. J Biogeogr 29: 835-849.

Manger R, Beukema J (2006) De libellen van de Grafelijkheidsduinen. De invloed van het waterpeil op de libellenfauna. Brachytron (in press).

Parmesan C, and 12 co-authors (1999) Poleward shifts in geographical ranges of butterfly species associated with regional warming. Nature 399: 579-583.

Purse BV, Hopkins GW, Day KJ, Thompson DJ (2003) Dispersal characteristics and management of a rare damselfly. J Appl Ecol 40: 716-728.

Schutte G, Reich M, Plachter H (1997) Mobility of the rheobiont damselfly *Calopteryx splendens* (Harris) in fragmented habitats (Zygoptera: Calopterygidae). Odonatologica 26: 317-327.

Stettmer C (1996) Colonisation and dispersal patterns of banded (*Calopteryx splendens*) and beautiful demoiselles (*C. virgo*) (Odonata: Calopterygidae) in south-east German streams. Eur J Entomol 93: 579-593.

Thompson DJ (1991) Size-biased dispersal prior to breeding in a damselfly: conflicting evidence from a natural population. Oecologia 87: 600-601.

Utzeri C, Carchini G, Falchetti E, Belfiore C (1984) Philopatry, homing and dispersal in *Lestes barbarus* (Fabricius) (Zygoptera : Lestidae). Odonatologica 13: 573-584.

Watts PC, Rouquette JR, Saccheri IJ, Kemp SJ, Thompson DJ (2004) Molecular and ecological evidence for small-scale isolation by distance in an endangered damselfly, *Coenagrion mercuriale*. Mol. Ecol. 13: 2931-2945.

Wigley TML, Raper SCB (2001) Interpretation of high projections for global-mean warming. Science 293: 451-454.

Odonata : Biology of Dragonflies
Ed.: B.K. Tyagi
Scientific Publishers (India), 2007, pp. 183 - 199

13

LOCAL ASSEMBLAGE PATTERNS OF ODONATES IN CENTRAL CHOCÓ, COLOMBIAN PACIFIC

L. PÉREZ, D. MONROY and E. REALPE

Zoology and Ecology Aquatic Laboratory LAZOEA
Biologic Sciences Department, Universidad de los Andes
Bogotá, Colombia
e.mail: le-perez@uniandes.edu.co

ABSTRACT

The dragonfly community in Central Chocó region was sampled with the aim of detecting patterns of local distribution and its relationship with the habitats' configuration. The Chocoan odonatofauna shows strong relationship with the presence of pluvial forest and it is postulated as a landscape condition for this ecological distribution pattern. The community shows relationship with variables like vegetation architecture associated to classified habitats, characteristics of aquatic environment, habitat's maturity and human impact. The species adaptability categorization is presented and features of specificity related to the habitats. Pluvial forest fragmentation, technified mining, the excessive logging constitute the main threats for the stenotopic species (*Archilestes* nov. sp., *Leptobasis* sp., *Metaleptobasis* sp., *Philogenia cristallina, Palaemnema dentata, Heteragrion erythrogastrum, Perissolestes remotus, Neocordulia batesi*). These species were related with the presence of mature forest, and their abundances were always the lowest. The euritopic species (*Erythrodiplax andagoya, E. famula lativittata, Micrathyria dictynna, Zenithoptera americana, Ischnura hastata*) are related with lentic systems with strong anthropic intervention and their abundances showed to be much higher. For the first time, a preliminary list of 38 species for the odonatofauna of Central Chocó is presented.

INTRODUCTION

Chocó is a landscape where a good cover of primary and secondary pluvial forest is still appreciated. Also, like in many other countries, they are losing between 500 and 1000 km2 of rain forest per year. In Salvador, for example, it is considered that primary forests do not exist (IUCN, 1990), being as well as one of the more diverse geographic zones of the planet but one of the most impacted areas by human activities. The neotropical region constitutes a greater diversity in dragonflies (Paulson, 2004). For example, in other taxonomic groups, Panama lodges much more diversity of birds (> 700 species) than entire North America (IUCN, 1990).

Ponds or streams are those environments where it is possible to detect a community of dragonflies since usually the diversity of species is considered like an emergent property of this ecological category (Begon *et al.*, 1995). It is relevant in ecology to understand the distribution patterns of the organisms and the pressures that determine them (Apodaca & Chapman, 2004). The distribution of biota must be studied taking care of the criterion of scales. For this reason, it turns out a high-priority to define the temporal, spatial, geographical and ecological context on which we want to make inferences on detected patterns (Rodriguez & Domínguez, 2003). In addition, the habitat is an important element to predict the ocurrence of certain species and that they are altogether the variables that define the patterns in which the populations appear (Apodaca & Chapman, 2004).

It is considered that the odonates show low correlations with environmental variables in microhabitats scale, however, on biotopes scale the correlation is so strong that they have been postulated like good indicators of particular conditions of the habitat that they occupy (Samways *et al.,* 1996). The need for a better knowledge about the odonatofauna of a specific site as well as its environmental conditions and requirements is very important in order to make approaches in bioindication with Dragonflies. Schmidt (1985) gathered information concerning to inventories of dragonflies and their habitats in a period of 25 years and proposed a Representative Spectrum of Odonata species (RSO) to qualify habitats. Some associations with environmental variables has been proposed by mean of abundance and dominance data in lentic habitats for dragonfly species, particularly in aquatic environments with high carbon concentrations (Osborn, 2005).

The present study proposes a methodology that allows quantifying and qualifying the odonate populations in a replicable way, with the purpose of characterizing odonate communities and their possible local distribution patterns, taking into account the biotic and abiotic variables in the habitat that they occupy.

It is proposed to some a typifying study of species according to their presumable adaptative rank and their abundances, frequencies and characteristics of habitat that were established along this study. This proposal of typifying is useful for the establishment of sustainable criteria, in order to categorize the vulnerability of species (Samways *et al.*, 1996).

STUDY SITES AND METHODS

The present study was developed in two sectors of the central zone of Chocó district, Colombian Pacific (Fig. 1), in the localities of Tutunendo and Salero (Unión Panamericana) (Table 1).

Table 1. Some physical attributes of localities in this study

Aspect	** Salero	Tutunendo
Georeference	N 5° 22' 95" W 76° 36' 74"	N 5° 42' 38" W 76° 37' 07"
Temperature	28. 6 °C	26. 9 °C
Relative humidity	90%	92%
Elevation	100- 150 m	0- 60 m
Life Zone	Tropical Pluvial Forest (bp- T).	Tropical pluvial
Forest (bp- T).	Tropical humid forest (bmh- T)	Tropical humid forest (bmh-T)
Precipitation year	6775 mm/ year	11000 mm/ year
Landscape unit	low/ médium hills*	Alluvial plain*

*Poveda *et. al.*, (2003). ** Palacios *et. al.*, (2003 a)

In the locality of Salero, the samplings were made at the *Parcela Permanente de Investigación en Biodiversidad* (PPIB). The human activities in this sector are centered on the artisan and technified mining, and the logging constitutes an activity of smaller intensity (Palacios *et al.,* 2000a). The vegetation is characterized by primary and secondary pluvial forest with different degrees of conservation.

The floristic composition is dominated by the Sapotaceae, Lecythidaceae, Arecaceae, Euphorbiaceae, Myristicaceae, Chrysobalanaceae and Moraceae families (Palacios *et al.*, 2003 b).

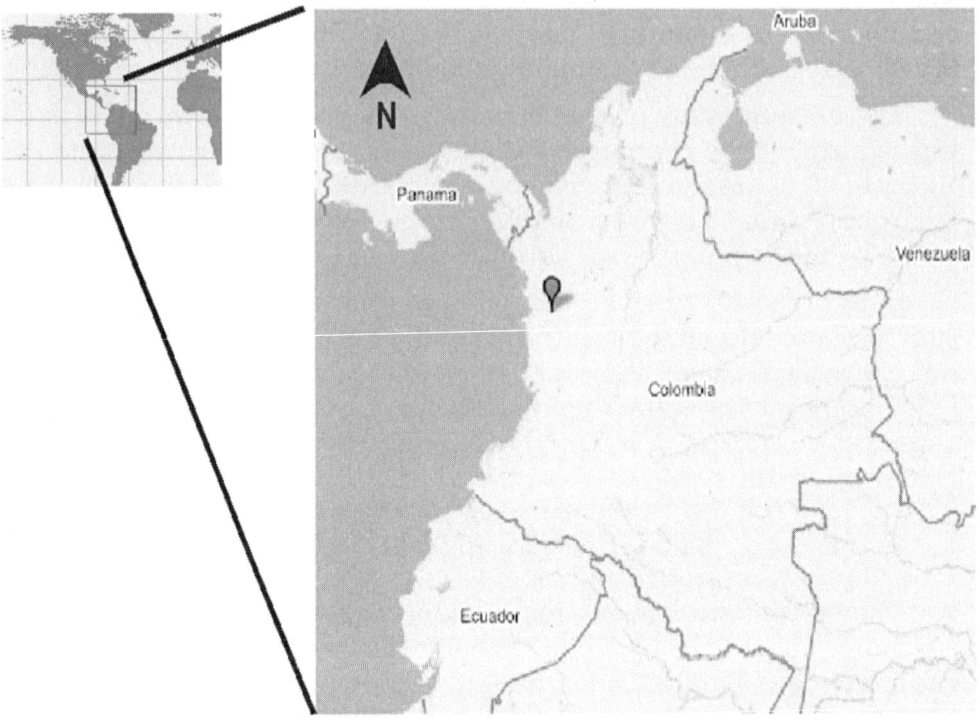

Fig. 1. Study sites in Choco district, Columbian Pacific.

Variables like type of aquatic environment, associated vegetation, intervention degree and land physiography were considered for the samplings (Clausnitzer, 2003; Samways *et al.*, 1996). Sampling units (SU) (Fig. 2), consisted on the route of triangular parcels (30 m by each side). The location of each SU, was determinated according to habitat heterogeneity available within each locality, determining a total of 37 SU.

The SU were visited several times (at least 2) during a period of 50 days, and intensively sampled in periods of 10 hours of collection.

The sampling was made with the collaboration of two people, collecting all the units sighted during the route of each parcel (90 ms). Collections were made between 09:00 - 16:00 hrs, with the purpose of including the hours of greater activity of the resident species.

The purpose was to obtain a greater heterogeneity space; the SUs in lotic habitats were located with the triangle base arranged throughout the creek and its other sides directed towards the surrounding vegetation. The SUs in lentic habitats were located with the triangle base along the littoral zone and the resting sides through the water associated vegetation. The anthropic intervention degree was categorized by the presence of human settlements, local activity and habitat deterioration. The statistical analysis was developed with MVSP 3.1 and PAST 1.12 software.

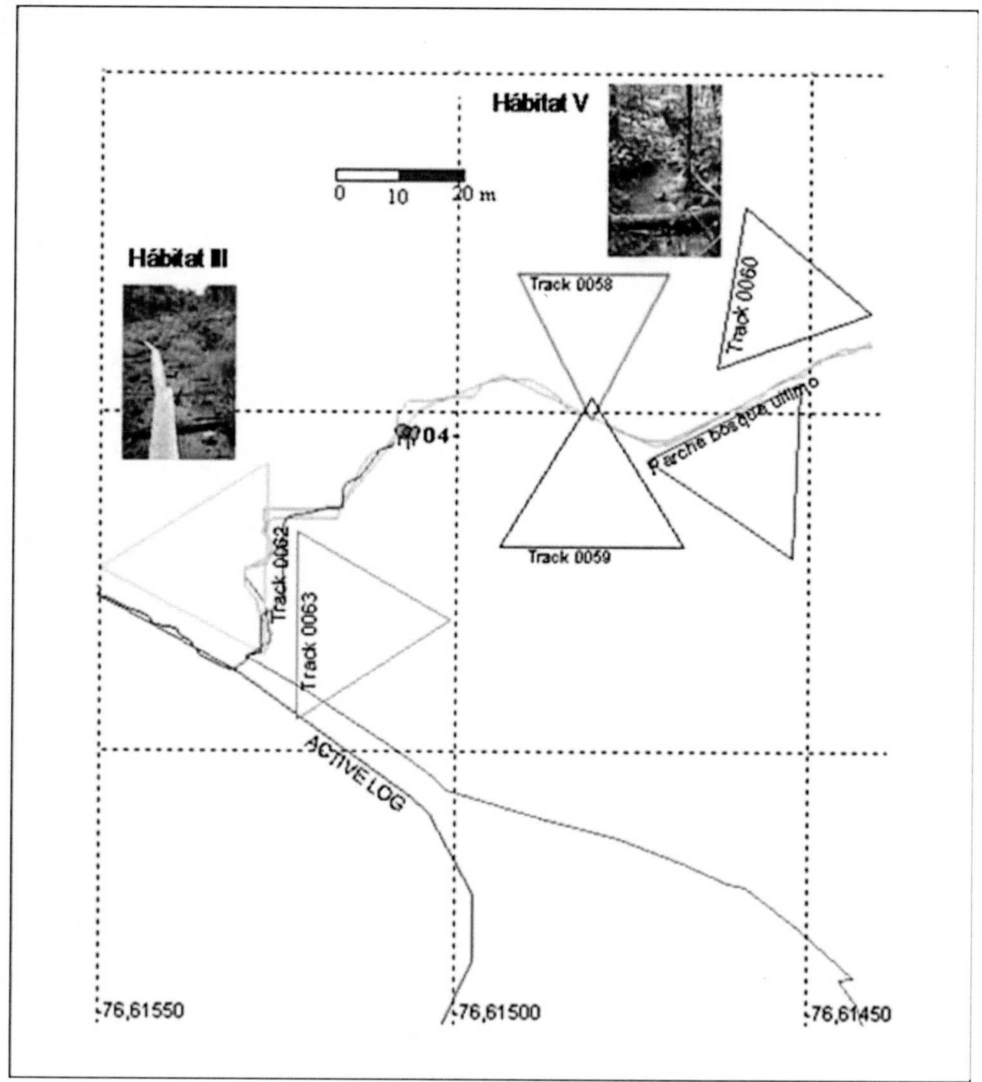

Figure 2. Sampling design scheme, Sample Units (SU) and their location in some habitats in Salero and Tutunendo localities.
The triangles have a 30 m side.

The collected specimens rest in the Entomological collection of the Museum of Natural History Universidad de los Andes.

The habitats were characterized according to the degree of intervention and some remarkable characteristics in the configuration of surrounding vegetation (Clausnitzer, 2003) (Table 2).

Table 2. Habitats classified according to different criteria of associated vegetation and anthropic intervention

Convention	Habitat type	Remarkable biotic Characteristics	Cause and Intervention degree
I	Ponds in forest patch	Vegetation riparian arboreal and shrubs, bottom with abundant leaves.	Medium, logging
II	Riparian forest 2nd order stream	Vegetation predominantly arboreal, herbaceous transition.	Medium, logging
III	Swamp zones	Vegetation predominantly herbaceous, shrubs transition.	High, logging
IV	Secondary forest	Vegetation arboreal, mature forest or in recovery.	Medium, logging mining
V	Riparian forest in 1st Order stream	Vegetation Arboreal mature forest.	Low, none
VI	Forest patch	Secondary forest with remarkable fragmentation	Medium, logging mining
VII	Forest gaps	Vegetation predominantly herbaceous, surrounding vegetation arboreal and shrubs.	High, logging
VIII	Lagoons	Surrounding vegetation predominantly herbaceous and shrubs, vegetation free floating and emergent aquatic in littoral vegetation.	High, human activities
IX	Backwater in 2nd Order stream	Arboreal Vegetation, secondary forest.	Medium, mining and logging

RESULTS AND DISCUSSION

The total of 38 species (Table 3) recorded in this study were distributed in 11 families between Zygoptera and Anisoptera according to Bridges (1994) classification.

Table 3. Species list and their hypothetic adaptive category

Taxa	Adaptive category
Libellulidae	
Cannaphila mortoni	Estenotopic
Erythemis credula	Euritopic
Erythrodiplax andagoya	Euritopic
E. castanea	Euritopic
E. connata fusca	Euritopic
E. famula lativittata	Euritopic
E. sp.	Euritopic
Idiataphe longipes	Estenotopic
Micrathyria dictynna	Euritopic
Nephepeltia phryne	Euritopic
Orthemis cultriformis	Euritopic
Perithemis tenera	Euritopic
Tramea binotata	Euritopic
Uracis imbuta	Estenotopic
U. fastigiata	Estenotopic
Zenithoptera americana	Euritopic
Coenagrionidae	
Argia indicatrix	Euritopic
A. cupraurea	Euritopic
Argia sp. 1	Euritopic
Argia sp. 2	Estenotopic
Helveciagrion sp.	Euritopic
Ischnura hastata	Euritopic
Leptobasis sp.	Estenotopic
Metaleptobasis sp.	Estenotopic
Acanthagrion trilobatum	Euritopic
Protoneuridae	
Protoneura amatoria	Estenotopic
Psaironeura remissa	Estenotopic
Megapodagrionidae	
Heteragrion erythrogastrum	Estenotopic
Heteragrion sp.	Estenotopic
Phillogenia cristalina	Estenotopic
Aeshnidae	
Gynacantha membranalis	Estenotopic
Lestidae	
Archilestes nov. sp.	Estenotopic
Calopterygidae	
Hetaerina miniata	Estenotopic
Perilestidae	

Perissolestes remotus	Estenotopic
Corduliidae	
Neocordulia batesi	Estenotopic
Pseudostigmatidae	
Megaloprepus coerulatus	Estenotopic
Mecistogaster sp.	Estenotopic
Platystictidae	
Palaemnema dentata	Estenotopic

In order to demonstrate differences between selected habitats and considering the total abundances of the general sampling a Kruskal-Wallis test was made, which showed a highly significant difference between the median of each treatments (habitat), with a confidence interval of 95% (P=0.009) (Fig. 3).

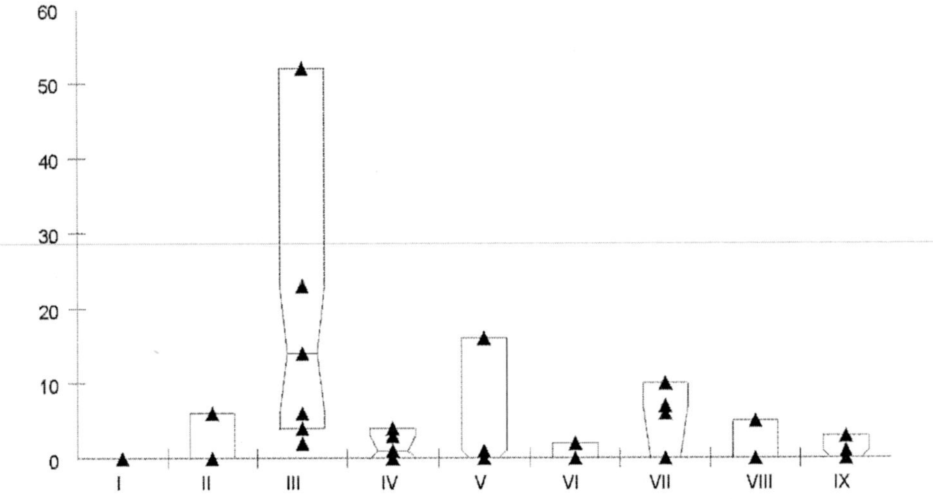

Figure 3. Abundance dispersion of the species in habitats I-IX. I (Ponds in forest patch), II (Riparian forest 2nd order stream), III (Swamp zones), IV (Secondary forest), V (Riparian forest in 1st Order stream), VI (Forest patch), VII (forest gaps), VIII (Lagoons), IX (Backwater in 2nd Order stream)

The differences are mainly determinated by the registered abundances in the habitat III (swamp zones), whose data dispersion is wider in comparison to the other habitats. The observed differences are product of certain species which are more dominant than others (Fig. 4), e.g., *Erythrodiplax andagoya, E. famula*, *Ischnura hastata* and *Zenithoptera americana*.

Jaccard similarity analysis performed, showed four grouping patterns with a distinguished species assemblage among them (Fig. 5). These groups represented as A, B, C and D, are composed of species proper of each habitat (lentic and lotic); groups A and C corresponding to lotic habitats and groups B and D to lentic habitats.

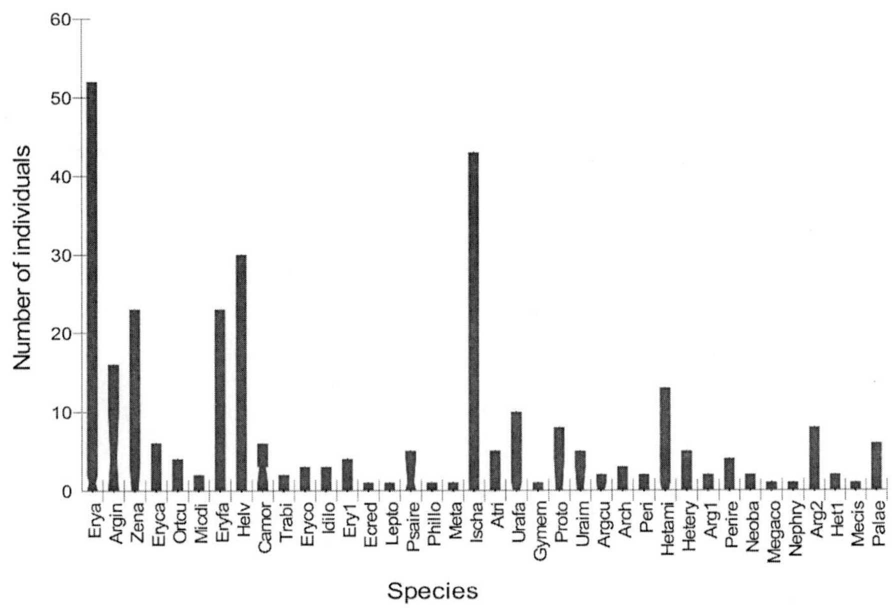

Figure 4: Schematic boxes analysis (Box & Wisker) according to abundances of the 38 Dragonfly species registered in SU in this study. Erya (*Erythrodiplax andagoya*), Argin (*Argia indicatrix)*, Zena (*Zenithoptera americana*), Eryca (*Erythrodiplax castanea*), Ortcu (*Orthemis cultriformis*), Micdi (*Micrathyria dictynna)*, Eryfa (*Erythrodiplax famula lativittata),* Helv *(Helveciagrion* sp.), Canmor (*Cannaphila mortoni), Trabi* (*Tramea binotata*), Eryco (*Erythrodiplax connata fusca*), Idilo (*Idiataphe longipes*), Ery1 (*Erythrodiplax* sp. 1), Ecred (*Erythemis credula*), Lepto (*Leptobasis* sp.), Psaire (*Psaironeura remisa*), Phillo (*Phillogenia* sp.), Meta (*Metaleptobasis* sp.), Ischa (*Ischnura hastata*), Atri (*Acanthagrion trilobatum*), Urafa (*Uracis fastigiata*), (*Gynacantha membranalis), Proto (Protoneura amatoria), Uraim (Uracis imbuta), Argcu (Argia cupraurea), Peri (Perithemis* sp), *Hetami (Hetaerina miniata), Hetery (Heteragrion erythrogastrum), Arg1 (Agia* sp. 1), *Perire (Perissolestes remotus), Neoba (Neocordulia batesi), Megaca (Megaloprepus caerolatus), Nephry (Nephepeltia phryne), Arg2 (Argia* sp. 2), *Het1 (Heteragrion* sp 1)*, Mecis (Mecistogaster* sp), *Palae (Palaemnema dentata).*

Group A characterized by forest gaps habitats and riparian forest in 2nd order streams, gathered stenotopic species like *Hetaerina miniata, Neocordulia batesi, Megaloprepus caerulatus,* and *Mecistogaster* sp. which show preference for sunny habitats but with primary and secondary riparian forest cover. Group C shows species with lower abundances and corresponding to specialist species (stenotopic species) which are associated to low light penetration habitats represented in classification habitats like riparian forest in 1st order streams and secondary forest. *Palaemnema dentata*, *Heteragrion* sp., *Phillogenia cristalina*, *Perissolestes remotus*, *Metaleptobasis* sp. are typical species in this ecological adaptability. In Salero locality, a new *Archilestes* species was discovered, which evidenced arboreal behavior, fast and agile flight; this behavior has not been commonly recorded to this genus (Corbet, 1999; Westfall & May, 1996) and pluvial forest presence has been a key in this behavioral evolution.

Lentic habitats represented by B and D groups (Fig. 5) are remarkably occupied by eurytopic species which generally show very high abundance; to this ecological adaptability classification belong the species like *Cannaphila mortoni*, *Erythemis credula*, *Erythrodiplax andagoya*, *E. connata fusca*, *E. castanea*, *E. famula lativittata*, *Micrathyria dictynna*, *Uracis imbuta*, *Zenithoptera americana*, *Argia indicatrix*, *Ischnura hastata* y *Acanthagrion trilobatum*. These recorded habitats at high anthropical intervention zones were classified as swamp zones and lagoons.

Group B harbours species like *Psaironeura remissa* and *Protoneura amatoria* which were classified under stenotopic adaptability because they are strongly associated with little permanent ponds into forest gaps and backwater second order streams characterized with primary and secondary riparian forest presence.

Also it was shown that the coexistence of some species constitute certain specific patterns (Clausnitzer, 2003; Samways *et al.*, 1996), which indicate that some characteristics in its habitational requirements promote the coexistence. *P. remisa*, *P. amatoria*, *H. erythrogastrum* and *P. remotus* were recorded in a typical forest environment with low light. The diverse thermoregulatory requirements of dragonflies are a determinant variable for presence of some particular assemblages (Samways *et al.*, 1996).

In order to find associations that shows any other pattern about abundances of recorded species, a principal component analysis was made (PCA) among 38 species and their abundances in each of their 9 classified habitats and a mayor sense the patterns or groups suggested were supported by Jaccard index (Rodríguez & Domínguez, 2003).

PCA analysis shows the three standardized principal components explaining over the 95% of the data structure. So, with the first component is retained the 86.9% of the total variability (principal eigen- value) and second retaining the 5.3% (Table 4).

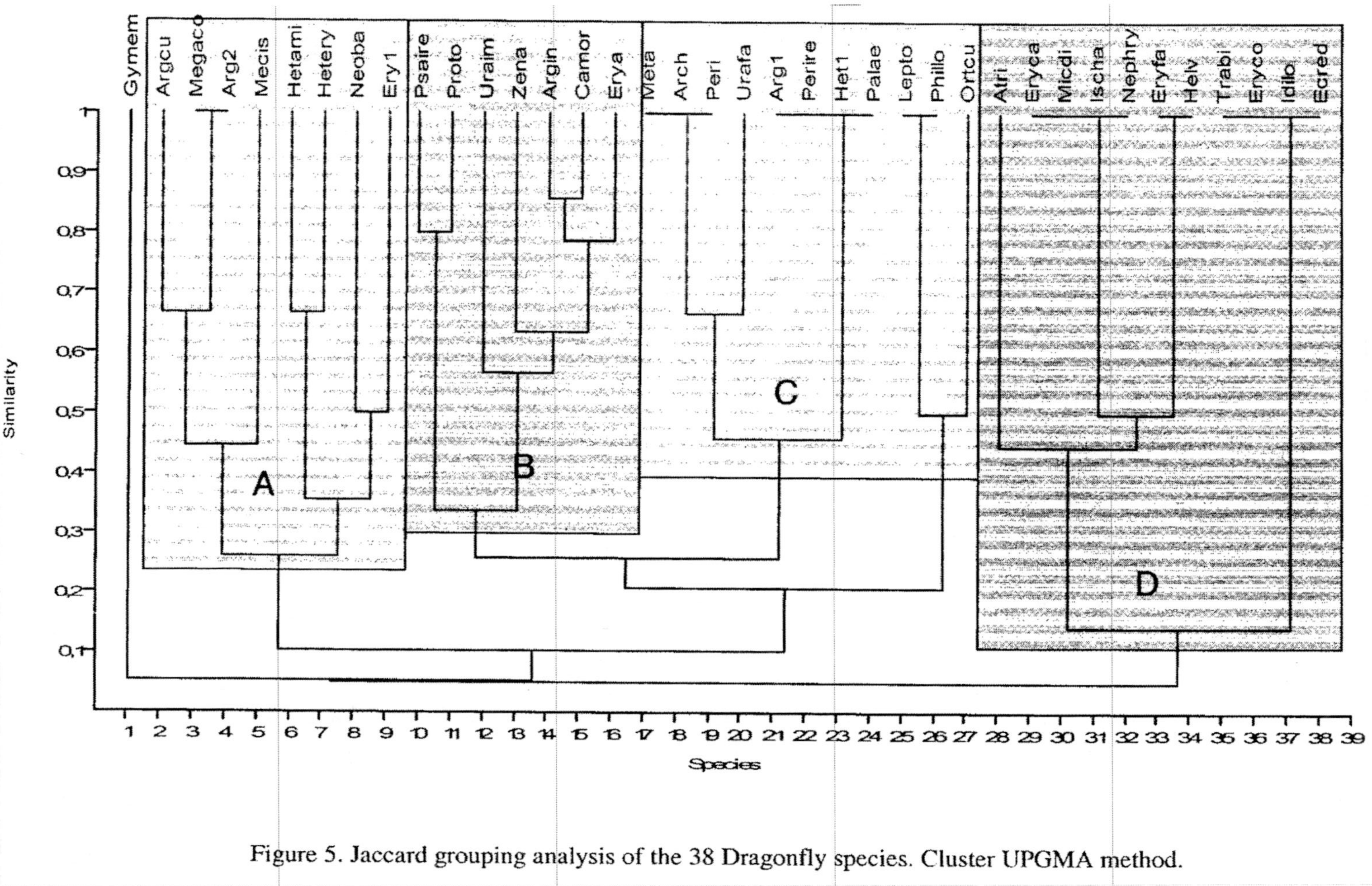

Figure 5. Jaccard grouping analysis of the 38 Dragonfly species. Cluster UPGMA method.

Table 4. Partial and accumulative percentages of the 3 components established in PCA analysis

Component	1	2	3
Eigen- values	722,508	44,631	28,690
Partial Percentage	86,959	5,372	3,453
Accumulative Percentage	86,959	92,331	95,784

Figure 6 reveals the mask generated by components 1 and 2 over the species associations less abundant Vs. habitat, explaining that species like *H. erythrogastrum, P. dentata, P. remotus, Heteragrion* sp, *Neocordulia batesi, Protoneura amatoria, M. caerulatus* are associated to habitat characterized by forest areas with canopy presence (secondary forest, riparian forest in 1st order streams and backwaters of 2nd order streams) (Fig. 7).

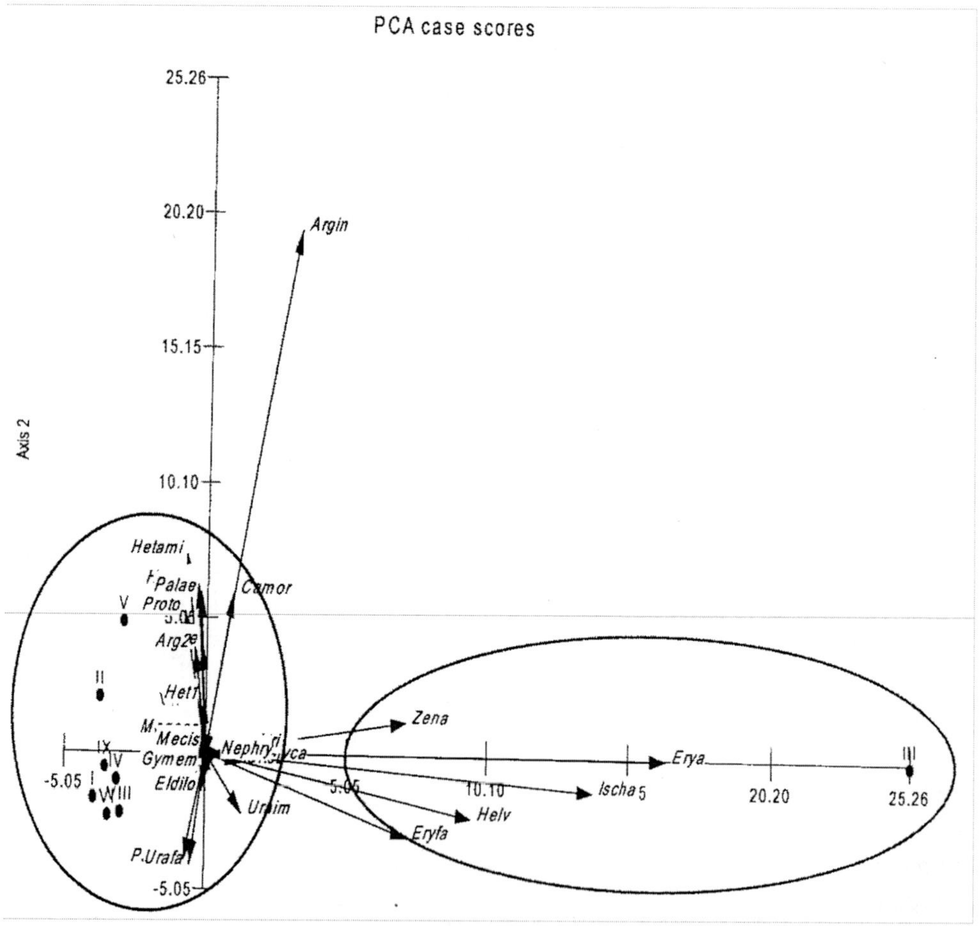

Figure 6. Principal component (1 and 2) diagram (93,33% of total variability).

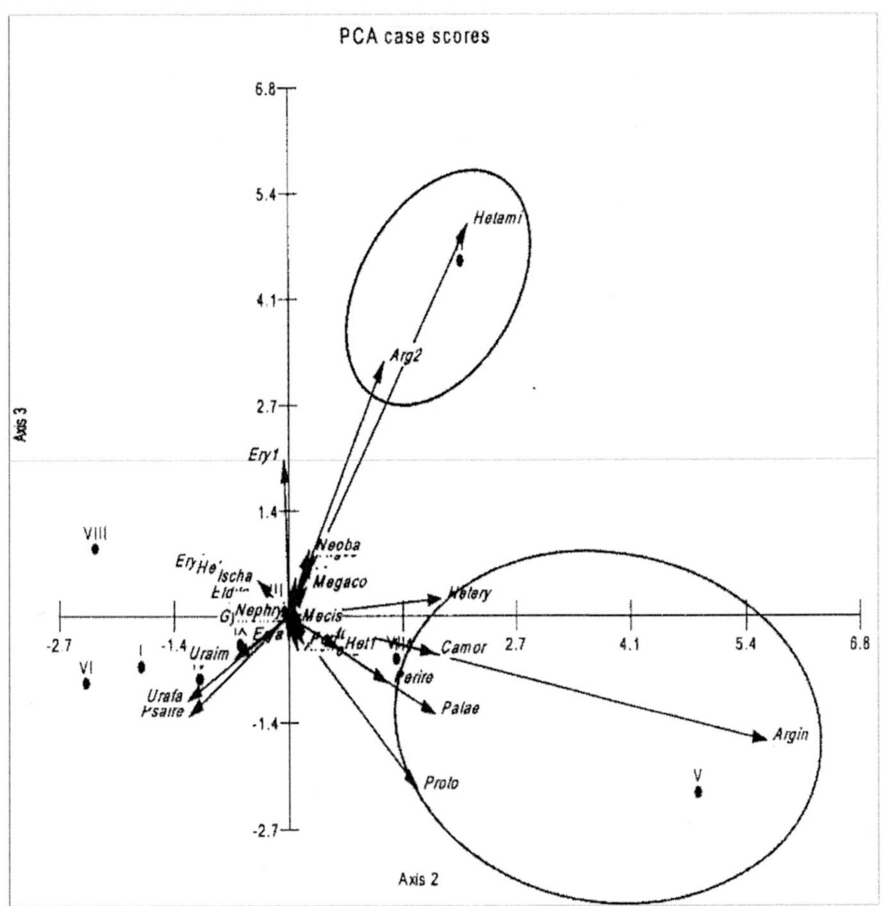

Figure 7. Principal component (2 and 3) diagrams (8,82% of total variability).

There is a strong habitational demarcation in the central Chocoan odonatofauna. Obviously data gives initial evidence in order to say that stenotopic species are clearly threatened, due mainly to the pluvial forest fragmentation, advances in technified mining and excessive logging.

This study also provides data suggesting that populations of this species are small. This, would sustain the necessity of digging in population kind of works and submit to test this hypothesis.

One of the most evident conclusions is that habitat heterogeneity in terms of associated vegetation (Corbet, 1999) (Riparian forest in lotic habitats), (Surrounding vegetation, littoral vegetation, aquatic vegetation in lentic habitats) is the best variable in order to explain diversity, richness and abundances of species (Buchwald, 1992; Buchwald & Schmidt, 1990). In case of *I. hastata* its greatest abundance in swamp habitats is stimulated by the dominance of herbaceous plants in the inner of lagoons. This species only occurred in environments that showed an advanced process of terrestrialization. This is a common phenomenon in the swampy habitats where some species like *E. andagoya* and *E. famula lativittata* also showed high abundances in these habitats.

The low rates of abundance of species like *H. erythrogastrum*, *P. remotus* and *H. miniata* can be related to low resource availability to the larval stages. In localities under study, the minor order streams (1st and 2nd Order) are characterized by having bottoms with low heterogeneity because they have been the result of volcanic activity and probably their little roughness limit the presence of gravel, pebbles, cobbles and other substrates that are necessary for establishment of higher population of larvae (Hawkings *et al.,* 1982; Horne & Goldman, 1994; Rosi-Marshall & Wallace, 2002). Thus, we suppose that the result of the emergence rates of images are lower too.

Osborn (2005) related high abundances of dragonflies in lentic habitats with very high concentrations of carbon. So it would be an explanation of what was observed in habitat III (*i.e.* swampy zones and lagoons) where species show very high abundances (40 ind/UM). On the other hand, the pluvial forest and its lotic habitats could be reflecting a contrary phenomenon since the main fraction of available carbon is represented in the biomass (Murphy & Lugo, 1986) and the reduced capability of the current waters to retain the organic matter (leaf litter, detritus etc.) restrict the maintenance of greatest odonata population.

Dragonfly community, considering specific condition that offers its habitats in a geographic context, exhibits changes in its structure according to some variables mostly affected by natural conditions or human intervention (Sahlen, 1999; Sahlen & Ekestubbe 2001; Apodaca & Chapman 2004; Osborn, 2005). We suppose that different groups of species can reflect the biotic and abiotic conditions from related habitats from which can be explained possibly the routes of specific species' degradation or, on the contrary, its own recovery. Paulson (2004) remarks that the main factor to the odonata population decline is the deterioration of habitat, especially the surrounding vegetation. In this context dragonflies become a potential determinant for evaluation of environmental deterioration and can be profitably exploited in detecting the inner change of the habitat and how the biocenosis responds to them (Carle, 1979).

Colombia is *terra incognita* (Paulson, 2004) about dragonfly diversity and there are areas that remain unknown for dragonfly dispersal and abundance (Donelly, 1992). This vacuum in knowledge makes this a potential study object of high priority in order to comprehand the association of dragonfly diversity with the habitat condition.

In this study it is proposed that the determination of RSO (Schmidt, 1985) and the habitat characterization according to its biotic structure (*i.e.* associated vegetation and anthropic intervention degree) can be a useful tool for detecting the state of trophic maturity trophic in aquatic habitats and for developing the short, half and long time monitoring in several temporal and spatial scales.

ACKNOWLEDGEMENTS

Investigators are thankful to all in the Faculty of Sciences, Universidad de los Andes, Zoology and Aquatic Ecology laboratory, LAZOEA, Univer-sidad de los Andes, Limnology group, Universidad Tecnológica del Chocó. Local communitarian councils in the municipalities Salero and Tutunendo, Chocó district.

REFERENCES

Apodaca, C. K. & Chapman, L.J., 2004. Zygoptera Adult of National Kibale Park, Uganda: Seasonal habitat associations and occurrence. Odonatologica 33(2): 129- 146.

Begon, M., Harper, L.J. & Townsend, R.J., 1995. Ecology: individuals, populations and communities. Editions Omega, S.A. Barcelona.

Bridges, C.A., 1993. The catalogues of family-group, genus-group and species-group names of the Odonata of the world (Second Edition). C. To Bridges, Urban, Illinois.

Buchwald, R., 1992. Vegetation and dragonfly biocenological fauna characteristics and examples of field studies. Vegetation 101: 99- 107.

Buchwald, R. & Schmidt, B., 1990. Der kleine Blaupfeil (Orthetrum coerulescens, Odonata) in Südbaden: Spezielle Untersuchungen zu ökologischen Ansprüchen, Populationsdynamik und gefährdung. Mitteilungen DES Badischen Landesvereins für Naturkunde und Naturschutz and V. Freburg im Breisgau 15: 109-144.

Carle, F. L., 1979. Potential Environmental monitoring of the Odonata, with to list of rare and endangered Anisoptera of Virginia, United states. Odonatologica 8(4): 319- 323.

Clausnitzer, V., 2003. Dragonfly communities in coastal habitats of Kenya: indication of biotope quality and the need of conservation measures. Biodivers. Cons. 12: 333- 356.

Corbet, P.S., 1999. Dragonflies: Behavior and ecology of Odonata. Comstock Publishing Associates, Cornell University Press. Ithaca, New York.

Donnelly, T.W. 1992. The Odonata of central Panama and to their position in the neotropical odonate fauna, with to checklist, and descriptions of new species. In: Quintero, D. & Aiello, A., 1992. Insects of Panama and Mesoamerica. Selected studies. Oxford University Press.

Hawkins, Ch. P., Murphy, M. L. & Anderson, N. H. 1982. Effects of Canopy, Substrate Composition, and Gradient on the Structure of Macroinvertebrate Communities in Cascade Range Streams of Oregon. *Ecology*, 63(6): 1840-1856.

Horne, A.J. & Goldman, C.R., 1994. Limnology. 2nd edition. McGraw-Hill. New York, the U.S.A. 576 pp.

IUCN, 1990. Las ultimas selvas tropicales. Ediciones Folio. España.

Murphy, P.G. & A.E. Lugo , 1986. Tropical Ecology of dry Forest. Ann. Rev. Ecol. Syst. 17: 67-88.

Osborn, R. 2005. Odonata ace indicators of habitat quality at lakes in Louisiana, United states. Odonatologica 34(3): 259- 270.

Palacios, Ll., J., García C, F. & Ramos P, Y. A. 2003 a. Características generales del corregimiento de Salero, Unión Panamericana, Chocó. En: SALERO Diversidad Biológica de un Bosque Pluvial Tropical (bp- T). Universidad Tecnológica del Chocó "Diego Luís Córdoba"

Palacios Ll., J., Ramos P, Y.A, Mosquera V., A. K., Castro C., F., García C, F., Arroyo V., J.E., Cogollo P., A. 2003 b. Estructura de un Bosque Pluvial Tropical (bp- T) en Salero, Unión Panamericana, Chocó. En: SALERO Diversidad Biológica de un Bosque Pluvial Tropical (bp- T). Universidad Tecnológica del Chocó "Diego Luís Córdoba"

Paulson, D., 2004. Critical odonata in Middle & South America. Guardians of the watershed. Global status of dragonflies: critical species, threat and conservation IUCN Regional report. International journal of odonatology 7(2): 163- 188.

Poveda, I. C., Rojas, C. A., Rudas, A., 2003. Unidades de paisaje con base en la topografía para la región del Chocó biogeográfico colombiano. Mapa. Universidad Nacional de Colombia. Instituto de Ciencias Naturales. En: Colombia Diversidad biótica IV: El Chocó biogeográfico/ Costa Pacifica/ ed. J. Orlando Rangel- Ch. – Bogotá: Instituto de Ciencias Naturales, 1024 p.: il., + 8 mapas.

Poveda, M. I. C., Rojas- P, C., Rudas- Ll, A, & Rangel, J. O., 2004. El Choco biogeográfico: ambiente físico. En: Colombia Diversidad biótica IV: El Chocó biogeográfico/ Costa Pacifica/ ed. J. Orlando Rangel- Ch. – Bogotá: Instituto de Ciencias Naturales, 1024 p.: il., + 8 mapas.

Rodríguez, P. & Domínguez, V. E., 2003. Escalas y diversidad de especies. En: Morrone, J. J. & Bousquets, LL. J. 2003. Una perspectiva latinoamericana de la biogeografía. Facultad de Ciencias, UNAM. Mexico.

Rosi-Marshall, E. J . & Wallace, J. B., 2002. Invertebrate food webs along a stream resource gradient. Freshwater Biology 47, 129- 141.

Sahlén, G. 1999. The impact of forestry on dragonfly diversity in central Sweden. International journal of Odonatology. 2(2): 177- 186.

Sahlén, G. & Ekestubbe, K., 2001. Identification of dragonflies (Odonata) as indicators of general species richness in boreal forest lakes. Biodivers. Cons. 10: 673- 690.

Samways, M.J., 1996. Workability of the New "IUCN categories of threat" with odonata in South America. Odonatologica 25(4): 347- 354.

Samways, M.J., Caldwell, P.M. & Osborn, R., 1996. Spatial patterns of dragonflies (Odonata) as indicators for design of a conservation pond. Odonatologica 25(2): 157- 166.

Schmidt, E., 1985. Habitat inventarization, characterization and bioindication by a "Representative spectrum of odonata species (RSO)". Odonatologica 14(2): 127- 133.

Westfall, M.J. & May, M.L. 1996. Damselflies of North America. Scientific Publishers, Gainesville. 649 pp.

Odonata : Biology of Dragonflies
Ed.: B.K. Tyagi
Scientific Publishers (India), 2007, pp. 201- 222

14

THE EXPANSION OF *CROCOTHEMIS ERYTHRAEA* (BRULLÉ, 1832) IN GERMANY – AN INDICATOR OF CLIMATIC CHANGES[1]

J. OTT

L.U.P.O.GmbH, Friedhofstrasse 28, D – 67705 Trippstadt, Germany,
e.mail: L.U.P.O.GmbH@t-online.de

(Dedication : Prof. B. Kiauta, whom I had promised a paper on Crocthemis for a long time, on his 70th birthday)

ABSTRACT

In this paper a detailed account of the northward expansion of *Crocothemis erythraea* in Germany is presented. While only a few decades ago the species was rare even in southern Germany, it is now found in nearly every federal state, in most of them autochthonous. The species conquered Germany from south to north, parallel to climatic changes in the country; similar expansions of this species are registered also in other European countries, as well as northward expansions of other southern species. The main reason for this range expansion are climatic changes, some consequences of which are discussed.

---

[1] Preparation of the paper funded by the EC-ALARM-project (code GOCE-CT-2003-506675)

INTRODUCTION

The Scarlet Darter (*Crocothemis erythraea*) (Brullé, 1832) (see Fig. 1) is called in German "Feuerlibelle", meaning "fire dragonfly" because of the intensive red colour of the male's body and the upper parts of the eyes (the female is olive green or brownish). The scientific species name "erythraea" also refers to the red colour (Greek "erythros" means red, see Fliedner, 1997). The range of this Orthetrum-sized libellulid covers the mediterranean (Brullé described specimens from Greece) and the whole of Africa, from there stretching eastwards via Afghanistan to India (Tsuda, 1991). It belongs now to the German odonatofauna (Müller & Schorr, 2001).

In Germany it was discovered for the first time already in 1918 in the federal state of Brandenburg, near Berlin, Northern Germany, by Bollow (1919), but in the following decades it was registered only very rarely. According to Schiemenz (1953), during the first half of the last century the original area of this species was still restricted to Southern Europe, from where it flew only occasionally to Middle Europe (Belgium, Vienna, Munich, Baden), but it never established autochthonous populations there. Also Robert (1959) described its range in Europe in that time span as "Italy, Spain, Greece and southern France", from where it would fly "in hot summers" to southern Germany, Belgium, Hungary and Switzerland, but "only single specimens". Without doubt it is a mediterranean element, and as such it was classified by St. Quentin (1960).

Dumont (1967) still postulated that individuals observed in the north originated from big populations in the Camargue, Southern France, and had followed the rivers as guiding lines.

Consequently the species was generally regarded as a "wanderer" and classified as an "invading species" from the south: Jurzitza (1978) called her "Wanderlibelle" ("wandering dragonfly" or "travelling dragonfly") and in his book he showed a photograph of a male which had reached Karlsruhe in southern Germany in the hot summer of 1976. Lohmann (1980) in his "Odonate fauna list of the German Federal Republic" also spoke of an "invasive species" and "wandering species", to be met with in Germany only in climatically favoured regions with high water temperatures in summer, and he classified the species as a "Vermehrungsgast" (a "guest sometimes reproducing"), as already a very few populations were known then from Baden-Württemberg.

A few years later Dreyer (1986) and Bellmann (1987) still classified *Crocothemis erythraea* as a species belonging to the

Mediterranean, entering northern countries occasionally with warm air streams, but they also pointed out that the species was already autochthonous since several years in the climatically favoured Rhine valley of the southern German federal state of Baden-Württemberg.

In the following decades the species dramatically expanded its range northwards in Germany and also up to higher altitudes. This range expansion is analysed here in detail, being a model also for other European countries as well as other mediterranean species expanding their range to the north.

CROCOTHEMIS ERYTHRAEA IN GERMANY – AN ONGOING "SUCCESS-STORY"

BIOLOGY AND ECOLOGY OF THE SPECIES – A SHORT INTRODUCTION

The following description is based on Robert (1959), Ott (1988), Sternberg (1989), Schorr (1990), Burbach (1998) and Sternberg & Höppner (2000).

Crocothemis erythraea is a medium sized libellulid and because of its intensive red colour (males) very striking when present at a water body or nearby, where it perches on the ground or in the vegetation.

In general the species prefers habitats (standing waters, in the mediterranean also running waters) with open zones at the shoreline, like sand- and gravel pits, but it occurs also at lakes with reeds and emerse vegetation (where the males often fly fast and with some "hovering stops" over the *Nymphaea-* or *Potamogeton*-zones and return to the reed-banks for perching). In the south it is very abundant in rice fields. It has been regarded as a "lowland species", but meanwhile it is found also at altitudes higher than 1600 meters a.s.l. in the Alps (Landmann, 1984) and even up to 2.000 meters a.s.l. in Spain (Anselin & Hoste 1996). In Germany it was found first of all at low altitudes only, but later on also higher up: in Bavaria it emerged even at a small garden pond (Bönisch in Burbach (1998) at an altitude of 550 meters a.s.l.. The species has a preference for clear water with good transparency (see Ott 1988), which may be the reason for its preference in northern Germany for *Leucorrhinia caudalis*-waters (Mauersberger, 2003). Fish are not problematic if there is enough water vegetation and if the number of carp (which cause turbidity and by this a reduction of submerse vegetation) and of grass carp (which reduce the submerged vegetation directly, thus increasing predation pressure by other fish upon dragonfly larvae) is not too high (Ott, 1993).

The males appear rather early in the morning at the water, a dependence on high temperatures could not be verified (Ott, 1988), but at hot temperatures the males tend to take up the so called "obelisk-position" (thermoregulation - or is it threat display ?). The males are aggressive against conspecific males and also against other species of similar size. Pairing – generally lasting only a few seconds during flight, with a tendency to finish it while seated – takes place mostly near the water, but was also observed in some cases away from the water. Males show a typical threat display behaviour against other *Crocothemis*-males, but a real territorial behaviour (as described by Falchetti & Utzeri, 1974 or Sternberg, 1989) is not the general behaviour - maybe only at low densities (different strategies ?). The female lays eggs guarded by its male, but sometimes alone (*e.g.* when the guarding male pursued another male and then lost its female).

By marking individuals in Germany a high site fidelity to the particular habitat (a gravel pit) was detected. After about 10 days of maturing, juveniles returned to the water where they had emerged. In the Camargue Aguesse (1959) also carried out marking experiments, but he could not register site fidelity in this wetland, so the type of biotope or the general density of the species in a habitat may have an influence. The minimum life span of *Crocothemis erythraea* is about 29-32 days (Ott, 1988).

The phenology in Germany lasts from May until September, with main activity from June to August, but pairing activity was occasionally observed even in September.

The larvae – oval and without dorsal spines – are found in standing water bodies of different sizes (always containing vegetation), even sometimes in brackish water, where they seem to prefer the shallow and warmer zones down to a depth of 50 cm; they are univoltine, but in the mediterranean and in warmer climates also bivoltine (Heidemann & Seidenbusch, 1993, Sternberg & Höppner, 2000). Exuviae are generally found close to the water in the vegetation.

CROCOTHEMIS ERYTHRAEA IN GERMANY: CONQUERING GERMANY IN A FEW DECADES

Here the continuous expansion of the species in Germany is documented, using 10-years time spans and referring to the German federal states (see Fig. 2).

FROM THE SEVENTIES OF THE LAST CENTURY TO PRESENT TIME

Before 1970

As mentioned above, *Crocothemis erythraea* was discovered already in 1918 in Brandenburg (Brodowin in the Mark Brandenburg, north-east of Berlin, Bollow 1919) and then near Konstanz by Rosenbohm (1922), in Bavaria in the Osterseen area in 1952 and 1954 (Burbach 1998), by Jurzitza (1963) near Karlsruhe in Baden-Württemberg, and then in Rhineland-Palatinate in 1967 (Friedrich et al., 1976). Always only single specimens were found, no autochthonous population was ever registered.

Between 1971 and 1979

In the seventies the species was again observed in Bavaria near Munich by Bilek (1978) and also again in Baden-Württemberg at several localities (Heidemann, 1977; Lohmann, 1980), as well as in Rhineland-Palatinate (Friedrich et al. 1976, Weitzel 1978). In this decade Heidemann (1977) for the first time in Germany reported reproduction of *C. erythraea* in Baden-Württemberg in several years (1974-76). Lohmann (1980), when reviewing the up to then existing literature from western Germany, mentioned only Heidemann`s observations of indigenous populations; in eastern Germany (GDR, German Democratic Republic) the species was not found at all during this decade (Donath, 1984). A single specimen was registered in 1977 (Kikillus & Weitzel, 1981) near Cologne in Northrhine-Westfalia, this being the sixth federal state with a record.

Between 1980 and 1989

During the next decade the species expanded its range in southern Germany, mainly in the two southern federal states of Baden-Württemberg and Rhineland-Palatinate. According to Buchwald et al. (1984), the species had become autochthonous in Baden-Württemberg in “several waters and during several years in the Rhine valley”. In Rhineland-Palatinate the number of sightings increased significantly (Niehuis 1984), and finally *C. erythraea* was found being autochthonous near Ludwigshafen for the first time for this federal state by Ott (1988), where it continued to reproduce from 1985 onwards (Ott 1990; unpubl. data). No data are available for Hessen, but without doubt the species was present also in this federal state, for the same biotope types exist there at similar altitude as in Rhineland-Palatinate. In the Saarland the species was registered for the first time in 1986 (Röhlinger 1988) and became indigenous from 1990 onwards (Trockur & Didion, 1994). In Northrhine-Westfalia in 1989 another single male was noticed by Schmidt (1990), which was the second record for this federal state.

Between 1990 and 1999

During this decade the populations in the Saarland and in Northrhine-Westfalia became indigenous and also much more numerous (see the compilation in Ott 1996, 2000, and in Rudolph, 1998; Bauhus, 2001). In Northrhine-Westfalia the first record of a reproducing population had been made in 1995 near Dortmund by Bauhus (1996). In Rhineland-Palatinate the species left the Rhine-valley (ca. 95 meters a.s.l.) and colonised also higher altitudes in the Palatinate (ca. 300 meters a.s.l., Ott, 2000). First confirmed records came now also from Hessen, where the species without doubt was also indigenous then in some areas (*e.g.* in the Rhine valley). After its "rediscovery" in 1991 in Bavaria – where the species had not been registered in the previous decade – the number of observation sites increased from single ones to several dozens (Burbach, 1998; *e.g.* in the Danube valley and near Freising), some of them also with autochthonous populations. After its long absence from Brandenburg since 1918, *Crocothemis erythraea* was registered there again in 1995 by Günther (see the addendum in Mauersberger, 2003), and in Saxony it was recorded as new for this federal state in1997 (and again in 1998, see Xylander, 2005).

From 2000 to present time (2006)

The number and sizes of populations in the southern federal states showed further increase, *e.g.* in the Saarland *Crocothemis erythraea* is known now from more than 50 sites, being autochthonous in more than 20 out of these (Trockur, pers. comm. 2006). In Baden-Württemberg an ongoing increase of population numbers was stated as well as colonisation of areas known from previous investigations as definitely uninhabited (Hunger et al., 2006): so a real expansion could be documented by these data. These new populations inhabit also regions at higher altitudes (up to 700 meters a.s.l.) outside the Rhine-valley. The same happened in Bavaria: also in this state the species "climbed" up to higher altitudes, *i.e.* distinctly over 700 meters (Voith, pers. comm. 2005).

Some more observations came from Saxony, where only two observation sites did exist until 2004, and autochthony was not verified (Xylander, 2005), but in 2005 and 2006 the numbers of observations increased again (Xylander, pers. comm. 2006). In the southern parts of Lower Saxony *Crocothemis erythraea* finally became a new member of the federal fauna in 2000 and was continuously observed in the following years (Lohr, 2003), in 2006 it was one of the most abundant dragonflies at some gravel pits near the river Weser (Lohr, pers. comm. 2006) and it is also now autochthonous in many waters. In 2001 the species was recorded in Thuringia for the first time by Mey (2003), as well as for Saxony-Anhalt in 2003 (Heidecke & Lindemann, 2004; Buttstedt & Zimmermann, 2005). In both federal states it is now also indigenous. Also in 2003 the first observation in Mecklenburg-Vorpommern was made by Mauersberger (2003), the northernmost observation (53°4` N) up to that date – yet this was topped in 2006 by Frank (2006), who found the species near Schönberg (53° 51`N); another observation was mentioned by Ritter (pers. comm. 2006) for the eastern part of the federal state. In Brandenburg the number of sightings increased (*e.g.* Donath, 2003), and also in this federal state the species is now indigenous (Schneider et al., 2005), having flourishing populations with some thousands of emerging individuals. Beutler & Beutler (2005) found 170 exuviae along 10 meters of shoreline and from that number calculated a total population of > 5000 individuals. Presently the species is known from more than 30 localities (Beutler, 2006, and pers. comm. 2006). In 2006 it was finally registered for the first time for Bremen (ARLT 2006) and also in northern Lower-Saxony (Jödicke, 2006: 53° 22`N (Friesland), Finch, pers. comm. 2006: 53,18° N (Hahn-Lemden)), the two latter observations being made at nearly the same latitude as the one in Mecklenburg-Vorpommern.

Actually, records of *C. erythraea* are missing from only three northern German federal states, but very probably it is already present there, as it has reached and even transgressed the geographical latitude of these states, namely Berlin and Hamburg and even Schleswig-Holstein.

During this present decade for the first time in Germany a probable second generation of this species – generally being univoltine in the north – was observed in Hessen and southern Lower Saxony in the hot summer of 2003 at shallow waters (Horn, 2003; Böhm, 2004).

Table 1. Distribution and expansion of the Scarlet Darter (*Crocothemis erythraea*) in the German federal states (o = observation(s) of single individuals, • = indigenous populations, ▲ = increasing / expanding, in ( ): probably; see text for details)

German federal state	< 1970	1971-1979	1980-1989	1990-1999	2000-2006
Baden-Württemberg (BW)	O	•	• ▲	•▲	•▲
Bavaria (BY)	O	O	O	•▲	•▲
Rhineland-Palatinate (RP)	O	O	•▲	•▲	•▲
Saarland (SL)			O	•▲	•▲
Hessen (HE)			(O)	O (•)	•
Thuringia (TH)					•
Saxony (SN)				O	O
Saxony-Anhalt (ST)					•
Northrhine-Westfalia (NW)		O	O	•	•▲
Lower Saxony (NI)					•
Brandenburg (BB)	O			O	• ▲
Bremen (HB)					O
Hamburg (HH)					(O)
Schleswig-Holstein (SH)					(O)
Mecklenburg-Vorpommern (MV)					O
Berlin (BE)					(O)

CROCOTHEMIS ERYTHRAEA EXPANDING IN THE REST OF EUROPE

Not only in Germany, but also in other Central and Middle European countries *Crocothemis erythraea* expanded its range in the recent decades. Only a few examples are presented here, as a detailed review will be published later (Ott, 2007). Already in the 18th century the species was known from Austria, where it had populations near Vienna, but in the 19th century it expanded its range in all Austrian federal states and is now, except for Tyrol, autochthonous. There it occurs even up to an altitude of 1,640 m a.s.l., and the highest breeding population was found at 570 m a.s.l. (Raab et al., 2006). In Switzerland it was indigenous only in very few cases in former times, but from 1980 onwards the number of populations increased continuously, combined with range expansion. It was registered up to 2090 meters a.s.l. and breeding populations were confirmed as high as 663 meters a.s.l. (Wildermuth et al., 2005). In The Netherlands only 3 individuals had been recorded in 1959, 1967 and 1968, respectively, but in 1993 a whole population was discovered in the southern part of that country, and in the following years several additional populations were established with an increasing tendency (NVL, 2002). Meanwhile *C. erythraea* has reached the UK, where it was first seen in 1995 and now has also established the first population (Smallshire & Swash, 2004).

Already these few examples show that range expansion of this species is a general process in Europe and not limited to Germany (and some of the above named authors also mention the climatic changes as a reason for this, see below).

DISCUSSION

Ways of expansion – a clear sign ?

As shown before, the species colonised the whole of Germany in about four decades, starting in the southern Rhine-valley and proceeding step by step to the north. At a first glance the species seems to have followed the rivers, at least it was found first of all in the river valleys – from where it expanded into adjacent regions and ascended from low altitudes to higher ones. Dumont (1967) supposed the species to use the large rivers as "corridors" or "guiding lines", entering Germany via the gate of Burgundy (junction of the rivers Saône and Rhine) and/or the river Mosel and then continuing down the river Rhine.

However, as the species needed nearly 80 years to proceed to areas more northerly than its first observation site in Germany (Bollow, 1919), this assumption sounds a bit too simple or generalising.

From comparing the map (Fig. 2) with Table 1 we may conclude that the species moved northwards on a rather "broad front". If it had strictly followed only the rivers as expansion corridors, it should have been expected to conquer the northern German federal states beginning at the lower course of the Rhine, from there heading eastwards through the lowlands of the northern federal states. But this was obviously not the case, as the species was not discovered on this route via Hannnover, Braunschweig *etc.* (especially along this route in Lower Saxony the "observation intensity of GdO-members" is very high, so it certainly would have been discovered). It was found, on the contrary, in that thoroughly investigated region for the first time as late as 2005 in the area of Celle (Clausnitzer, 2005).

So we may rather assume that *Crocothemis erythraea* obviously passed through the central German low mountain region as a whole and arrived simultaneously in several federal states of the north in the mid nineties (see here also Fig. 3). We may also plausibly consider the populations of the northern federal states being of different origin. Populations *e.g.* in the north-eastern federal state of Brandenburg, possibly have been established by animals having proceeded on the route via Austria, Hungary, through the gate of Hungary (Danube valley near Hainburg/Vienna) and through the Czech Republic and Poland – this assumption is supported by the fact that also in these countries the species was rediscovered and expanding around the mid nineties (Czekaj, 1994; Mocek, 1998; Bernard, 2000). At the moment genetic analyses of different populations are carried out (Ott & Sanchez Guillén, in prep.) to decide on the question, whether the northern *Crocothemis*-populations are of homogenous or different origin.

THE EXPANSION OF *CROCOTHEMIS ERYTHRAEA* IN GERMANY – WAS THE SPECIES ONLY OVERLOOKED OR IS THE EXPANSION A RESULT OF NEWLY CREATED WATERS?

The period before its expansion

It is hard to assess the real situation of the species during the decades before and after its first discovery in Germany in 1918. On one hand, of course less investigations than at present have been carried out on the whole in that phase, but on the other hand it has to be mentioned that especially in those particular regions where the species was found much later, several thorough odonatological investigations have been performed even in these first decades of the last century (*e.g.* Rhine-valley, see the compilation in Lohmann, 1980, or Niehuis, 1984). Moreover, in Germany Odonatology has a long tradition, leading to the detection of other very rare and much more cryptic species in this period (*e.g. C. hylas, O. curtisii*).

So we may conclude that in this time span *Crocothemis erythraea* very probably was not at all present or at the most in very small numbers, in any case there were no noticeable populations.

The period during its expansion

In Germany we saw a strong increase in the number of odonatologists during the last three decades: in the seventies there were only a few dozens, in the early eighties, when the GdO (German Odonatological Society, see: www.libellula.org) has been founded, there were about 100, and presently the GdO comprises more than 500 members, this number being a reliable indicator of the general "odonatological activity".

Also in the eighties environmental issues and nature protection became more and more important in Germany; the federal states started mapping and monitoring programmes, most often including Odonata – in the same period other private initiatives started to document the distribution of dragonflies (*e.g.* Niehuis, 1984 for Rhineland-Palatinate).

Without doubt our detailed knowledge of the species distribution as well as the density of observations increased enormously during this period, though already at the beginning of the expansion of *C. erythraea* the number of odonatologists had been pretty high. Moreover, the species is of such a striking appearance that even amateurs or beginners in odonatology do recognise it immediately when present at a water. In addition, in a lot of areas we do have reliable data for the situation before its expansion or intrusion into these previously unsettled areas; so it is rather impossible that this striking species – in particular an indigenous population – has been overlooked during this time (see the increase of faunistic literature during the second half of the last century in Schorr (1990) as another "indicator" of high odonatological activity).

Another aspect may be the increase of the number of water bodies in Germany. In particular garden ponds were created in huge numbers, and along the rivers the number of gravel pits increased – the latter function as good and typical *Crocothemis erythraea*-habitats, if not used too intensively for fishing or recreation purposes. In total Germany's land surface is presently covered by 2, 3% of different waters (BFN, 1997, 2004; BFS, 2006) with an increasing tendency during the last decades. Consequently in the beginning of the expansion this was assumed to be the main reason, as no other factor could be recognised then (*e.g.* Ott, 1988).

From old and historic maps it is easily seen that already in former times a big number of waters existed, in general more natural waters existed in those times (running waters with morphodynamic; less exploitation of land !), and today the species is present in waters, which do exist already since decades. In addition, the increasing creation of garden ponds and exploitation of gravel pits along the rivers during the decades after WW 2nd was occurring more or less simultaneously all over Germany and not at first in the south and then moving northwards.

Finally there is no good reason, why the number of waters should have been a limiting factor during the first half of the last century and why those which already existed for a long time should not have been populated earlier. As shown before, the species colonised Germany continuously from south to north, but being a vigorous flier and with regard to its ability to cross higher altitudes and to reproduce also in small water bodies, colonisation should have happened much quicker: so the creation of new waters can't be the (only) reason for the documented way of expansion of *Crocothemis erythraea.*

ANOTHER REASON FOR THE EXPANSION: CLIMATIC CHANGE

At the beginning of the nineties of the last century the problem of climatic changes and the first registered effects on flora and fauna became a theme in science (*e.g.* Gates, 1993; BFN, 1995; Graves & Reavey, 1996). For dragonflies – and in particular for *Crocothemis erythraea* – this theme was mentioned for the first time by Ott (1996 a) at the SIO-conference in Essen, where it was discussed controversially, as many odonatologists interpreted the first steps of its expansion in Germany as a normal oscillation which they expected to stop soon.

But the expansion of *Crocothemis erythraea* continued and similar processes were shown also for several other species of Mediterranean origin in Germany, all over Europe, the United States and Japan (Ott, 1996 b, 2000, 2001, 2007; Parmesan & Yoke, 2003; Hickling et al., 2005). In this context it must be pointed out that – with only very few exceptions – other species, in particular those with a typical Euro-Siberian distribution, did not expand their range.

So today the well documented range expansion of southern dragonfly species to the north is accepted as a fact (*e.g.* Hickling et al., 2006).

If species with a previous distribution restricted to warmer southern areas extend their range to northern areas, the reason may be (or should be) related to temperature. Looking at the history of its expansion, *Crocothemis erythraea* first of all populated the warmer regions in the Rhine valley (according to its bio-climatic envelope as a mediterranean species) and then spread to the north and up to higher altitudes.

Parallel to this process we had in Germany an increase of temperature: the warmer regions (*e.g.* Rhine valley) became even warmer and colder regions also became warmer, a still continuing phenomenon.

From 1988 onwards, in Germany the mean temperature of every single year except 1996 exceeded the long-term medium temperature of 8,3°C. (period: 1961-1990): in 1994 it was 9,7°C, in 2000 it was even

Fig. 1. Male Scarlet Darter (*Crocothemis erythraea*) (Brullé, 1832) (Pescara, Italy, 2006)

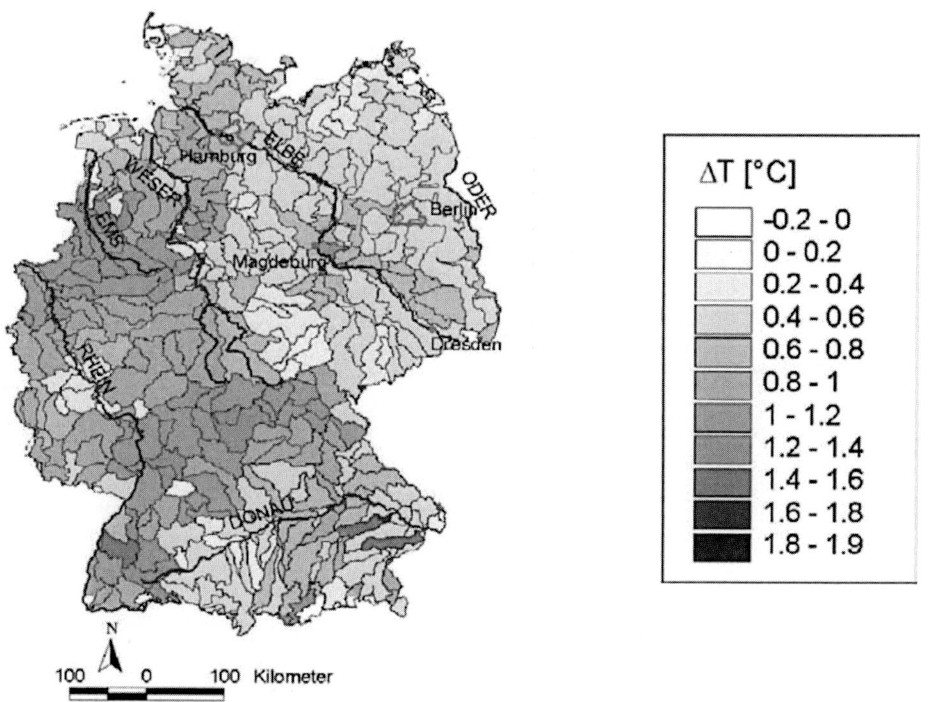

Fig. 3: Change of mean temperature in Germany from 1901-1998 (data source: DWD; Fig. prepared by F. Badeck)

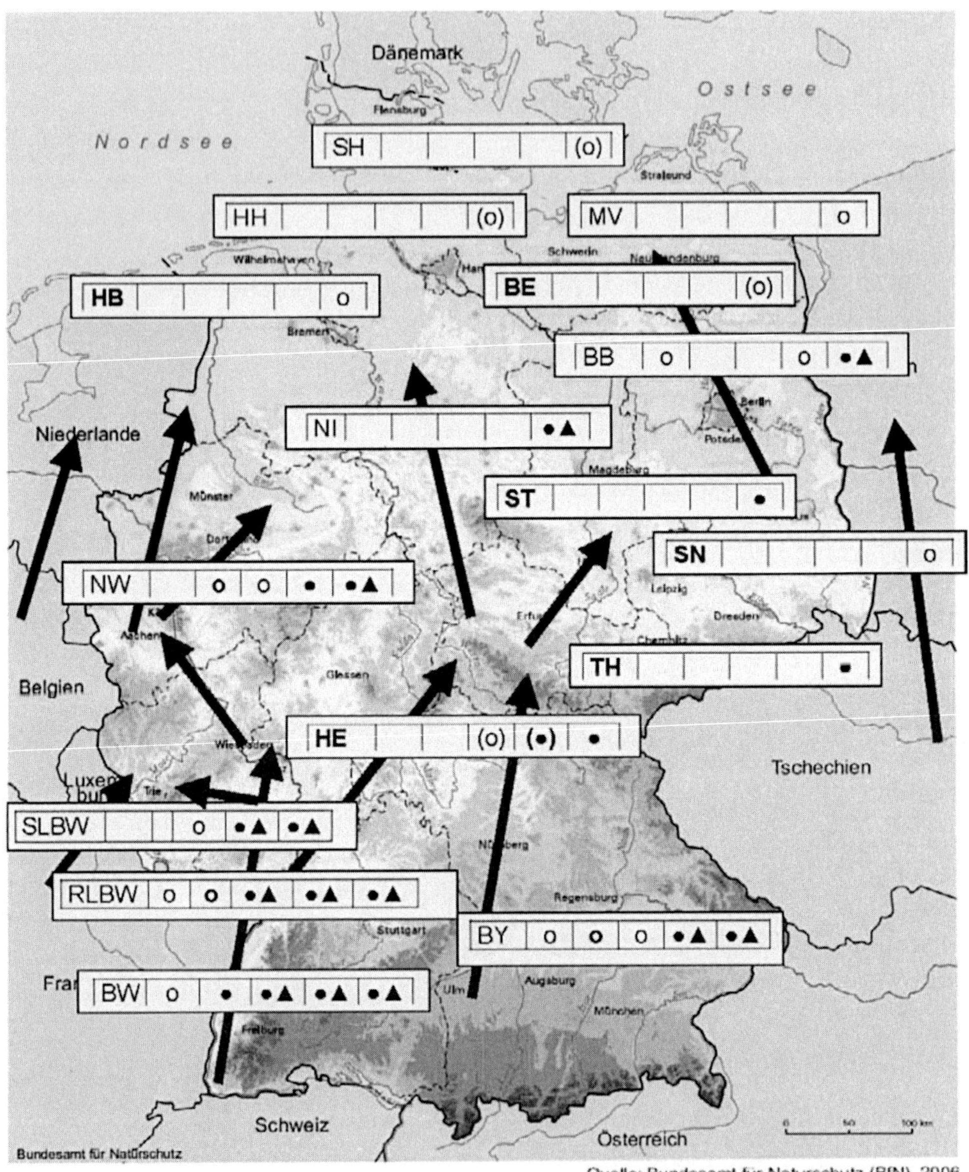

Quelle: Bundesamt für Naturschutz (BfN), 2006

Fig. 2: Expansion of the Scarlet Darter (*Crocothemis erythraea*) in Germany (codes for the federal states see Table 1) — arrows indicate the known or possible ways of expansion

9,9°C, and in 2003 again 9,4°C. In 2003 the duration of sunshine in Germany reached the extremely high amount of 2014 hours, which is 32% more than medium duration (DWD 2000-2004). In addition we are experiencing an increasing number of so called "summer days" and "hot days", and a decreasing number of "cold days" and "frost days" (DWD, 2004, see also Rapp & Schönwiese, 1996).

These data given here for Germany as a whole of course vary in the single federal states, but the general increase is found everywhere. For example, for Rhineland-Palatinate an increase of about one degree was shown already by Ott (1996 b) for various climatic stations, and this process continued in the recent years (Am, 2006).

The increase of the mean temperature in Germany from 1991-1998 is shown in the following figure and here we see that the previously demonstrated ways of expansion of *Crocothemis erythraea* reflects very well the trend in the temperature increase in the different areas or federal states.

So expansion of *Crocothemis erythraea* reflects the change of the abiotic climatic conditions in Germany, and this is in concordance with a lot of other investigations and results stating obvious changes in plant and animal phenology (*e.g.* earlier breeding of birds, earlier flowering of trees etc., *e.g.* Walther et al., 2002; Parmesan & Yohe 2003; Root et al., 2003; Leuschner & Schipka, 2004; Schröder et al., 2005) related to climatic changes, although in detail some questions of course remain as yet unanswered (see below) as well as some controversial discussions on the species level (*e.g.* Schmidt, 2004).

Still remaining is the important question, whether the species reacts to key factor "temperature" (or another one ?) via the adult stage, the larval stage or via both of them, and also the question whether it is influenced by a combination of several factors – and, moreover, whether the northern populations are already adapted in any way (*e.g.* differing body sizes and behaviour of the adults, differing behaviour of the larvae (ice-avoiding strategy), altered physiology).

Various possibilities may be considered as key factors:

- increase of mean temperature per annum (for the species in general)
- increase of mean summer temperature, influencing activity period of the adults (for the adult stage – higher activity and mobility)
- increase of water temperature (for the larval stage – lower mortality)

- increase of the sunshine duration during summer (for the adult stage – higher activity and mobility)
- increase of so called "summer days" and "hot days" (for the adult stage – higher activity and mobility) or
- decrease of so called "frost days" (for the larval stage – lower mortality)

As the mean air temperature increased during the last decades, the water temperatures-especially in shallow waters-must have risen as well (unfortunately there are no long term data *e.g.* from gravel pits in Rhineland-Palatinate, available; Wendling, *pers. comm.* 2006), so that a combination of these two factors might be effective. According to Lohmann (1980) the species is "depending on high water temperatures in summer", but he did not give any proof or reference for this.

A "non freezing over" of the waters, as postulated by Jurzitza (2000), is obviously not an essential factor: larvae of the species survive unharmed a freezing over of the waters, as shown by Ott (1988: *C. erythraea* emerged after the water bore an ice cover as thick as 30 cm in the previous winter) and also emphasised by Burbach (1998). It goes without saying that long periods of higher temperature definitely increase the species` mobility (as in most other odonata / insects) and its ability to expand its range by colonising new waters.

OUTLOOK

Consequences for the biotopes

As shown before, *Crocothemis erythraea* has invaded a lot of waters and also different biotope types, mainly sand- and gravel-pits of different successional stages, but also alluvional waters, ponds, lakes with a rich reed and submersed vegetation, and obviously even moorland biotopes (Ott, 2007; Wildermuth, 2006; Jödicke, 2006). As no experiments have been carried out up to now, it is unclear whether the larvae of *C. erythraea* do have a negative impact on other larvae in the water, either by direct predation or indirectly via the food chain. But as they do co-exist with a lot of other dragonfly species over a long time span (Ott, 1990; Ott, 2007) it is more probable that they occupy their special niche within the whole coenosis, exerting no severe and direct effect on other (protected) Odonata species. However, a decline of another species (*Orthetrum cancellatum*) during an increase of *Crocothemis erythraea* at a water has been reported (Lohr, pers. comm. 2006).

The appearance of *Crocothemis erythraea* may indicate altered conditions in the waters, in particular if it settles in an untypical biotope like moorland bogs, and here it should be considered a disturbance indicator revealing some negative impact (*e.g.* decrease of typical moorland species like *Aeshna juncea, Somatochlora arctica, Coenagrion hastulatum*: see Ott, 2001, 2006, 2007). So the presence of *C. erythraea* indicates changes on the biotope-level (*e.g.* after heat waves: Mouthon & Daufresne, 2006), and without doubt it indicates also changes in the structure of the food-web, the whole landscape or ecological system (*e.g.* Emmerson et al., 2005, Schröter et al., 2006). As all recent scenaria (*e.g.* IPCC) expect an increase of greenhouse gas emissions and an ongoing global warming, we have to expect effects on the waters and a continuing range expansion of southern species (Ott, 2007).

CONSEQUENCES FOR THE RED LISTS

When *Crocothemis erythraea* was seen more often in Germany and had established its first populations, all odonatologists were happy and proud to have this "new member". Consequently the species was included in the red lists, as in the early years it did of course not appear in big numbers. *E.g.* in the federal state of Rhineland-Palatinate it was classified at first in 1983 as a "vulnerable guest" (Itzerott et al., 1988) and then in 1993 – now having autochthonous populations – as "vulnerable" (Eislöffel et al., 1993), in Northrhine-Westfalia (Schmidt & Woike, 1986) as a "Vermehrungsgast" (a guest sometimes reproducing), and in Baden Württemberg still in 1994 even as "endangered" (SGL, 1994).

In the first national list of 1984 (Clausnitzer et al., 1984) it was also classified as a "Vermehrungsgast" (a guest sometimes reproducing).

The problems of including this continuously spreading species in the red list were pointed out by Ott (1996) for the first time, and consequently it was excluded from the following national red list of 1998 (Ott & Piper, 1998). Why some authors – *e.g.* Jurzitza (2000) – still today classify the species as "vulnerable" for the south western parts of Germany remains unclear; for Baden-Württemberg Hunger & Schiel (2006) classify *C. erythraea* as "not endangered", a seemingly correct classification, as the species is still extending its range, in particular in the south, and so it has to be excluded from the red list according to the IUCN-criteria (IUCN 2006).

ACKNOWLEDGEMENTS

Many thanks to Dr. B.K. Tyagi for his kind invitation to give a contribution to this book ! Thanks also to Dr. E. Schroeder and U. Euler (BfN), who made the basic map for figure 2 available, Dr. F. Badeck who provided figure 3 and to A. Ritter, M. Voith, Dr. B. Trockur, M. Lohr, Dr. K. Wendling,. Dr. H. Beutler, Dr. O.D. Finch, Dr. W. Xylander for unpublished data. Very special thanks finally to Dr. R. Rudolph, who made several suggestions to the manuscript and helped to improve the English.

REFERENCES

Aguesse, P., 1959. Notes sur l`accouplement et la ponte chez Crocothemis erythraea Brullé (Odonata, Libellulidaee). *Vie et Milieu* 10: 176-184.

AM [Agrarmeteorologie], 2006. www.agrarinfo.rlp.de (climatic data for Rhineland-Palatinate).

Anselin, A. & Hoste, 1995. Dragonfly records from the Sierra de la Demanda and the Sierra de Urbión (Spain) with notes on habitat and altitudinal range. In: R. Jödicke, 1996. *Studies on Iberian Dragonflies.* Adv. Odonatol. Suppl. 1: 9-12.

Arlt, J., 2006. GdO-Mailingliste 10/2006.

Bauhus, S., 1996. Funde von Crocothemis erythraea (Brullé) und Aeshna affinis (VanderLinden, 1823 In der Lippe-Aue (Anisoptera: Libellulida, Aeshnidae). *Libellula* 15 (1/2): 79-84.

Bauhus, S., 2001. Vorkommen und Status der mediterranen Libellenarten Aeshna affinis Vander linden und Crocothemis erythraea (Brullé) in Westfalen (Odonata). *Natur und Heimat* 61(3): 73-82.

Bellmann, H., 1987. *Libellen.* Neumann-Neudamm, Melsungen.

Bernard, R., 2000. An interesting record of Crocothemis erythraea (Brullé) in midwestern Poland (Anisoptera: Libellulidae). *Notul. Odonatol.* 5 (5): 65-66.

Beutler, D. & H. Beutler, 2005. GdO-Mailingliste 11/2005.

Beutler, H. (2006). Libellenmonitoring: Aktuelle Ausbreitung der Feuerlibelle in Brandenburg. In : Landesumweltamt Brandenburg (Eds.) : Umweltdaten aus Brandenburg – Bericht 2006. pp. 102-103, Potsdam-Groß Glienicke.

Biesmeijer, J.C. et al., 2006. Parallel Declines in Pollinators and Insect-Pollinated Plants in Britain and the Netherlands. *Science* 313: 351-354.

Bilek, A., 1978. Zur Faunistik europäischer Libellen. *Articulata* 1(7): 47-49.

BFN [Bundesamt Für Naturschutz], 1995. *Klimaänderungen und Naturschutz.* Angewandte Landschaftsökologie Vol. 4. Landwirtschaftsverlag, Münster-Hiltrup.

BFN [Bundesamt Für Naturschutz], 1997. *Erhaltung der Biologischen Vielfalt – Wissenschaftliche Analyse deutscher Beiträge.* Landwirtschaftsverlag, Münster-Hiltrup.

BFN [Bundesamt Für Naturschutz], 2004. *Daten zur Natur.* Landwirtschaftsverlag, Münster-Hiltrup.

BFS [Bundesamt Für Statistik], 2006. www.destatis.de (statistical data for Germany).

Böhm, K., 2004. Zur Entwicklung und Phänologie von Crocothemis erythraea in Nordrhein-Westfalen: Nachweis einer zweiten Jahresgeneration? (Odonata: Libellulidae). *Libellula* 23 (3/4): 153-160.

Bollow, C., 1919. Crocothemis erythraea Brullé in der Mark. *Deutsche Entomol. Zeitschr.*, p. 191.

Buchwald, R., B. Gerken, K. Siedle, & K. Sternberg, 1984. Übersicht über die Libellenvorkommen in Baden-Württemberg mit kurzer Charakteristik des Fortpflanzungsgebietes und Angaben zur Verbreitung. *Libellula* 3 (3/4): 101-110. 1984.

Burbach, K., 1998. Feuerlibelle Crocothemis erythraea (Brullé 1832). In: K. KUHN & K. BURBACH, 1998. *Libellen in Bayern*, pp. 172-173. Ulmer, Stuttgart

Buttstedt, L. & W. Zimmermann, 2005. Über Entwicklungsnachweise der Feuerlibelle Crocothemis erythraea (Brullé, 1832), in Thüringen und Sachsen-Anhalt. *Entomol. Nachr. Ber.* 49 (3/4): 171-178.

Czekaj, A., 1994. New records of Crocothemis erythraea (Brullé) and Tarnetrum fonscolombiii (Sel.) from Poland (Anisoptera, Libellulidae). *Notul odonatol.*4 (3): 53.

Clausnitzer, H.J., 2005. GdO-Mailingliste 14/2005.

Clausnitzer, H.-J., P. Pretscher & E. Schmidt, 1984. Rote Liste der Libellen (Odonata). In: J. Blab, E. Nowak, W. Trautmann & H. Sukopp, 1984. *Rote Liste der gefährdeten Pflanzen und Tiere in der Bundesrepublik Deutschland*, pp. 116-118, Kilda, Greven.

Donath, H., 1984. Situation und Schutz der Libellenfauna in der Deutschen Demokratischen Republik. *Ent. Nachr. Ber.* 28 (4): 151-158.

Donath, H., 2003. Verstärktes Auftreten südlich verbreiteter Libellenarten. *Biologische Studien Luckau* 32:100-102.

Dreyer, W., 1986. *Die Libellen.* Gerstenberg, Hildesheim.

Dumont, H., 1967. A possible scheme of the migration of Crocothemis erythraea Brullé - populations from the Camargue (Odonata. Libellulidae). *Biologisch Jaarboek* 35: 222-227.

DW [Deutscher Wetterdienst], 2000-2004. *Klimastatusbericht*e (www.dwd.de) (climatic data for Germany).

Eislöffel, F., Niehuis, & M. Weitzel - unter Mitarbeit von M. & U. Braun, J. Ott, H. Schausten & L. Simon, 1993: *Rote Liste der bestandsgefährdeten Libellen (Odonata) in Rheinland-Pfalz.* Mainz.

Emmerson, M., M. Bezemer, M.D. Hunter & T.H. Jones, 2005. Global change alters the stability of food webs. *Global Change Biology* 11: 490-501.

Falchetti, E. & C. Utzeri, 1974. Preliminary observation on the territorial behaviour of Crocohemis erythraea Brullé. *Fragmenta Entomologica* 10: 295-300.

Fliedner, H., 1997. Die Bedeutung der wissenschaftlichen Namen europäischer Libellen. *Libellula Supplement* 1: 1-111.

Frank, M., 2006. Erstfund der Feuerlibelle (Crocothemis erythraea) in Mecklenburg-Vorpommern (Odonata, Libellulidae). *Virgo, Mitteilungsblatt des entomologischen Vereins Mecklenburg* 9(1): 69-70.

Friedrich, E., M. Niehuis & S. Ohliger, 1976. Beitrag zur Libellenfauna der Südpfalz und angrenzender Gebiete (Insecta: Odonata). *Mitt. Pollichia* 64: 153-163.

Gates, D., 1993. *Climatic Change and Its Biological Consequences.* Sinauer Associates, Sunderland

Graves, J. & D. Reavey, 1996. *Global Environmental Change – Plants, Animals & Communities.* Longmann, Essex.

Heidecke, F. & K. Lindemann, 2004. Erster Reproduktionsnachweis von Crocothemis erythraea (Odonata: Libellulidae) in der Goitzsche bei Bitterfeld in Sachsen-Anhalt im Jahre 2003. *Naturwiss. Beitr. Mus. Dessau* H. 16: 63-64.

Hickling, R., D.B. Roy, J.K. Hill, & C.D. Thomas, 2005. A northward shift of range margins of British Odonata. *Global Change Biology* 11: 1-5.

Hickling, R., D.B. Roy, J.K. Hill, R. Fox & C.D. Thomas, 2006. The distribution of a wide range of taxonomic groups are expanding polewards. *Global Change Biology* 12: 450-455.

Horn, R., 2003. Eine zweite Jahresgeneration bei Crocothemis erythraea in Deutschland während des extrem heißen Sommers 2003 (Odonata: Libellulidae) ?. *Libellula* 22 (3/4):139-142.

Hunger, H., Schiel, F.-J. & B. Kunz, 2006. Verbreitung und Phänologie der Libellen Baden-Württembergs. *Libellula Supplement* 7: 15-184.

IPCC [International Panel On Climate Change]: www.ipcc.org

Itzerott, H., Niehuis, M. & M. Weitzel unter Mitarbeit von R. Kikillus, S. Ohliger & E. Schmidt, 1988. *Rote Liste der bestandsgefährdeten Libellen (Odonata) in Rheinland-Pfalz* (Stand: April 1983). Mainz

IUCN [International Union For The Conservation Of Nature], 2006. Criteria for the red lists: www.redlist.org

Jödicke, R., 2006. GdO-Mailingliste 10/2006.

Jurzitza, G. 1963. Libellenbeobachtungen in der Umgebung von Karlsruhe/Baden, 3. *Mitteilung. Beitr. Naturk. Forsch. SW-Deutschl.* 22 (2): 107-111.

Jurzitza, G. 1965. Libellenbeobachtungen in der Umgebung von Karlsruhe/ Baden. 5. *Mitteilung. Beitr. naturk. Forsch. SW-Deutschl.* N.F. 24 (1): 37-39.

Jurzitza, G.,1978. *Unsere Libellen.* Kosmos, Stuttgart.

Jurzitza, G., 2000. *Der Kosmos Libellenführer.* Franckh-Kosmos, Stuttgart.

Kikillus, R. & M. Weitzel., 1981. Grundlagenstudien zur Ökologie und Faunistik der Libellen des Rheinlandes. *Pollichia Buch* Nr. 2, Bad Dürkheim.

Landmann, A., 1983. Zum Vorkommen und Status der Feuerlibelle (Crocothemis erythraea Brullé, 1832) in Österreich (Insecta: Odonata, Libellulidae). *Ber. Naturwiss.-Medizin. Vereins in Innsbruck* 70: 105-110.

Landmann, A., 1984. Die Libellenfauna eines subalpinen Hochmorkomplexes in den Salzburger Zentralalpen. *Libellula* 3 (1/2): 65-74.

Landmann, A., G. Lehmann, F. Mungenast & H. Sonntag, 2005. *Die Libellen Tirols.* Berenkamp, Innsbruck.

Leuschner, C. & F. Schipka, 2004. *Vorstudie Klimawandel und Naturschutz in Deutschland.* BfN-Skripten 115, Bonn-Bad-Godesberg.

Lohr, M., 2003. Crocothemis erythraea auch in Niedersachsen (Odonata: Libellulidae). *Libellula* 22 (1/2): 35-39.

Lohmann, H., 1980. *Faunenliste der Libellen (Odonata) der Bundesrepublik Deutschland und Westberlins.* SIO Rapid Communications No. 1., Utrecht.

Mauersberger, R., 2003. Crocothemis erythraea im Nordosten Deutschlands (Odonata: Libellulidae). *Libellula* 22 (1/2): 55-60.

Mey, D., 2003. Vorkommen und Beobachtungen zur Verhaltensweise der Feuerlibelle Crocothemis erythraea Brullé, 1832 (Odonata: Libellulidae) in Thüringen *Veröffentlichungen Naturkundemuseum Erfurt* 22/2003: 137-148.

Mocek, B., 1998. Contribution to the knowledge of the dragonflies (Odonata) in Eastern Bohemia with the findings of the species Crocothemis erythraea (Brullé, 1832). *Acta Musei Regionaehradecensis s.*A. 26: 27-37.

Mouthon, J. & M. Daufresne, 2006. Effects of the 2003 heatwave and climatic warming on mollusc communities of the Saône: a large lowland river and ist two main tributaries (France). *Global Change Biology* 12: 441-449.

Müller, J. & M. Schorr, 2001. Verzeichnis der Libellen (Odonata) Deutschlands. *Entomofauna Germanica* 5: 9-43.

Niehuis, M., 1984. Verbreitung und Vorkommen der Libellen (Insecta: Odonata) im Regierungsbezirk Rheinhessen-Pfalz und im Nahetal. *Fauna und Flora in Rheinland-Pfalz* 3 (1): 1-203.

NVL [Nederlandse Vereniging Voor Libellenstudie], 2002. *De Nederlandse Libellen (Odonata).* Nederlandse Fauna 4., Leiden.

Ott, J., 1988. Beiträge zur Biologie und zum Status von Crocothemis erythraea (Brullé, 1832). *Libellula* 7 (1/2): 1-25.

Ott, J., 1991. Die Libellenfauna der Kiesgrube "Schleusenloch" bei Ludwigshafen. Fauna *Flora Rheinland-Pfalz* 6: 609-645

Ott, J., 1993 a. Die Veränderung der Libellenfauna einer Kiesgrube nach dem Besatz mit Graskarpfen. *Artenschutzreport* 3: 3-7.

Ott, J., 1993 b. Die Libellenfauna des Stadtgebietes von Kaiserslautern - Ergebnisse einer Stadtbiotopkartierung und planerische Konsequenzen. *Fauna Flora Rheinland-Pfalz* 7: 103-146.

Ott, J., 1995 a. Zum Einfluß intensiver Freizeit- und Angelnutzung auf die Fauna von Sekundärgewässern und Konsequenzen für die Landschaftsplanung - dargestellt am Beispiel der Libellen (Odonata). *Fauna Flora Rheinland-Pfalz* 8: 147-185.

Ott, J., 1995 b. Die Beeinträchtigung von Sand- und Kiesgruben durch intensive Angelnutzung Auswirkungen auf die Libellenfauna und planerische Lösungsansätze. *Limnologie aktuell* 7: 155-170.

Ott, J.,1996 a. Crocothemis erythraea – an indicator for climatic change ? In: Schmidt, E., R. Rudolph, J. Fey & J. Kaminski (1995): *Abstact Booklet of the 13th International Symposium on Odonatology 1995*: pp. 42. Essen.

Ott, J., 1996 b. Zeigt die Ausbreitung der Feuerlibelle Crocothemis erythraea Brullé in Deutschland eine Klimaveränderung an ? *Naturschutz und Landschaftsplanung* 2/96: 53-61.

Ott, J., & W. Piper, 1998. Rote Liste Libellen (Odonata). In: BFN (Ed.)(1998): *Rote Liste gefährdeter Tiere Deutschlands*: pp. 260-263. Bonn-Bad Godesberg.

Ott, J., 2000. Die Ausbreitung mediterraner Libellenarten in Deutschland und Europa - die Folge einer Klimaveränderung ?. *NNA-Berichte* 2/2000: 13-35.

Ott, J., 2001. Expansion of mediterranean Odonata in Germany and Europe – consequences of climatic changes. In: Walter, G.R. et al. (Eds.) (2001): *„Fingerprints“ of Climate Change*: pp. 89-111. Kluwer Academic Publishers, New York et al.

Ott, J., 2006. Die Arktische Smaragdlibelle – Somatochlora arctica (Zetterstedt, 1840) – in der Pfalz: übersehen oder kurz vor dem Aussterben? (Insecta: Odonata: Corduliidae). *Fauna Flora Rheinland-Pfalz* 10 (4): 1323-1338.

Ott, J., 2007. The dragonfly fauna of the “Schleusenloch” – results of a long term monitoring of a gravel pit (in prep.).

Ott, J., 2007. The expansion of dragonflies in Europe – recent changes and their ecological consequences. *In*: J., Ott (Ed.), 2007. *Monitoring and climate change – the example of dragonflies*. pp. §§-§§. Pensoft Publishers, Sofia.

Ott, J. (Ed.), 2007. *Monitoring and climate change – the example of dragonflies*. Pensoft Publishers, Sofia.

Ott, J. & R.A. Sanchez-Guillén, 2007. On which route arrived Crocothemis erythraea in northern Germany during its northward expansion? (in prep.).

Parmesan, C. & G. Yohe, 2003. A globally coherent fingerprint of climate change impacts across natural systems. *nature* 421: 37-42.

Raab, R., A. Chovanec & J. Pennerstorfer, 2006. *Libellen Österreichs*. Springer, Wien & New York.

Rapp, J. & C.D. Schönwiese, 1996. *Atlas der Niederschlags- und Temperaturtrends in Deutschland 1891-1990*. Frankfurter Geowissens-chaftliche Arbeiten, Serie B - Meteorologie und Geophysik, Vol. 5, Frankfurt.

Robert, P.A.,1959. *Libellen*. Kümerly & Frey, Bern.

Röhlinger, H., 1988. Zum Vorkommen und zur Verbreitung der Libellen im Saarland nach Untersuchungen in den Jahren 1985 und 1986. *Dendrocopus* 15: 135-144.

Rosenbohm, A., 1922. Weitere Beobachtungen zur Libellenfauna des Oberrheins und Bodensees. *Mitt. bad. Landesver. Naturk. Naturschutz*, N.F. 1 (10): 248-251.

Root, T.R., J.T. Price, K.R. Hall, S.H. Schneider, C. Rosenzweig & J.A. Pounds, 2003. Fingerprints of global warming on wild animals and plants. *nature* 421: 57-60.

Rudolph, R., 1998. Südliche Libellenarten in Westfalen. *Natur- und Landschaftskunde* 34: 114-116.

Schiemenz, H., 1953. *Die Libellen unserer Heimat*. Urania, Jena.

Schmidt, E. & M. Woike, 1986. Rote Liste der in Nordrhein-Westfalen gefährdeten Libellen (Odonata). In: Löbf [Ed.] (1986): *Rote Listen der in Nordrhein-Westfalen gefährdeten Pflanzen und Tiere*. Schriftenreihe der Landesanstalt für Ökologie, Landschaftsentwicklung und Forstplanung Nordrhein-Westfalen Vol. 4: pp. 199-204.

Schmidt, E., 1990. Die Odonatenfauna eines ehemaligen Bleisandabsetzbecken (Buchholzweiher bei Mechernich, Nordeifel). *Mitt. Pollichia* 77: 383-393.

Schmidt, E., 2004. Klimaerwärmung und Libellenfauna in Nordrhein-Westfalen – divergierende Fallbeispiele. *Entomologie heute* 16: 71-82.

Schneider, T., Brauner, O. & A. Reichling, 2005. Entwicklungsnachweis von Crocothemis erythraea und Funde von Aeshna affinis im Odertal Südostbrandenbugs (Odonata: Libellulidae, Aeshnidae). *Libellula* 24 (1/2): 73-82.

Schröder, W., G. Schmidt & J. Hasenclever, 2005. Korrelation meteorologischer und pflanzenphänologischer Indikatoren des Klimwandels. *Z. Umweltchem. Ökotox.* 17 (2): 94-105.

Schröter, D. et al., 2005. Ecosystem Service Supply and Vulnerability to Global Change in Europe. *Science* 319: 1333-1336.

SGL [Schutzgemeinschaft Libellen Baden-Württemberg] (1994): *10. Sammelbericht (1994) über Libellenvorkommen (Odonata) in Baden-Württemberg.* Karlsruhe.

Schorr, M., 1990. *Grundlagen zu einem Artenhilfsprogramm Libellen der Bundesrepublik Deutschland.* Ursus, Bilthoven.

Smallshire, D. & A. Swash, 2004. *Britain`s Dragonflies.* WildGudes, Hampshire.

Sternberg, K., 1989. Beobachtungen an der Feuerlibelle (Crocothemis erythraea) bei Freiburg im Breisgau. *Veröff. Naturschutz Landschaftspflege Bad.-Württ.* 64/65: 237-254.

Sternberg, K. & B. Höppner, 2000. Crocothemis erythraea (Brullé) Feuerlibelle. *In:* Sternberg, K. & R. Buchwald (2000): *Die Libellen Baden-Württembergs.* Vol 2.: pp. 374-384. Ulmer, Stuttgart.

ST. Quentin, D., 1960. Die Odonatenfauna Europas, ihre Zusammensetzung und Herkunft. *Zool. Jb. Syst. Ökol. Geograph. Tiere* 87 (4/5): 301-316.

Trockur, B. & Didion, A. 1994. Bemerkenswerte Libellenfunde für das Saarland aus den Jahren 1988 bis 1993. *Faunistisch-floristische Notizen aus dem Saarland* 26 (2): 329-344

Tsuda, S., 1991. *A Distributional List of World Odonata 1991.* Osaka.

Walter, G.R. et al. 2002. Ecological responses to recent climate change. *nature* 416: 389-395.

Weitzel, M., 1978. Funde der Wanderlibelle Crocothemis servilia DRURY im Rheintal (Odonata, Libellulidae). *Atalanta* 9 (2):179-180.

Wildermuth, H., 2006. GdO-Mailigliste 11/2006.

Wildermuth, H., Y. Gonseth & A. Maibach, 2005. *Fauna Helvetica Odonota – Die Libellen der Schweiz.* Fauna Helvetica 12., CSCF, Neuchâtel.

Xylander, W.E.R., R. Stephan & R. Franke, 1998. Erstnachweise und Wiedernachweise von Libellen (Odonata) für den Freistaat Sachsen und für die *Oberlausitz. Abhandl. Ber. Naturkundemuseum Görlitz* 70: 33-42.

Xylander, W.E.R., 2005. Feuerlibelle Crocothemis erythraea (Brullé, 1832). In: Brockhaus, T. & U. Fischer, 2005. *Die Libellen Sachsens*: pp. 247-249. Natur & Text, Rangsdorf.

Zimmermann, W., F. Petzold & F. Fritzlar, 2005. Verbreitungsatlas der Libellen (Odonata) im Freistaat Thüringen. *Naturschutzreport* Vol. 22: 1-224.

Odonata : Biology of Dragonflies
Ed.: B.K. Tyagi
Scientific Publishers (India), 2007, pp. 223-240

15

ADAPTATIONIST APPROACH OF REPRODUCTIVE BEHAVIOUR IN LIBELLULIDAE : A CASE REPORT ON *DIASTATOPS OBSCURA* FABRICIUS

J.B. IRUSTA and A. ARAÚJO

Programa de Pós-Graduação em Psicobiologia,
Departamento de Fisiologia - Centro de Biociências,
Universidade Federal do Rio Grande do Norte.
Caixa Postal 1511 – Campus Universitário, 59078-970 Natal, RN, Brasil.
e.mails: banuelos@ufrnet.br; arrilton@cb.ufrn.br

ABSTRACT

In this chapter we discuss three topics related to reproductive behaviour in Libellulidae, specifically theoretical aspects based on data obtained from the species *Diastatops obscura* (Fabricius, 1775) in its natural environment in Northeast Brazil. First, we studied reproductive strategies, with emphasis on intrasexual competition among males in territorial areas and the interconnection and synchronism of the behaviours of both sexes. Second, we analysed the females' choice of mate, a subject of lively discussion today among odonatologists throughout the world. Finally, we studied the possible relation between male body size and reproductive success. It was shown that males compete for areas of greatest access to sexually mature females, and that females select dominant males over the satellites. In competition among males, it was observed that larger individuals are more likely to achieve more copulations than smaller and medium-size males. It was also shown that larger individuals accomplish oviposition of the females inseminated by them more often and in greater numbers, and are territorial for more days than their smaller counterparts. Accordingly, we discuss aspects related to female choice and stabilizing selection.

INTRODUCTION

The evolutionary success of living beings depends on their adaptation to the environment. This adaptation over the course of their history manifests itself through natural selection, which selects aspects that, on the one hand, favour the survival of individuals and, on the other, the propagation of their genetic endowment to future generations. Charles Darwin (1859) himself described these two components of evolutionary success in his famous treatise, "On the Origin of Species". These two environmental pressures became known as natural selection (*sensu strictu*) and sexual selection, respectively.

Sexual selection occurs mainly through the preference of females for certain male attributes. Over the course of its evolution, this female attitude gave rise to morphological structures and behaviours in males who, by being attractive to their mates, may represent relative disadvantages for the survival of these individuals. Thus, sexual selection and its extravagant manifestations are submitted to or limited by natural selection, as if these two evolutionary forces were acting in opposite directions. So, sexual selection, modulated by natural selection, is largely responsible for system diversity and reproductive strategies currently found in the animal kingdom; it determines the genetic, morphological and behavioural aspects of individuals, populations and species.

Reproductive strategies in Libellulidae

Odonata have been the object of a great deal of scientific study to bring to light diverse aspects of their reproductive behaviour and have proved to be one of the most fascinating orders in the insect world. Studies emphasize the reproductive tactics of males and females, whose copulation is unique among insects and who display interesting behaviours such as selective mate choice, sexual mimicry, sperm competition and female guarding.

In the Libellulidae Family, this peculiar reproductive behaviour can be summed up by the following general considerations: sexually mature males are usually territorial, staying near bodies of water that are suitable as breeding grounds. They compete with one another in these areas, both ritualistically and with real aggression, to conquer and defend a territory attractive to females. These fights may result in chasing, expulsion, injury and, in some situations, to the death of some of the participants. These general rules, however, are not universal and new studies have shown an increasing number of exceptions and variations.

The females of species with territorial males only go to reproductive areas to copulate and lay eggs. Oviposition usually occurs by mass release directly into the water or into suitable substrate (Corbet, 1999) and is more successful when a dominant male guards the female during the entire process. Post-copulation guarding of the female may be with or without contact and apparently depends on the territorial strategies adopted and on environmental aspects such as competitor density.

Some species develop secondary characters (distinctive colours, for example) and behaviours associated to female courtship. The role of females in mate choice is currently under discussion, mainly in species without defined courtship behaviour. Several opinions attribute most of the participation in this selection exclusively to males and to the quality of oviposition sites, while other investigations, including this chapter, opt for more active female participation.

Female choice

Two components that act in sexual selection can be distinguished: on one hand, constant search and competition, normally by males, for mates susceptible to being inseminated (intrasexual selection), and on the other, the development of the capacity to choose, normally practiced by the females, the best suited mates (epigamic or intersexual selection). This is the basis for understanding the “conflict of interests” between the sexes, which has evolved into elaborate reproductive behaviours, as a result of an evolutionary race that seeks to optimize this conflict in favour of both sexes. Males maximize their reproductive efforts by fertilizing the greatest number of females, while females choose a male that provides material or genetic elements that enable offspring success, besides producing more and better eggs.

The optimization relation for males has long been known: Bateman (1948), in an experiment with *Drosophila*, showed that male heredity is positively related to the number of fertilizations that is achieved during their life. The active participation of females in mate choice is not as evident for most of the insects that have already been studied.

The reproductive effort of an organism, understood as the proportion of energy available that is used in reproduction, can be divided into mating effort and parental effort (Alexander & Borgia, 1979). In many animals, the first component is more typical in males, and the second more common in females.

Parental effort is rare in Odonata, with egg and larval development occurring with no assistance from the parents, who channel practically all their reproductive effort to mating. Thus, males dedicate most of their energy to intrasexual selection, which manifests itself in a series of pre and postcopulatory activities. Before coupling, males must search for or attract females while simultaneously competing with other males for potential mates. The copulations may be long or short, depending on the bursa copulatrix of the females and the male penis, which is adapted to remove or put aside the preceding sperm, resulting in different and very interesting sperm competition and sperm precedence systems (Waage, 1984; Córdoba-Aguilar *et al.*, 2003). After coupling, the males must also avoid other likely fertilizations of this female by other competitors, by practicing mate-guarding behaviours.

For their part, female choice is theoretically guided by males with genetic characteristics that provide competitive advantages, which are usually indicated by morphological or genetic characteristics known as secondary sexual characters, or by males who offer some interesting material resource. In the field of behavioural entomology, female choice is still not totally accepted for many species that develop polygenic reproductive systems. The behaviours shown by males and females at reproductive sites and during reproductive periods often seem to respond both to intrasexual evolutionary pressures and, to a lesser degree, intersexual pressures. It is always difficult to differentiate these two conditioning behaviours and interpretations in both directions continue to feed this discrepancy. While some authors attribute all sexual initiative to males who try to copulate with every female within their reach, others recognise certain female behaviours that would clearly influence the selection of a mate that will fertilise most of its eggs (Thornhill & Alcock, 1983).

Among the criteria that must be evaluated in sexually mature females during the reproductive phase, those related to suitable sites and opportune moments become mixed with those involving mate choice itself. Thus, when a reproductive event is interrupted, resulting in failure for the male involved, it may be difficult to determine if the cause is the male himself (his physical or hierarchical condition) or other circumstances (territorial characteristics, environmental conditions) (Thornhill & Alcock, 1983). In any case, some incontrovertible evidence of female choice has already been shown in other insects. Thus, in species where the males offer material resources (nuptial presents), it was verified that females choose mates that offer better (in this case bigger) resources (Thornhill & Alcock, 1983).

In territorial Odonata, when a female enters the reproductive area, the males compete to grab and hold her to copulate and fertilise the eggs, which will be oviposited and guarded by the male if there is no other impeding circumstance. This competition among males in reproductive areas makes it difficult to determine whether females choose only the most suitable sites (territories), thus favouring their owner, or if they exercise some type of mate choice as a criterion differentiated from other choices associated with reproduction, even when there is no courtship.

This polyandry is usually more intense in reproductive areas with high male density, which could be interpreted as a more difficult choice for females, or as a succession of forced copulations to which they were submitted by more aggressive males. This last idea, known as convenience polyandry by forced copulations, is defended by various authors (Koenig, 1991; Conrad & Pritchard, 1992), in that female choice may be eclipsed by male behaviour.

However, from a theoretical point of view, female choice should be universally understood, since this capacity to discriminate among mates is the engine of evolutionary pressure that resulted in the appearance of secondary characters found in insects and other animals.

Size and reproductive success

The phenotype characters of the individual often determine reproductive success, since they usually indicate genotype advantages susceptible to being transferred to offspring, and play an important part in inter- and intra-specific competitions. On the other hand, investment in secondary characters, such as size advantage or more intense coloration often associated with sexual selection, are not accompanied indefinitely by natural selection, since basic species format is preserved generation after generation. This suggests that, once a certain threshold is reached, sexual selection and natural selection act in opposite directions, setting limits to the supposed advantages. This is the concept of stabilising selection studied by Fincke (1988), Neems *et al.* (1998), Stoks (2000) and Thompson & Fincke (2002).

In insects, reproductive success is usually explained as a function of phenotype characters such as size (Thornhill & Alcock, 1983; Blanckenhorn *et al.,* 2000) and coloration (Van Dyck *et al.,* 1997). Among Odonata, it has been described that male mating success may vary with maturity, flying agility, stored fat reserves, longevity, territorial characteristics and body size (Corbet, 1999; Sokolovska *et al.,* 2000).

The relation between male body size and reproductive success has been previously studied in various Odonata, with different conclusions. In studies with damselflies, Harvey & Walsh (1993) found a positive correlation between body size and fitness, while Fincke (1986), Richardson & Baker (1997), Stoks (2000) and Purse & Thompson (2005) found no correlation. Andersson (1994) found advantages for smaller scrambling damselfly males while Fincke (1988) found size stabilising selection in relation to lifetime mating success, but not for mating efficiency or life span.

Wing coloration in Odonata seems to be related to sexual selection. The implication of wing pigmentation in reproductive success has been previously studied in diverse species, especially Calopterygid damselflies (Córdoba-Aguilar, 2002; Grether, 1996; Siva-Jothy, 1999).

This chapter discusses the repercussion of body size and the parameters that influence reproductive success in male *Diastatops obscura*. We also look for evidence of the influence of wing coloration and degree of wing damage on male fitness.

Despite the long tradition of studying the reproductive systems of Libellulidae, little work has been done with Neotropical species. Very few behavioural studies deal with South American members of the Palpopleurinae Subfamily: *Perithemis mooma* (Wildermuth, 1991, 1992), *Diastatops intense* (Wildermuth, 1994), *D. obscura* (Irusta & Araújo, 2006 & accepted).

METHODOLOGY

The genus *Diastatops* (Rambur, 1842) belongs to the Libellulidae Family and is included in the Palpopleurinae Subfamily, along with genera *Palpopleura, Zenithoptera* and *Perithemis.* They are usually small dragonflies with short and pigmented wings, found in tropical and subtropical regions. *Diastatops* are neotropically distributed and include 8 species in Brazil (Paulson, 2000).

D. obscura is a small dragonfly (average male body size: 26.5 mm ± 1.1, females: 24.9 mm ± 1.8) characterized by dark wings and reddish abdomen. Males and females exhibit sexual dimorphism, especially evident in the intense coloration (Figure 1). Its short wings impede the rapid flight characteristic of other dragonflies, a fact which largely conditions its behaviour.

From the ecological point of view, *D. obscura* chooses clean, moderately lotic bodies of water with abundant aquatic and marginal vegetation as reproductive sites. Males establish contiguous territories, whose conquest and defence require innumerable fights and persecutions. Contrary to other Palpopleurinae, there is no male courtship of the female.

MATERIAL AND METHODS

The reproductive behaviour of *Diastatops obscura* was studied in the middle stretch of the Pitimbu River (S: 05° 56' 04.6", W: 35° 10' 50.0"), Rio Grande do Norte State, Brazil, for 92 hours (continuous focal observation) between September, 2002 and February, 2004. A total of 135 males were measured and had their wings marked with numeric codes to enable individualized monitoring. They were released immediately at the capture site. Another 39 non-captured males were observed and recognized individually for morphological singularity (usually tears in the wings).

In the pilot study, we also began to capture females arriving in the reproductive area; however, we discontinued this practice because marked females fled and did not return to the research area, making it impossible to observe any mating behaviour. So, we decided to proceed without any manipulation whatsoever of the females. In spite of their lacking marks, the low density at rendezvous enabled, in most cases, a differentiated observation of individual female reproductive behaviour. A total of 161 females were observed entering the study area, resulting in 187 reproductive events (attempts at copulation, copulation and ovipositions).

All of these mating behaviour aspects were associated to each of the different tactics used by the males in the reproductive areas: territorial males at the river margin or territorial males at mid-river and satellite males in riverbank vegetation, close to the marginal territories. Margin, in this paper, corresponds to the lateral strip of river next to the riverbank, permanently under water and with abundant aquatic and semi-aquatic vegetation. Satellite males were considered to be the non-territorial males that remained near, but outside the aquatic territorial area, on the side traversed by incoming mature females. For each observation day territorial males were considered to be those that dominated an aquatic territory (suitable for oviposition) in the reproductive area for 30 minutes or more.

The influence of male size on reproductive success was studied. Total body length (head-abdomen) and fore wing length and width (from wing joint to wing tip) were measured to the nearest 0.2 mm. It was verified that the two variables are strong and positively correlated (Pearson: $r_{0.61}$=0.98). Three size categories were established: small (body size < 25.5 mm), medium (25.5 -27.5 mm) and large (>27.5 mm).

Two other phenotype characters were observed: the brilliance and degree of wing damage of males in the reproductive area. These criteria were judged visually and therefore must be interpreted with caution. According to wing brilliance, observed under sunlight at different angles, two categories were established: "very brilliant" and "moderately brilliant" wings. To classify wing damage two categories were created: "clearly torn" (missing one-third or more of the entire distal edge of any of the wings) and "basically preserved" (less deterioration than described above).

Body size and, to a lesser extent, the other two phenotype characters analysed, were contrasted with four variables related to male reproductive success: territorial/satellite behaviour, mating success, oviposition success of the inseminated female and territory tenancy (number of days the male is territorial in the study area).

The following statistical tests were used: Chi-square test (χ^2) (to compare males of different categories with and without reproductive activity), Two-way ANOVA and Student's t-test (to compare the number of reproductive events among the three size categories), and Pearson's correlation coefficient (r) (to correlate wing length and body length). The confidence level was set at 95% for all tests.

RESULTS

Reproductive behaviour in *Diastatops obscura*

Males start arriving at the river two hours after dawn (after 7:15h on sunny days) to establish their territories, after an entire repertoire of agonistic interactions lasting around two hours. Three types of strategies were defined during this period: riverside territorial strategy, mid-river territorial strategy and satellite males. After this agitated stage, a calmer phase succeeded in which the territories were defined and both displays and visual communication between them became predominant.

After 11:00 h. male density in the reproductive area decreased significantly until a few minutes after 12:00 h., when no male guarded territory. On rainy days no males appeared at the rendezvous.

The females usually started arriving after 7:45 h., when male density was high. Around 90% reached the reproductive area in the next two hours.

While males and females do not interact directly with one another in the foraging areas, the mere sight of a female in the reproductive areas immediately provokes the males, who try to grab her and form a tandem to copulate. Territorial males usually take females to their territories, and satellites to sites beyond the reach of other males, to be able to complete copulation. The males engage in no-contact female guarding during oviposition; this usually occurs a few minutes after copulation.

Visualisation likely defines important reproductive behaviour aspects of this species. The males occupy their territories at sites and in ways that highlight their chromatic brilliance. They show a preference for territories near the margin, from which approaching females are more visible (Irusta & Araújo, 2006). Significantly more copulations take place in these marginal territories (67.1% of the observed copulations) than those recorded in areas farther away from the margin (12.7%). There is also a much higher oviposition rate (77.9% of observed ovipositions and 52.0% of copulations by males at the margin) than that observed for satellite males (5.7% of copulations by satellite males) (Irusta & Araújo, accepted). So, weaker males can at most opt for territories some distance away from the riverside, where they have fewer opportunities to catch a receptive female. This explains the strong competition among males to obtain a territory next to the river margin.

In the postcopulatory stage, visual contact with the female is equally important and when this is lost, the males abandon mate guarding.

Female passive choice in *Diastatops obscura*

The males follow strategies derived from intrasexual competition that are modulated by female behaviour and by male density in the reproductive areas. Thus, when male density is low or moderate, there are practically no satellite males present, a situation in which females avoid interaction with other males after being inseminated, and if males persist the females often stop ovipositing and leave the reproductive area. In these cases a second successive copulation is infrequent in *D. obscura* females (Irusta & Araújo, 2006).

Higher male density gives rise to a line of satellites on the riverbank, resulting in a significant change in female behaviour. When females approach territorial areas and are grabbed by satellite males, they are taken to a dry area to complete copulation, because these males do not occupy any river territory and try to avoid interaction with other males. This seems to alert the female that her mate is not a dominant male, causing in her a change of attitude. Thus, of the 35 copulations achieved by male satellites, which represent 20.2% of total copulations observed, only on two occasions (5.7%) did females manage to oviposit for a few seconds. This was followed in the vast majority of cases by a second copulation by another male or by spontaneous flight from the reproductive area. On the contrary, females that copulated with territorial males managed to lay their eggs immediately after copulation in 54.3% of the cases (Irusta & Araújo, accepted).

Thus, the majority of copulations with satellite males were followed by second copulations (or copulation attempts) of the female with other males (56.7%) or by spontaneous flight from the area when the female was released by the satellite male (36.7%) after copulation; both situations led to failed oviposition. Females that copulated with territorial males only copulated a second time in 22.5% of the cases in high male density areas and 9.3% in low density areas.

Male body size and reproductive success

With respect to the relation between body size and male reproductive strategy (territorial or satellite behaviour), a larger number of small males exhibited satellite behaviour: 33.33% compared to 14.29% and 26.92% for medium-size and large individuals, respectively; however, these differences were not significant ($\chi^2 = 3.23$, $p = 0.199$).

Larger size also resulted in more prolonged territoriality for these males in the reproductive areas: 2.88 ± 2.97 days (mean ± sd) compared to 1.54 ± 1.65 (mean ± sd) for the medium-size and 1.70 ± 1.44 (mean ± sd) for the small ($F_{2,85} = 37.77$, $p < 0.001$); with half of the large males defending a territory in the same area for at least two consecutive days, 26.92% for at least 5 days and 11.54% for 10 or more days (Table 1).

Table 1. Relation between male size and territory tenancy. Larger males remained territorial for longer periods than medium-size and small males

Male body size	% of males with territorial behaviour for ≥2 days	% of males with territorial behaviour for ≥5 days	% of males with territorial behaviour for ≥10 days
small	25.9	14.8	0.0
medium	25.7	5.7	2.9
large	50.0	26.9	11.5

The proportion of males that managed to mate was significantly higher for large males than for small ones ($\chi^2 = 6.95$, $p = 0.031$). Medium-size males had slightly more mating success than the small males ($p = 0.138$) (Table 2). The number of copulations achieved by large males was significantly higher than those of medium-size and small males (Anova, $F_{2,32} = 45.79$, $p < 0.001$) (Table 2; Figure 2).

Table 2. Relation between male size and mating success. Larger and medium-size males had more likelihood of copulation, and larger males achieved a higher mean number of copulations

Male body size	% of copulating males	Mean no. of copulations per copulating male
small	22.2	1.3
medium	40.0	1.7
large	57.7	2.5

Large males were significantly more likely to achieve oviposition of the females they inseminated and guarded ($\chi^2 = 7.18$, $p = 0.028$) (Table 3). Larger size also resulted in a significantly higher mean number of ovipositions achieved by their mates ($F_{2,13} = 12.49$, $p = 0.024$) (Table 3).

Table 3. Relation between male size and female ovipositing success. Larger males achieved greater ovipositing success, demonstrating more efficient postcopulatory guarding

Male body size	% of copulating males that achieved female oviposition	Mean no. of ovipositions achieved by mates
small	7.4	1.0
medium	14.3	1.8
large	34.6	2.1

Figure 1. Male (left) and female *Diastatops obscura.* Distinctive wings and obvious sexual dimorphism. Photos by J.B. Irusta

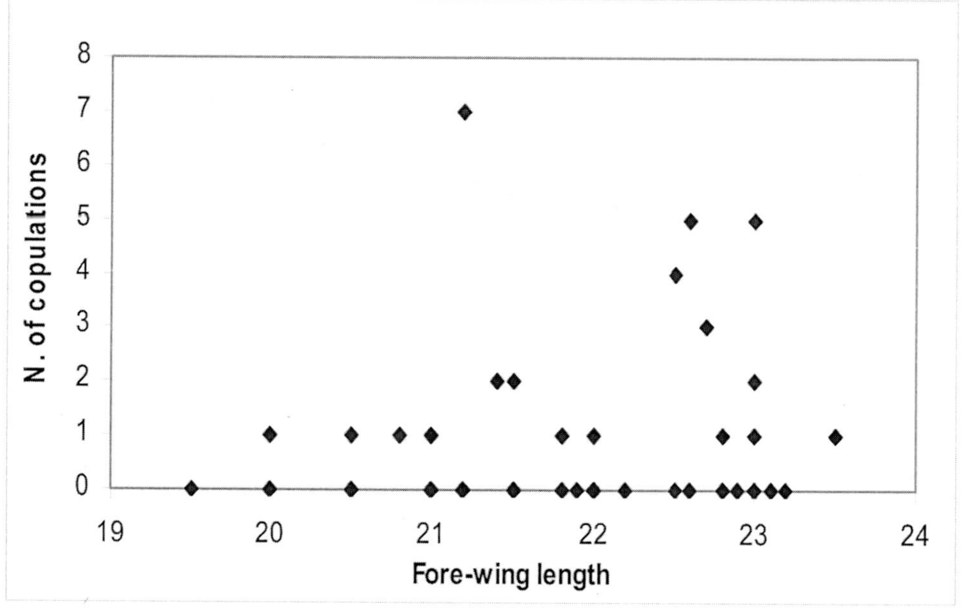

Figure 2. Correlation between number of male copulations and wing size. Among those who achieved copulation, the relation is direct.

The analysis of wing brilliance, carried out on 100 individuals, showed slight differences in mating probability ($\chi^2 = 2.55$, $p = 0.279$), in favour of the more brilliant males. Their copulations were slightly more frequent and they remained some more days in the study area behaving as territorial males, but in both cases the differences were not significant. Even though these results show a tendency of the possible importance that wing coloration may have in sexual selection, more systematized studies must be conducted to corroborate these findings. Despite the differences in wing pruinosity of males caused by incident light, we verified that wing brilliance remains practically unchanged once males reach sexual maturity.

In regards to wing damage, among a sample of 108 males, those with torn wings had a greater number of copulations (Student's t-test: $t_{34} = 2.083$, $p < 0.001$) with 3.15 ± 0.99 copulations observed compared to 1.83 ± 0.89 for males with well preserved wings. The conclusions of this characteristic are not immediate, since young males who did not take part in many interactions with other competitors showed marks similar to those of dominant males who managed to preserve wing integrity for long periods of time.

DISCUSSION AND CONCLUSIONS

Male strategies

In order to understand the reproductive behaviours exhibited by *Diastatops obscura* we must analyse simultaneously the strategies adopted by males and females. The males compete daily for suitable territory that allows them to attract and grab females. Other strategies such as stealing females and satellite tactics are used to achieve a greater number of copulations and ensure the siring of most of the eggs. The hierarchical order resulting from male competition for the best territories offers all the information that a receptive female needs and dispenses with the courtship behaviours found in other paired species, such as *Perithemis mooma* (Wildermuth, 1991).

The reproductive tactics followed by the males in this study were analysed according to reproductive success, through daily observations of copulations and ovipositions and by the reoccurrence of males in reproductive areas over successive days. Given that this study was undertaken under natural environmental conditions, there were potentially important variables that could not be controlled (climactic conditions, density *etc.*). However, similar methodologies have been used by various authors such as Banks & Thompson (1985), Corbet (1999) and Thompson & Fincke (2002), who also report that individual male reproductive success can be evaluated by the achievement of a combination of longevity and mating success. Longevity is important since it is related to the number of days the male can act territorial, a fundamental aspect, given that the oviposition rate of *D. obscura* females fertilized by satellite males is very low (5.7% of copulations) and is always interrupted by territorial males (Irusta & Araújo, accepted).

Daily mating success includes not only copulations, but also intrasexual competitive capacity in reproductive areas, in order to obtain a suitable territory (that is with better access to arriving females), to capture females (grab them first or steal them from another male) and to guard them until their postcopulatory egg-laying, ensuring the siring of most of the eggs.

Female strategies

The females wait for male hierarchical stratification in the reproductive areas to facilitate an encounter with a high quality mate to fertilize her eggs. There seems to be a combination of tactics among males and females, from precopulatory behaviours to oviposition. The competition itself among territorial males causes the best individuals to locate themselves in areas that enable them to grab the greatest number of females, who, if caught by the best positioned males, are sure to be inseminated by dominant specimens. When other situations arise, such as the capture by a satellite male, the usual female tactic, which involves mating with the first male she encounters, does not seem advantageous. In this case, females change strategy, by fleeing or allowing another copulation that enables sperm displacement by another higher status male (territory owner). Rejection behaviours of satellite males have been suggested for other female dragonflies such as: *Pachydiplax longipennis* (Sherman, 1983), *Erythemis simplicicollis* (Mcvey, 1988), *Leucorrhinia intacta* (Wolf *et al.*, 1989), *Libellula luctuosa* (Moore, 1989) and *Plathehis lydia* (Koenig, 1991).

Thus, not only do male dragonflies exhibit a previously known flexibility in reproductive strategies that varies according to environmental conditions, but the females also modify their mate selection strategy according to the density and status of the males. Although other studies must be developed, we can affirm, using Thornhill & Alcock's (1983) argument that *D. obscura* females "exercise their control of the reproductive process in such as way as to reject the sperm of inferior males and use the sperm of superior ones", which shows the existence of female choice, albeit in a passive form.

Body size

Sokolovska *et al.* (2000), after analysing 33 studies on the relation between Odonata body size and reproductive fitness, concluded that for territorial males, larger size was correlated with greater longevity, higher mating rate, greater lifetime mating success and more efficient territory conquest and defence success. However, none of these studies included South American species, and of the seven Libellulidae investigated, none belonged to the Subfamily Palpopleurinae. In relation to our species, these results should be looked upon with caution, both geographically and phylogenetically. Furthermore, several criticisms of these authors' data have been made by Thompson & Fincke (2002), raising doubts about the fitness advantages for larger male and female Odonata. More accepted are size advantages in species where males are larger than females (Thompson & Fincke, 2002), as occurs, albeit slightly, in *D. obscura*.

Nevertheless, some species of Zygoptera showed that advantages when defending a territory were due to the amount of fat accumulated and not to the weight of the individual (Mardem & Waage, 1990), while other studies with dragonflies found significant positive correlation between body width (but not wing length) and territorial defence (Kasuya *et al.,* 1997).

From the results of this study we can conclude that larger body size is a positive factor for aspects related to daily mating success such as: greater likelihood of copulating, greater number of copulations and a higher rate of immediate ovipositions of the fertilised females. The same is seen for aspects of longevity such as: greater number of days behaving as territorials. The smaller males, despite managing to defend their territories, had lower values in all these parameters, whereas medium-size individuals generally had intermediate values. This indicates a clear sexual selection in favour of larger males, as described by Sokolovska *et al.* (2000).

Since the males of this species are not significantly larger in each successive generation, size advantages in adults must have their drawbacks in pre-imago stages, when size increment must generate high ecophysiological costs (late eclosion, higher food demands, higher predation rate, *etc.*). A scenario likely exists in which stabilising selection limits growth in the larval stage or makes survival difficult, but not necessarily in adults, whose fitness improves with size.

Observations on coloration intensity of male wings showed that males with little brilliance seldom manage to defend a reproductive territory for more than two consecutive days. This situation suggests, as occurs in other Odonata, that wing coloration reflects the amount of energy reserves accumulated in prematurity stages (Hooper *et al.,* 1999). If we consider that dark wings in this species have a thermo-regulatory function in addition to the intraspecific communication role suggested by Irusta & Araújo (2006) and by other authors for other Odonata with distinctive wings (Córdoba-Aguilar, 2002), wings with little coloration, badly torn or smaller in size may indicate an inferior condition. This could be caused by more frequent aggressions toward these individuals, resulting in diminished fitness.

In relation to wing damage, although the results cannot be considered conclusive, shorter consecutive territorial occupation was observed in males with wings damaged by agonistic interactions with other congeners. This could indicate that wing deterioration created disadvantages that resulted, in the middle or long term, in shorter longevity or less competitive capacity than that observed in the other males. However, it did not seem to have a negative impact on daily reproductive success.

Obviously, many other extrinsic and intrinsic characters influence reproductive success: genetic load, experience, maturity, amount of stored fat, food ingested during the larval and teneral phases, climactic and environmental conditions, and so on. Other studies will be needed to better understand the complex and fascinating field of reproductive behavioural ecology in dragonflies.

ACKNOWLEDGEMENT

This study was supported by Conselho Nacional de Desenvolvimento Científico e Tecnológico – CNPq (grant no. 140004/ 03-3).

REFERENCES

Alexander, R.D. & G. Borgia 1979. On the origin and basis of the male-female phenomenon. In: M.S. Blum & N.A. Blum, Eds., *Sexual selection and reproductive competition in insects*. New York: Academic Press.

Andersson, M. 1994. *Sexual selection*. Princeton, New Jersey: Princeton University Press.

Banks, M.J. & D.J. Thompson 1985. Lifetime mating success in the damselfly *Coenagrion puella*. *Animal Behaviour* 33: 1175-1183.

Bateman, A.J. 1948. Intra-sexual selection in *Drosophila*. *Heredity* 2: 349-368.

Blanckenhorn, W.U., C. Mühlhäuser, C. Morf, T. Reusch & M. Reuter 2000. Female choice, female reluctance to mate and sexual selection on body size in the dung fly *Sepsis cynipsea*. *Ethology* 106: 577-593.

Conrad, K.F. & G. Pritchard 1992. An ecological classification of odonate mating systems: the relative influence of natural, inter- and intra-sexual selection on males. *Biological Journal of the Linnean Society* 45: 255-269.

Corbet, P.S. 1999. *Dragonflies: Behaviour and Ecology of Odonata*. Harley Books, Essex. 830pp.

Córdoba-Aguilar, A. 2002. Wing pigmentation in territorial male damselflies, *Calopteryx haemorrhoidalis*: a possible relation to sexual selection. *Animal Behaviour* 63: 759-766.

Córdoba-Aguilar, A., E. Uhía-Castro & A. Cordero Rivera 2003. Sperm competition in Odonata (Insecta): the evolution of female multiple matings and rivals' sperm displacement. *Journal of Zoology* 261: 381-398.

Darwin, C. 1859. *On the origim of species*. Murray, London.

Fincke, O.M. 1986. Lifetime reproductive success and the opportunity for selection in a nonterritorial damselfly (Odonata: Coenagrionidae). *Evolution*, 40: 791-803.

Fincke, O.M. 1988. Sources of variation in lifetime reproductive success in a nonterritorial damselfly (Odonata: Coenagrionidae). In: Clutton-Brock, T.H. (Ed.), *Reproductive success*, 3: 24-43. Chicago, Univ. Chicago Press.

Grether, G.F. 1996. Sexual selection and survival selection on wing coloration and body size in the rubyspot damselfly *Hetaerina americana. Evolution* 50: 1939-1948.

Harvey, I.F. & K.J. Walsh 1993. Fluctuating asymmetry and lifetime mating success are correlated in males of the damselfly *Coenagrion puella* (Odonata: Coenagrionidae). *Ecological Entomology* 18: 198-202.

Hooper, R.E., Y. Tsubaki & M.T. Siva-Jothy 1999. Expression of a costly, plastic secondary sexual trait is correlated with age and condition in a damselfly with two male morphs. *Physiological Entomology* 24: 364-369.

Irusta, J.B. & A. Araújo 2006. Reproductive behaviour of *Diastatops obscura* (Fabricius, 1775) in a riverine environment (Anisoptera, Libellulidae). *Odonatologica* 35: 279-285.

Irusta, J.B. & A. Araújo, (accepted). Reproductive tactics of sexes and fitness in *Diastatops obscura* Fabricius (Odonata : Libellulidae). *Journal of Insect Science.*

Kasuya, E., K. Edanami & I. Ohno 1997. Territorial conflicts in males of the dragonfly *Orthetrum japonicum japonicum* (Odonata : Libellulidae): the role of body size. *Zoological Science* 14: 505-509.

Koenig, W.D. 1991. Levels of female choice in the white-tailed skimmer *Plathemis Lydia* (Odonata : Libellulidae). *Behaviour* 119: 193-224.

Marden, J.H. & J.K. Waage 1990. Escalated damselfly territorial contests are energetic wars of attrition. *Animal Behaviour* 39: 954-959.

Mcvey, M.E. 1988. The opportunity for sexual selection in a territorial Dragonfly, *Erythemis simplicicollis*. In : T.H. Clutton-Brock (Ed.), *Reproductive success*. The University of Chicago Press.

Moore, A.J. 1989. The behavioral ecology of *Libellula luctuosa* (Burmeister) (Odonata: Libellulidae): III. Male density, OSR, and male and female mating behavior. *Ethology* 80: 120-136.

Neems, R.M., J. Lazarus & A.J. Malachlan 1998. Lifetime reproductive success in a swarming midge: trade-offs and stabilizing selection for male body size. *Behavioral Ecology* 9: 279-286.

Paulson, D.R. 1998 (updated March 2006). *List of the Odonata of South America, by Country*. http://www.ups.edu/home.html. University of Puget Sound.

Purse, B.P. & D.J. Thompson 2005. Lifetime mating success in a marginal population of a damselfly, *Coenagrion mercuriale*. *Animal Behaviour* 69: 1303-1315.

Richardson, J.M. & R.L. Baker 1997. Effect of body size and feeding on fecundity in the damselfly, *Ischnura verticalis* (Odonata: Coenagrionidae). *Oikos* 79: 477-483.

Sherman, K.J. 1983. The adaptative significance of postcopulatory mate guarding in a dragonfly, *Pachydiplax longipennis*. *Animal Behaviour* 31: 1107-1115.

Siva-Jothy, M.T. 1999. Male wing pigmentation may affect reproductive success via female choice in a Calopterygid damselfly (Zygoptera). *Behaviour* 136: 1365-1377.

Sokolovska, N., L. Rowe. & F. Johansson 2000. Fitness and body size in mature odonates. *Ecological Entomology* 25: 239-348.

Stoks, R. 2000. Components of lifetime mating success and body size in males of a scrambling damselfly. *Animal Behaviour* 59: 339-348.

Thompson, D.J. & O.M. Fincke 2002. Body size and fitness in Odonata, stabilising selection and meta-analysis too far? *Ecological Entomology* 27: 378-384.

Thornhill, R. & J. Alcock 1983. *The evolution of insect mating systems*. Harvard University Press, Cambridge.

Van Dyck, H., E. Matthysen & A.A. Dhondt 1997. The effect of wing colour on male behavioural strategies in the speckled wood butterfly. *Animal Behaviour* 53: 39-51.

Waage, J.K. 1984. Sperm competition and the evolution of Odonate mating systems. In: Smith, R.L. (Ed.), *Sperm competition and the evolution of animal mating systems*. Acad. Press, N.Y., London, pp. 251-290.

Wildermuth, H. 1991. Behaviour of *Perithemis mooma* Kirby at the oviposition site (Anisoptera: Libellulidae). *Odonatologica* 20: 471-478.

Wildermuth, H. 1992. Visual and tactile recognition of the oviposition substratum in *Perithemis mooma* Kirby, 1889 (Odonata: Libellulidae). *Odonatologica* 21: 309-321.

Wildermuth, H. 1994. Reproductive behaviour of *Diastatops intensa* Montgomery (Anisoptera: Libellulidae). *Odonatologica* 23: 183-191.

Wolf, L.L, E.C. Waltz, K. Wakeley & D. Klockowski 1989. Copulation duration and sperm competition in white-faced dragonflies (*Leucorrhinia intacta*; Odonata: Libellulidae). *Behavioral Ecology and Sociobiology* 24: 63-68.

Odonata : Biology of Dragonflies
Ed.: B.K. Tyagi
Scientific Publishers (India), 2007, pp. 241-255

16

CAUSES AND COSTS OF LAMELLAE AUTOTOMY IN DAMSELFLY LARVAE: A REVIEW

R. STOKS and M. DE BLOCK

Laboratory of Aquatic Ecology, KULeuven, Deberiotstraat 32, B-3000 Leuven, Belgium
e.mail: robby.stoks@bio.kuleuven.be

ABSTRACT

Autotomy, the amputation of a body part at a breakage plane, is a well-known escape mechanism when animals are caught by a predator. Here, we review the ecological causes, field occurrence, and costs of lamellae autotomy in damselfly larvae. Lamellae autotomy is widespread in nature and is an important escape mechanism when caught by invertebrate predators and small fish, but has no survival advantage against large fish. However, associated with the other functions of lamellae, autotomy carries short-term costs in the form of a reduced ability to withstand low oxygen levels and high temperatures, and an increased vulnerability to conspecific and heterospecific predation. To deal with the latter, larvae show threat-sensitive antipredator behaviour after autotomy. The resulting reduced food intake together with their increased predator-induced stress, may explain negative effects on mass and size at emergence and wing asymmetry. Based on the known short-term costs, long-term costs on adult fitness and a regulatory role of autotomy in population regulation are likely, but await experimental proof.

INTRODUCTION

Autotomy, the amputation of a body part at a breakage plane, is a well-known escape mechanism both in vertebrates and invertebrates (*e.g.,* reviews in Mcvean, 1975; Maginnis, 2006). In damselfly larvae autotomy may include legs and caudal lamellae, but only the latter mechanism occurs frequently (Robinson *et al.*, 1991b; Stoks & De Block, 2000) and has been fairly well-studied. In fact, among invertebrates damselfly larvae are one of the few groups where ecological aspects of autotomy have been studied in great detail.

Damselfly larvae have three caudal lamellae which are connected with breaking joints to the abdomen. When grasped by a predator larvae autotomize these lamellae. A sphincter at the breaking joint is then closed to avoid hemolymph efflux or water influx (MacNeill, 1960). The preformed cleavage plane can speed up wound healing, reduce bacterial infection, and minimize the potential water-borne cues that could signal the presence of a wounded animal (Juanes & Smith, 1995).

In contrast, these remediating effects do not occur when lamellae are removed by cutting (Eriksen, 1986). As a result, no short-term mortality could be observed in *Lestes disjunctus* when lamellae were removed at the breaking joint, but several died due to the cutting treatment.

When lost, lamellae can usually be replaced completely within two or three moults (MacNeill, 1960; Baker & Dixon, 1986). This suggests a low cost of regeneration or a selective advantage to possess a full set of lamellae (Burnside & Robison, 1995). The role of lamellae as a source of oxygen uptake and as a body part that serves in autotomy has long been questioned and poorly motivated (see e.g., MacNeill, 1960). Here, we want to bring together the recent information obtained on the ecological causes, occurrence and costs of lamellae autotomy in damselfly larvae and to identify gaps in our knowledge and promising avenues for future research.

ECOLOGICAL CAUSES

Autotomy has a clear survival advantage in damselfly larvae: once caught by a predator, damselfly larvae almost entirely rely on autotomy to escape predation. They rely much more on lamellae autotomy than on leg autotomy to escape once caught. In a study on *Ischnura posita* leg damage did not occur at a density of 8 larvae/aquarium where 23% of the individuals showed lamellae loss (Robinson *et al.,* 1991b). Although at a density of 32 larvae/aquarium leg loss increased to 11%, lamellae loss was still much more important with 74% of the larvae missing lamellae. Similarly, during staged encounters 14% of *L. sponsa* larvae could escape a capture by stickleback *Gasterosteus aculeatus*, *Notonecta* and *Aeshna* and 90% of these did so by lamellae loss, while no larva did so by leg loss (Stoks & De Block, 2000).

Lamellae loss can be due to escape predation by conspecifics and heterospecifics and there has been some debate about which of these two mechanisms is the most important. The former view was supported by Baker & Dixon (1986), who even suggested using wounding (lamellae loss and leg damage) as an index of aggressive interactions between damselfly larvae. They found that in laboratory trials the frequency of wounding in *I. verticalis* and *Enallagma ebrium* was positively correlated with the frequency of aggressive interactions reaching up to 38% during 4d lab trials. Two other studies showed that conspecific larvae may cause considerable lamellae loss. During 24h laboratory trials with *I. posita* larvae, 72% of the final instar larvae and 24% of the F-1 larvae lost one or more caudal lamellae due to aggressive interactions (Shaffer & Robinson, 1993). Final instar larvae lost on average 1.6 lamellae with most larvae losing 1 lamella, but some larvae losing all three lamellae. F-1 larvae lost on average 1.2 lamellae, with most larvae still having all three lamellae and none having lost all lamellae. Similarly, in another study on *I. posita* autotomy to escape attacks by conspecifics was quite common (Robinson *et al.*, 1991b). Moreover, these authors showed that this process was density dependent, ranging from 17% at the lowest density up to 73% at the highest density during 24h trials (Robinson *et al.*, 1991b). In contrast, in *I. verticalis* only 2.5% of the aggressive encounters resulted in autotomy (McPeek & Crowley, 1987). This led McPeek (1990a) to assume that most cases of autotomy in his study lakes are due to heterospecific predators.

Attacks by heterospecific predators may indeed be a major cause of autotomy. When caught by sticklebacks and the invertebrate predators like *Notonecta viridis* and *Aeshna cyanea*, *L. sponsa* larvae lost on average 1.5 lamellae during their escape (Stoks & De Block, 2000). In most cases, larvae lost one single lamella (62%), 30% of the larvae survived by autotomizing two lamellae and 8% lost all three lamellae during a single attack. These frequencies were not dependent upon which of these three predators attacked them.

Laboratory experiments learned that incidences of lamellae autotomy may depend upon damselfly species, size and predator environment. The main reason being that damselfly species differ in their escape strategy and its efficiency, and that predators differ in their attack strategy and efficiency. Large fish take prey in a gulp and are very efficient predators on damselfly larvae. Accordingly, they only rarely induce lamellae autotomy. Dixon & Baker (1986) reported that all *I. verticalis* and *E. ebrium* larvae attacked by pumpkinseed sunfish were invariable swallowed whole. Similarly, in staged encounters of *L. sponsa* larvae with pumpkinseed sunfish, larvae could escape only ca 5% of the attacks by swimming away, and when caught in no single occasion a larva could escape by autotomy (Stoks & De Block, 2000). In the case of a stickleback, a smaller fish, which are also fast swimmers only about 5% of the *Lestes* larvae could escape predation by swimming away. However, sticklebacks are more gape-limited than pumpkinseed and ca. 10% of F-2 larvae and up to 30% of F-0 larvae that were caught could escape by lamellae autotomy once caught (Stoks & De Block, 2000). In contrast, 30-40% of the larvae could escape by swimming when attacked by the much smaller invertebrate predators *N. viridis* and *A. cyanea*. The incidence of lamellae autotomy as an escape strategy did not differ between the stickleback and these two invertebrate predators. The opportunity of using it differed: the slower F-2 larvae were less likely to be able to escape once caught by autotomy than the faster F-0 larvae as they were less likely to deflect an attack toward the caudal lamellae. This was especially striking in larvae caught by *Notonecta* where none of the caught F-2 larvae could escape, while 40% of the F-0 larvae could escape by using autotomy.

Based on these laboratory studies it is clear that using lamellae loss as a measure of aggressive interactions between larval damselflies (Dixon & Baker, 1986; Stoks, 1998) or as a measure of aggressive encounters with heterospecific predators (McPeek, 1990a) should be done with caution. The former may be useful in ponds with large fish and relative low numbers of insect predators. The latter may give some idea about the intensity of encounters with heterospecific predators in ponds with relatively high numbers of insect predators and low densities of damselfly larvae.

FIELD OCCURRENCE

Several studies documented the occurrence of autotomy under natural conditions by reporting the frequencies of larvae with different number of lamellae in field surveys. In such studies, alternative explanations for lamellae loss, besides autotomy, can not be completely excluded. One cause for lamellae loss may be injuries not caused by predation including collection damage. *Enallagma* damselfly larvae autotomize lamellae that are artificially cut within 20 min

(McPeek *et al.,* 1996). Such injuries are probably rare under natural conditions and have not been reported in aquaria. Dixon & Baker (1986) specifically tested whether collection with a net caused lamellae loss and found this not to be the case. Another cause for limb loss in general in arthropods is moulting: appendages may become stuck in the old exoskeleton and these body parts must be shed if the organism is to survive (Maginnis, 2006). However, lamellae loss does occur very rarely during moulting: in *I. posita* in only 1 out of 117 monitored moults (Robinson *et al.,* 1991b). Therefore, it is reasonable to assume that missing lamellae may accurately reflect autotomy. Obviously, in cases of larvae with several missing lamellae one cannot be sure on the number of separate attacks that occurred, and therefore one may underestimate true incidences of autotomy. Nevertheless, field surveys showed us that the natural occurrence of autotomy under field conditions can be high. Moreover, one can combine the frequency of missing lamellae with those of regenerating (hence reflecting older autotomy events) to calculate incidences of autotomy.

Two studies only reported combined frequencies of lamellae loss and leg loss without any details on the lamellae status of the larvae. Dixon & Baker (1986) monitored incidences of combined lamellae and leg loss of *I. verticalis* and *E. ebrium* in a Canadian fishpond during ten months. Wounding frequencies in monthly samples varied between ca. 20% and ca. 45%, and were higher when larval were more evenly spread in the substrate. This may be explained by the fact that with more aggressive interactions there will not only be more wounding, but larvae will also spread out more evenly to avoid encounters. McPeek (1990a) reported a field survey of regenerated legs and lamellae in three *Enallagma* species. Incidences in *E. hageni* and *E. vesperum,* two fish lake specialists, were ca. 20-30% in two lakes dominated by sunfish, but much higher, ca. 50-60%, in *E. boreale* in two fishless lakes with large dragonfly larvae as top predators. This is consistent with the previously reported finding that fish are much less likely than large dragonfly larvae to remove lamellae.

Two other field survey studies reported lamellae autotomy in detail by also including information on the lamellae status of the larvae. Robinson *et al.* (1991b) reported that at least one missing or regenerating lamella was evident in 59% of field collected final instar *I. posita* larvae. These percentages were lower in F-1 (42%) and F-2 (25%) larvae and may reflect their lower swimming speed (see above). The lamellae status (including regenerated lamellae) in the final instar was as follows: 69% had three lamellae, 22% had two lamellae, 5% one lamella and 4% no lamellae. In a field collection of 80

Proischnura subfurcatum larvae in a swamp in Uganda the lamellae status was as follows: 44% had three lamellae, 45% had two or one lamella and 11% had no lamellae (Apodaca & Chapman, 2004).

Finally, Stoks (1998) determined the occurrence of autotomy in *L. viridis* using larval skins. During 57 emergences that were monitored, no animals lost lamellae, so lamellae status of the larval skin may be a valid alternative to indirectly study lamellae autotomy. As larvae with lamellae may have a lower survival (see further) this may underestimate real autotomy frequencies. Nevertheless, the study showed very high levels of lamellae loss with ca. 78% of the larval skins with missing lamellae, and ca. 90% when also including animals with regenerating lamellae (hence that has lost their lamellae before the final instar). The distribution of lamellae status was as follows: 22% had three lamellae, 35% had two lamellae, 24% had 1 lamella and 19% had no lamellae. These numbers did not differ between sexes.

SHORT-TERM COSTS

Two types of short-term costs directly linked to the dual function of the lamellae have been described. First, caudal lamellae function in oxygen uptake and lamellae autotomy therefore may seriously reduce oxygen uptake. In *L. disjunctus*, larvae without lamellae have a 20-30% lower oxygen uptake at high oxygen levels, and up to 60-88% at low oxygen levels (Eriksen, 1986). Larvae of *Argia vivida* without lamellae have a 27% decrease in oxygen consumption at 3.5 mg/l (Pritchard, 1991). Apodaca & Chapman (2004) showed that the rate of oxygen consumption was 51% lower in larvae of *Proischnura subfurcatum* without lamellae compared to larvae with lamellae.

This reduced oxygen uptake may shape a larva's ability to withstand low oxygen levels and high temperature and affect its microhabitat distribution. Eriksen (1986) showed that *L. disjunctus* larvae without lamellae perform worse in withstanding low oxygen levels; larvae with lamellae were able to survive water temperatures that were 13-14°C higher than larvae without lamellae can tolerate. Typically, larvae without lamellae migrate to the water surface when confronted with oxygen stress (Hagemeier, 1990; Robinson *et al.*, 1991b), prefer microhabitats with higher oxygen levels (Robinson *et al.*, 1991b), or show more pull-downs to ventilate their body (Eriksen, 1986). Therefore, the reduced oxygen uptake may not only cause direct lethal effects due to oxygen lack but may also indirectly make the larvae more vulnerable to predation at the water surface and may affect their foraging efficiency.

Secondly, lamellae can be used in several ways in avoiding predation, and as a result after autotomy larvae will become more vulnerable to predation : (1) Lamellae can be used to struck opponents which may avoid being attacked. Larvae of *I. posita*, for example use this behaviour in agonistic encounters with crayfish and conspecifics (Shaffer & Robinson, 1996). Obviously, when all lamellae are lost this behaviour can not longer be used, (2) Similarly, after autotomy, larvae can not longer rely on autotomy until lamellae are regenerated, and (3) Lamellae are used to generate thrust in swimming (Brackenbury, 2001) and several studies showed that damselfly larvae with removed or with experimentally reduced lamellae area have a lower swimming speed (Robinson *et al.,* 1991a; Burnside & Robinson, 1995; McPeek *et al.*, 1996; Stoks, 1999a; Gyssels & Stoks, 2005). Although the presence of three lamellae has been hypothesized to be a redundant system (Burnside & Robinson, 1995), startle performance decreased each time a lamella was removed in *L. sponsa* (Stoks, 1999a). Admittedly, the effect of removing yet another lamella increased quadratically, being relatively small when the first lamella was removed, and getting larger and larger with the removal of more lamellae. As swimming is a major strategy of damselfly larvae to escape a predator attack (Mcpeek, 1990b; Stoks & De Block, 2000; Stoks & McPeek, 2003), lamellae autotomy makes larvae more vulnerable to predation.

An increased mortality by predation after lamellae autotomy has been shown both when confronted with conspecific and heterospecific predators. Cannibalism can be widespread in damselfly populations (e.g., Anholt 1994; De Block & Stoks, 2004). Two studies showed that larvae without lamellae are more prone to cannibalism. *Ischnura posita* larvae without lamellae suffer twice as many cannibalism during 24h trials (Robinson *et al.,* 1991b). *Lestes sponsa* larvae without lamellae suffer more than three times higher predation by conspecifics during 48h trials (Stoks, 1998b).

Several studies showed a mortality cost of autotomy due to predation by heterospecific predators. During laboratory trials, *L. sponsa* larvae without lamellae suffered more mortality by predation from *N. glauca* (Stoks, 1998) and *A. cyanea* dragonfly larvae (Stoks, 1999b). Additionally, an increased mortality by *A. cyanea* predation of *L. sponsa* larvae was also shown under more realistic conditions in field enclosures (Stoks *et al.,* 1999). Interestingly, this mortality cost has been shown to be strongly depended upon prey species and predator type, and this in line with the differential survival value of swimming. While swimming is a good strategy to survive an attack by dragonfly larvae or boatmen, it is not when attacked by fish (Stoks &

De Block, 2000). As a result, *Enallagma* and *Lestes* species that occur strictly in fishless ponds with large dragonfly predators use swimming to evade predators while fishpond specialist of these genera do not (McPeek, 1990a,b; Stoks & Mcpeek, 2003). In accordance, in trials where the lamellae area of larvae was reduced, larvae with more reduced lamellae of the fishless pond specialist *E. aspersum* were eaten more by *A. canadensis* while this was not the case in the fishpond specialist *E. vesperum* (McPeek *et al.,* 1996). This also holds for lamellae autotomy: *I. elegans* larvae without lamellae suffered a higher predation by *Anax imperator* but not by pumpkinseed sunfish compared to larvae with lamellae (Gyssels & Stoks, 2005).

In accordance with the threat-sensitivity hypothesis that animals assess and adjust their behaviour flexibly in accordance with the magnitude of the threat imposed by a predator (Sih, 1986), the more vulnerable larvae without lamellae show less risky behaviour. Effects have been shown on propensity to swim away, thanatosis (death feigning), overall activity (as measured by position changes), and refuge use. Robinson *et al.* (1991b) showed that *I. posita* larvae without lamellae are more reluctant to release from a wooden stalk and swim when threatened (9% release) than are larvae with lamellae (29% release). Similarly, after autotomy *I. elegans* larvae from a fishless pond were less inclined to swim away after a simulated attack (3% vs 57%) (Gyssels & Stoks, 2005). Larvae without lamellae most often (80%) did not move when attacked, while this was much less frequent in larvae with lamellae (40%). Interestingly, in line with the differential survival value of swimming away when attacked in fishponds, larvae of *I. elegans* with and without lamellae from a fishpond were as reluctant to swim away (with lamellae: 78%, without lamellae: 80%) (Gyssels & Stoks, 2005). Further, in accordance with their higher vulnerability, *I. elegans* larvae without lamellae more frequently entered thanatosis after a simulated attack with a forceps than larvae with lamellae (85% vs 53%) (Gyssels & Stoks, 2005).

After lamellae autotomy, *L. sponsa* larvae decreased their activity in a fixed way, irrespective of the presence of predators (Stoks, 1998b). When confronted with a caged *Notonecta* predator larval activity was no further reduced. Swimming behaviour is an especially obvious behaviour, increasing detection by predators (Convey, 1988; Mcpeek 1990a). Autotomy did not affect swimming behaviour of *I. elegans* larvae from a fishpond: both larvae with and without lamellae decreased swimming in the presence of fish kairomones (Gyssels & Stoks, 2006). However, in larvae from a fishless pond, only larvae without lamellae reduced swimming behaviour. This pattern may be explained by the fact that all larvae (irrespective of their pond of origin) associate fish kairomones with an increased predation risk, and that larvae in the fishpond with lamellae do not reduce this behaviour as they rely on the fact that they have a rather high chance of escaping an attack by their syntopic predators. Another very conspicuous behaviour of damselfly larvae are rigid abdominal bends, which often attracts the attention of predators and generates an attack (R. Stoks, personal observation). *Ischnura elegans* larvae without lamellae showed this behaviour less frequent compared to larvae with lamellae (Gyssels & Stoks, 2006).

The increased vulnerability of larvae without lamellae may also determine their microhabitat choice between safe patches with no food and dangerous patches with food. When varying the foraging value (food present or absent) and predation risk (caged *A. cyanea* or no predator) in a microhabitat without cover, *L. sponsa* larvae with and without lamellae responded differentially to the imposed trade-off (Stoks, 1999b). All larvae spent more time in the microhabitat without cover when food was present than when food was absent. Larvae without lamellae, however, only sporadically left the safe microhabitat, irrespective of the presence of the predator, again indicating a fixed antipredator strategy. In contrast, larvae with lamellae shifted more frequently towards the risky microhabitat than those without lamellae, and more often in the absence than in the presence of the predator.

LONG-TERM COSTS

As a result of their reduced mobility and changed microhabitat use damselfly larvae without lamellae have a lower food intake than larvae with lamellae, and this both in the absence and presence of predators. Over a period of 10h *L. sponsa* larvae without lamellae ate 37% less *Daphnia* than larvae with lamellae in the absence of predators, and 26% less in the presence of a caged *Notonecta* predator (Stoks, 1998a). Similarly, in another study *L. sponsa* larvae without lamellae had a 44% lower intake of *Daphnia* prey over a period of 8h (Stoks, 1999b). There are indications that larvae may show both a fixed and a flexible (only shown in the presence of a predator) component in their antipredator behaviour when confronted with autotomy. Besides their fixed lower foraging rate, *L. sponsa* larvae without lamellae tended to decrease their foraging rate more in the presence of an *Aeshna* predator than larvae with lamellae (26% vs 19%) (Stoks, 1999b). In contrast, no reduction in foraging rate were observed in *I. elegans* larvae without lamellae compared to larvae with lamellae (Gyssels & Stoks, 2006). These trials were, however, much shorter (20 min), and effects may only become apparent over a longer observation period.

Reduced foraging intake should eventually result in a lower growth rate. In line with this, the only field enclosure study so far evaluating the long-term costs of autotomy reported a reduction of final body mass with 10% in *L. sponsa* larvae without lamellae compared to larvae with lamellae (Stoks *et al.,* 1999). No effect of autotomy on body size was found. This suggests that animals dealing with lower energy levels try to maintain size constant and relax body mass (see also Strobbe & Stoks, 2004). Such strategy makes sense as size is fixed at emergence, while damselflies can still compensate for low mass at emergence during the maturation period (Anholt *et al.,* 1991; Stoks, 2001b).

Intriguingly, negative effects on life history of autotomy may also occur independent of food intake. In a laboratory experiment with *L. sponsa* larvae where all larvae had the same food intake, larvae with lamellae had a 12% lower mass and a 2% smaller size than larvae with lamellae after 40 days (Stoks, 2001). Also, development time from their moult in F-1 until emergence was 7% longer, and their survival was 13% lower, although these effects did not reach statistical significance. As the larvae partly regenerated their lamellae when moulting into the final instar this can be explained by a cost of regeneration. Alternatively, larvae without lamellae may suffer fixed higher predator-induced stress levels and, as such, are less efficient in converting ingested food into body mass. Several studies in damselfly larvae showed such negative effects of non-lethal predator stress on digestive efficiency (McPeek *et al.,* 2001; Stoks & McPeek, 2003; Stoks *et al.,* 2005). In line with the hypothesis that predator stress underlies the lower mass, wing asymmetry increased in the adults under a combination of autotomy and predation risk imposed by a caged *Aeshna* (Stoks, 2001a).

Ultimately, it is to be expected that the negative effects of autotomy on mass (and to a much lesser extent on size) at emergence and wing asymmetry should translate into a reduced adult fitness. Animals with a lower mass at emergence have a lower chance of becoming sexually mature (Stoks, 1999c), and especially in females this is associated with a lower lifetime mating success (De Block & Stoks, 2005). Smaller and less heavy females tend to have a lower fecundity (Sokolovska *et al.,* 2000). Adult males with more asymmetrical wings have a lower mating success (e.g., Harvey & Walsh, 1993; Stoks, 1999c; De Block & Stoks, in review). Moreover, as animals with more asymmetrical wings probably have a lower flight performance, they may be less efficient in foraging, thereby accumulating less body mass, which may negatively affect female fecundity. However, despite the strong indirect information, we have no direct proof for a link between lamellae autotomy in the larval stage and adult fitness so far.

CONCLUSIONS AND FUTURE DIRECTIONS

There is convincing evidence that lamellae autotomy can be an efficient escape mechanism of larvae once caught by invertebrate predators or small fish, but it plays no role in escaping predation by large, not-gape limited fish predators. It can be very common in damselfly populations and be due both to intraspecific and interspecific aggressive encounters. Therefore, studies using it as an index of intraspecific or interspecific interactions may be misleading. Short-term costs of autotomy are well-studied in damselfly larvae and include increased oxygen uptake, hence a reduced ability to withstand low oxygen levels and high temperatures, and an increased risk of being killed during next encounters with predators. As for the efficiency of autotomy, also predation costs after lamellae loss are predator specific, being largest with invertebrate predators and nearly absent with fish predators. In line with this, larvae show threat-sensitive behaviour, which often takes the form of fixed lower activity levels. In the long term this results in a lower food intake. This, together with their higher perceived predation risk causes a decreased mass and, to a lesser extent, size at emergence and an increased wing asymmetry.

It is to be noted that studies on costs so far only compared larvae with three lamellae and larvae with no lamellae. So, we have no direct proof whether losing one or two lamellae also imposes costs. It has been suggested that the 3-lamellae system is redundant. However, studies on swimming speed do show, albeit small, reductions in swimming speed after the loss of one lamella. Such small differences may, however, be important as even small differences in total lamellae size may give a survival advantage against dragonfly predators (McPeek, 1997). Moreover, the fact that damselfly larvae regenerate also one missing lamellae may suggest there is selection to have a complete set (Robinson *et al.,* 1991b). More general, the discussed short- and long-term costs of autotomy are probably higher than the costs of lamellae regeneration and may explain why larvae do show regeneration of their lamellae. Costs of regeneration itself are, however, not quantified so far and are not easy to demonstrate. This is because larvae without lamellae probably not only have to allocate more food toward new lamellae tissue, but also suffer a lower food intake and probably also a lower digestive efficiency. One potential to solve this issue is by comparing individuals experiencing autotomy with regeneration and individuals experiencing autotomy without regeneration (Maginnis, 2006). This can for example be done, by comparing life history traits at emergence in three groups of larvae: (1) larvae that loose lamellae in F-2 and fully regenerate them, (2) larvae that loose lamellae in F-1 and will only partially regenerate them, and (3) larvae that loose lamellae in the final instar and will not regenerate them.

Based on the demonstrated costs of autotomy in the larval stage and at emergence, autotomy has the potential to influence adult fitness and to play a key role in population regulation. However, we lack direct proof for both linked contentions. Long-term costs have been less studied and no studies directly evaluated fitness costs in the adult stage. For this, we need longitudinal studies where lamellae status will be manipulated, and the lifetime mating success of the resulting adults monitored (De Block & Stoks, 2005). In a theoretical model, Harris (1989) showed that nonlethal injury could regulate population size if injury rates are density-dependent and injury reduced survival and/or reproduction. Current evidence shows that these assumptions are fulfilled for lamellae autotomy in damselfly larvae. Lamellae autotomy by conspecifics is density-dependent and may be very common (Baker & Dixon, 1986; Robinson *et al.,* 1991b). Moreover, there is solid proof that autotomy reduces survival and mass at emergence, and therefore may also negatively affect female fecundity. Studies looking at artificial larval populations in mesocosms with different densities and with manipulated lamellae status may therefore be rewarding in directly evaluating the assumed role of autotomy in regulating population size.

ACKNOWLEDGEMENTS

We acknowledge the significant contribution Prof. B. Kiauta made to the field of Odonatology. MDB is a postdoctoral researcher at the Fund for Scientific Research Flanders (FWO). This study was supported by research grants from FWO and the KULeuven Research Fund to Luc De Meester and RS.

REFERENCES

Anholt, B.R. 1994. Cannibalism and early instar survival in a larval damselfly. *Oecologia* 99: 60-65.

Anholt, B.R., J.H. Marden & D.M. Jenkins 1991. Patterns of mass gain and sexual dimorphism in adult dragonflies (Insecta, Odonata). *Can. J. Zool.* 69: 1156-1163.

Apodaca, C.K. & L.J. Chapman 2004. Larval damselflies in extreme environments: behavioural and physiological response to hypoxic stress. *J. Ins. Phys.* 50: 767-775.

Baker, R.L. & S.M. Dixon 1986. Wounding as an index of aggressive interactions in larval Zygoptera (Odonata). *Can. J. Zool.* 64: 893-897.

Burnside, C.A. & J.V. Robinson 1995. The functional morphology of caudal lamellae in coenagrionidf (Odonata: Zygoptera) damselfly larvae. *Zool. J. Linn. Soc.* 114: 155-171.

Brackenbury, J. 2003. Escape manoeuvres in damselfly larvae: kinematics and dynamics. *J. Exp. Biol.* 206: 389-397.

Convey, P. 1988.Competition for perches between larval damselflies - the influence of perch use on feeding efficiency, growth-rate and predator avoidance. *Freshw. Biol.* 19: 15-28.

Corbet, P.S. 1999. *Dragonflies. Behaviour and ecology of Odonata.* Harley Books, Colchester.

De Block, M. & R. Stoks 2004. Cannibalism-mediated life history plasticity to combined time and food stress. *Oikos* 106: 587-597.

De Block, M. & R. Stoks 2005. Fitness effects from egg to reproduction: bridging the life history transition. *Ecology* 86: 185-197.

Eriksen, C.H. 1986. Respiratory roles of caudal lamellae (gills) in a lestid damselfly (Odonata: Zygoptera). *J. N. Am. Benthol. Soc.* 5:16-27.

Gyssels, F.G.M. & R. Stoks 2005. Threat-sensitive responses to predator attacks in a damselfly. *Ethology* 111: 411-423.

Gyssels, F.G.M. & R. Stoks 2006. Behavioral responses to fish kairomones and autotomy in a damselfly. *J. Ethol.* 24: 79-83.

Hagemeier, D.H. 1990. Hypoxia induced vertical migration of zygopteran (Odonata) larvae and the role of caudal gills. M.S. Thesis, The University of Texas at Arlington.

Harris, R.N. 1989. Nonlethal injury to organisms as a mechanism of population regulation. *Am. Nat.* 134: 835-847.

Harvey, I.H. & K.J. Walsh 1993. Fluctuating asymmetry and lifetime mating success are correlated in males of the damselfly *Coenagrion puella* (Odonata, Coenagrionidae). *Ecol. Entomol.* 18: 198-202.

Juanes, F. & L.D. Smith 1995. The ecological consequences of limb damage and loss in decapod crustaceans: A review and prospectus. *J. Exp. Mar. Biol. Ecol.* 193: 197-223.

Maginnis, T.L. 2006. The costs of autotomy and regeneration in animals: a review and framework for future research. *Behav. Ecol.* 17: 857-872.

MacNeill, N. 1960. A study of the caudal gills of dragonfly larvae of the suborder Zygoptera. *Proc. R. Ir. Acad* 61B: 115-140.

McPeek, M.A. 1990a. Determination of species composition in the *Enallagma* damselfly assemblages of permanent lakes. *Ecology* 71: 83-98.

McPeek, M.A. 1990b. Behavioral differences between *Enallagma* species (Odonata) influencing differential vulnerability to predators. *Ecology* 71: 1714-1726.

McPeek, M.A. 1997. Measuring phenotypic selection on an adaptation: Lamellae of damselflies experiencing dragonfly predation. *Evolution* 51: 459-466.

McPeek, M.A. & P.H. Crowley 1987. The effects of density and relative size on the aggressive behavior, movement and feeding of damselfly larvae (Odonata, Coenagrionidae). *Anim. Behav.* 35: 1051-1061.

McPeek, M. A., A. Schrott & J.M. Brown 1996. Adaptation to predators in a new community: Swimming performance and predator avoidance in damselflies. *Ecology* 77: 617-629.

McPeek, M.A., M. Grace & J.M.L. Richardson 2001. Physiological and behavioral responses to predators shape the growth/predation risk trade-off in damselflies. *Ecology* 82: 1535-1545.

Mcvean, A. 1975. Mini-review: Autotomy. *Comp. Biochem. Physiol.* 51(A): 497-505.

Pritchard, G. 1991. Insects in thermal springs. *Mem. Ent. Soc. Can.* 155: 89-106.

Robinson, J.V., D.A. Hayworth & M.B. Harvey 1991a. The effect of caudal lamellae loss on swimming speed of the damselfly *Argia moesta* (Hagen) (Odonata: Coenagrionidae). *Am. Midl. Nat.* 125: 240-244.

Robinson, J.V., L.R. Shaffer, D.D. Hagemeier & N.J. Smatresk 1991b. The ecological role of caudal lamellae loss in the larval damselfly *Ischnura posita* (Hagen) (Odonata, Zygoptera). *Oecologia* 87: 1-7.

Shaffer, L.R. & J.V. Robinson 1993. Ontogenetic differences in intraspecific aggression of damselfly larvae. *Odonatologica* 22:311-317.

Shaffer, L.R. & J.V. Robinson 1996. Do damselfly larvae recognize and differentially respond to distinct categories of macroinvertebrates? *J. Ins. Behav.* 9: 407-419.

Sih, A. 1986. Antipredator responses and the perception of danger by mosquito larvae. *Ecology* 67: 434-441.

Sokolovska, N., L. Rowe & F. Johansson 2000. Fitness and body size in mature odonates. *Ecol. Entomol.* 25: 239-248.

Stoks, R. 1998a. Effect of lamellae autotomy on survival and foraging success of the damselfly *Lestes sponsa* (Odonata: Lestidae). *Oecologia* 117:443-448.

Stoks, R. 1998b. Indirect monitoring of agonistic encounters in larvae of *Lestes viridis* (Odonata: Lestidae) using exuviae lamellae status. *Aquat. Ins.* 3: 173-180.

Stoks, R. 1999a. The effect of lamellae autotomy and sexual size dimorphism on startle-response performance in larvae of a lestid damselfly (Odonata). *J. Zool.* 247: 269-273.

Stoks, R. 1999b. Autotomy shapes the trade-off between seeking cover and foraging in larval damselflies. *Behav. Ecol. Sociobiol.* 47: 70-75.

Stoks, R. 1999c. Natural and sexual selection in the damselfly *Lestes sponsa.* PhD thesis, University of Antwerp.

Stoks, R. 2001a. Food stress and predator-induced stress shape developmental performance in a damselfly. *Oecologia* 127: 222-229.

Stoks, R. 2001b. What causes male-biased sex ratios in mature damselfly populations? *Ecol. Entomol.* 26: 188-197.

Stoks, R. & M. De Block 2000. The influence of predator species and prey age on the immediate survival value of antipredator behaviours in a damselfly. *Arch. Hydrobiol.* 147: 417-430.

Stoks, R. & M.A. Mcpeek 2003. Antipredator behavior and physiology determine Lestes species turnover along the pond-permanence gradient. *Ecology* 84: 3327-3338.

Stoks, R., M. De Block, F. Van De Meutter & F. Johansson 2005. Predation cost of rapid growth: behavioural coupling and physiological decoupling. *J. Anim. Ecol.* 74: 708-715.

Stoks, R., M. De Block, H. Van Gossum, F. Valck, K. Lauwers, R. Verhagen, E. Matthysen & L. De Bruyn 1999. Lethal and sublethal costs of autotomy and predator presence in damselfly larvae. *Oecologia* 120: 87-91.

Strobbe, F. & R. Stoks 2004. Life history reaction norms to time constraints in a damselfly: differential effects on size and mass. *Biol. J. Linn. Soc.* 83: 187-196.

Odonata : Biology of Dragonflies
Ed.: B.K. Tyagi
Scientific Publishers (India), 2007, pp. 257-271

17

ENDEMIC ODONATES OF THE WESTERN GHATS: HABITAT DISTRIBUTION AND CONSERVATION

K.A. SUBRAMANIAN

Centre for Ecological Sciences, Indian Institute of Science, Bangalore- 560 012, India
e.mail: subbu@ces.iisc.ernet.in

ABSTRACT

The habitat distribution of endemic odonates of Western Ghats is studied based on field work and published literature. One hundred and seventy six species of odonates with 68 endemics are reported from the Western Ghats. The breeding habitats of 50 endemic species are known. The current field survey discovered new populations of three monotypic species from the region. Most of the endemic species are restricted to riverine habitats as compared to non-endemics. Streams flowing through evergreen forests and Myristica swamps support high diversity of endemics. Draining of Myristica swamps, diversion of streams for agriculture and structural alterations are major threats to the conservation of endemic species of the region. Long term conservation of endemic odonate fauna of region should focus on conservation of riverine habitats of the region.

INTRODUCTION

The Western Ghats running parallel to the west coast of India from the river Kanyakumari (8°N) to Tapti (21°N) is a prominent landscape feature of the peninsular India (Map 1). This 1600 Km long chain of mountain along with Sri Lanka is also one of the 25-biodiverisity hotspot regions of the world. It is well known for its unusually high diversity of angiosperms and amphibians (Myers *et. al*, 2000). Moreover, the Western Ghats is the catchment of 3 large, 13 medium and 17 minor rivers of the peninsular India (Rao, 1975). These 33 rivers and associated wetlands harbour high diversity of freshwater organisms. A recent study identified the Western Ghats region of Indian subcontinent as one of the important areas of freshwater biodiversity (WCMC, 2000).

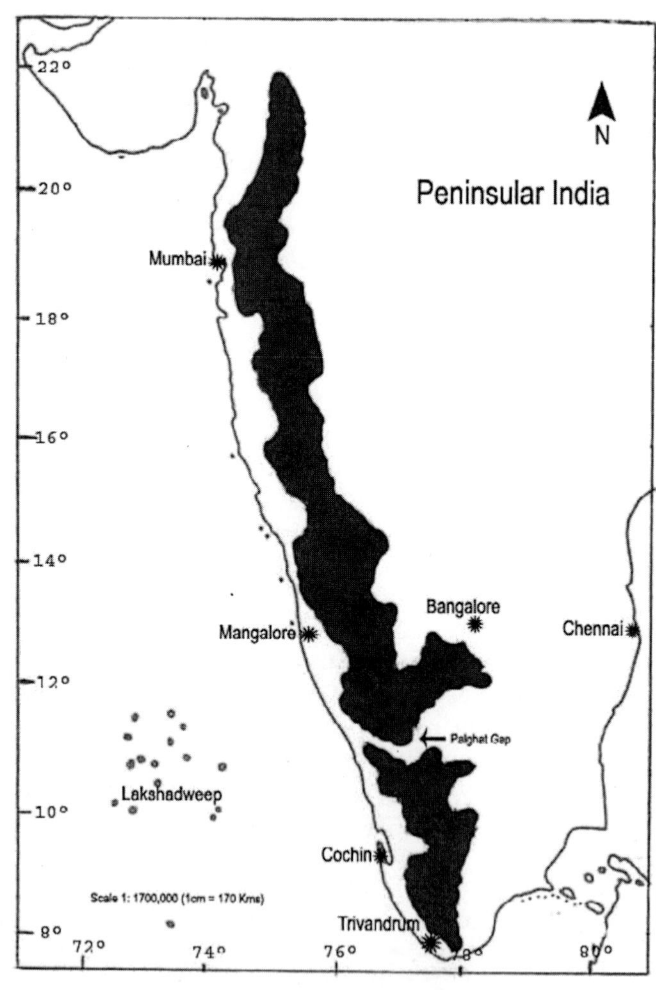

Map 1. The Western Ghats

The region is known for its high diversity of fresh water fishes (Daniels, 2001; Bhat, 2002). However, species level inventories are incomplete for most of the freshwater invertebrates. Exception to this is Odonata (dragonflies and damselflies), thanks to the works of Fraser (1924, 1932, 1933-36).

Like many other tropical regions of the world, the Western Ghats is densely populated. It also faces increasing degradation and loss of natural vegetation cover. Biologists have estimated that 12,450 km^2 of natural forest cover exist in the Western Ghats, which is about 6.8% of the original cover (Menon & Bawa, 1997).

Irrigation and hydroelectric projects, timber operations and agriculture expansion were the important factors for this loss of natural forest cover (Chandran, 1997). Between 1920 and 1990, forest cover declined by 40% with fourfold increase in the number of fragments and an 83% reduction in the size of forest patches (Menon & Bawa, 1997). Few areas remain in the Western Ghats with large continuous patch of natural vegetation; most areas are a mosaic of various natural and human modified landscape elements (Nagendra & Gadgil, 1998).

The riverine ecosystems and associated wetlands of the Western Ghats with its high diversity of odonates provide a unique opportunity to study tropical odonate communities. A study on odonate communities focusing on unperturbed and human impacted wetlands from the region is essential to understand the freshwater biodiversity of the Western Ghats. This study based on literature survey and field work focuses on the diversity and habitat distribution of endemic odonates of the region. It also addresses to the ongoing threats to the conservation of odonate habitats of the region. Results from this exercise, along with the ongoing studies on other freshwater plant and animal taxa will form the baseline dataset on the freshwater biodiversity of the region. Such a dataset is essential for developing reliable biodiversity monitoring tools and identifying areas for conservation of freshwater biodiversity.

METHODOLOGY

Literature survey: The odonate fauna of the Western Ghats is well known taxonomically. Taxonomy of adults is well worked out and descriptions are available for all the reported species (Fraser, 1924, 1932, 1933-36; Davis & Tobin, 1984,1985; Prasad & Varshney, 1995). Recent studies on the odonate fauna of the region are species' checklists based on field surveys (Peters, 1981, Rao & Lahiri, 1982;

Prasad, 1987, Mathavan & Miller, 1989; Radhakrishanan, 1997; Emiliyamma & Radhakrishnan, 2000, 2002; Jaffer *et al.,* 2002; Subramanian & Sivaramakrishanan, 2002; Mitra, 1994,1999, 2002, 2003; Radhakrishnan & Emiliyamma, 2003; Emiliyamma, *et al.*, 2005; Subramanian, 2005). Though partial, these published studies give valuable information on geographic and habitat distribution of odonates of the region.

Field survey: Odonates were surveyed in different regions of the Western Ghats since 1999. The details of study sites are given in Table 1. The wetland habitat of odonates was observed in the field. These wetlands were classified into different habitats following Dugan (1990). Current threats faced by different wetland types were also noted during the field survey.

Table 1. Study sites in the Western Ghats

S.N.	Site Name	District	State	Latitude (°N)	Longitude (°E)
1	Bondla	South Goa	Goa	15.40	74.06
2	Kigga	Chikkamagalur	Karnataka	13.44	75.18
3	Neralekoppa	Chikkamagalur	Karnataka	13.67	75.45
4	Makut	Coorg	Karnataka	12.08	75.75
5	Koyyur	Dakshina Kannada	Karnataka	12.95	75.29
6	Panaje	Dakshina Kannada	Karnataka	12.66	75.17
7	Gurupura	Dakshinkanda	Karnataka	12.55	74.55
8	Kuke-Subramanya	Dakshinkanda	Karnataka	12.68	75.63
9	Moodabadri	Dakshinkanda	Karnataka	13.04	74.59
10	Panaje Bridge	Dakshinkanda	Karnataka	13.02	74.47
11	Someswara Beach	Dakshinkanda	Karnataka	12.47	74.55
12	Surathkal	Dakshinkanda	Karnataka	13.00	74.57
13	Ullal Beach	Dakshinkanda	Karnataka	12.49	74.50
14	Haridravathi	Shimoga	Karnataka	14.10	75.13
15	Hosanagara	Shimoga	Karnataka	13.88	75.05
16	Nagodi	Shimoga	Karnataka	13.91	74.86
17	Nandihole	Shimoga	Karnataka	14.00	75.13
18	Nellibedu	Shimoga	Karnataka	13.94	74.85
19	Nittur	Shimoga	Karnataka	13.93	75.90
20	Gangoli	Udupi	Karnataka	13.39	74.39
21	Halladi	Udupi	Karnataka	13.35	74.51
22	Heru	Udupi	Karnataka	13.24	74.44
23	Kollur Town	Udupi	Karnataka	13.87	74.87
24	Kudremukh NP	Udupi	Karnataka	13.20	75.11
25	Mala	Udupi	Karnataka	13.20	75.11
26	Malpe Beach	Udupi	Karnataka	13.22	74.41

S.N.	Site Name	District	State	Latitude (°N)	Longitude (°E)
27	Malyadi	Udupi	Karnataka	13.34	74.42
28	Marvanthe	Udupi	Karnataka	13.43	74.38
29	Ramasamudra	Udupi	Karnataka	13.12	75.00
30	Someswara	Udupi	Karnataka	13.37	75.12
31	Badal	Uttara Kannada	Karnataka	14.36	74.79
32	Abgeri	Uttarakannada	Karnataka	14.64	74.58
33	Andhalli	Uttarakannada	Karnataka	14.67	74.80
34	Angdibail	Uttarakannada	Karnataka	14.62	74.53
35	Chittagri	Uttarakannada	Karnataka	14.86	74.98
36	Dabguli	Uttarakannada	Karnataka	14.85	74.66
37	Devanalli	Uttarakannada	Karnataka	14.63	74.66
38	Gundubale	Uttarakannada	Karnataka	14.63	74.43
39	Hemmadi	Uttarakannada	Karnataka	14.75	74.86
40	Hulidevarakodlu	Uttarakannada	Karnataka	14.44	74.64
41	Kalleswara	Uttarakannada	Karnataka	14.76	74.61
42	Kamani	Uttarakannada	Karnataka	14.71	74.60
43	Karadrolli	Uttarakannada	Karnataka	14.99	74.84
44	Karikan	Uttarakannada	Karnataka	14.35	74.51
45	Kathlekan	Uttarakannada	Karnataka	14.28	74.75
46	Kodigathe	Uttarakannada	Karnataka	14.35	74.65
47	Malemanne	Uttarakannada	Karnataka	14.28	74.73
48	Manchikeri	Uttarakannada	Karnataka	14.89	74.79
49	Mavinahole	Uttarakannada	Karnataka	13.97	75.10
50	Mundgod	Uttarakannada	Karnataka	15.05	74.94
51	Necti stream	Uttarakannada	Karnataka	14.64	74.61
52	Ramanguli	Uttarakannada	Karnataka	14.80	74.59
53	Muthapan Puzha	Calicut	Kerala	11.43	76.08
54	Thatekkad	Ernakulam	Kerala	10.10	76.72
55	Eravikulam	Idukki	Kerala	10.13	77.09
56	Aralarm	Kannur	Kerala	12.09	75.79
57	Dhoni	Palakkad	Kerala	10.88	76.63
58	Meenvallom	Palakkad	Kerala	10.90	76.60
59	Silent Valley	Palakkad	Kerala	11.5	76.27
60	Aghastyamalai	Trivandrum	Kerala	8.67	77.15
61	Bonnacaurd	Trivandrum	Kerala	8.68	77.20
62	Neyyar	Trivandrum	Kerala	8.58	77.22
63	Shendurani	Trivandrum	Kerala	8.55	77.03
64	Radha Nagari WLS	Kholapur	Maharastra	16.50	73.85
65	Triambak	Nasik	Maharastra	19.94	73.53
66	Dongarvadi	Pune	Maharastra	18.48	73.42
67	Tamhini	Pune	Maharastra	18.46	73.44
68	Amba	Raigad	Maharastra	16.98	73.78

S.N.	Site Name	District	State	Latitude (°N)	Longitude (°E)
69	Koyana	Satara	Maharastra	17.57	73.76
70	Topslip	Coimbatore	Tamil Nadu	10.48	76.86
71	Karandhamalai	Madurai	Tamil Nadu	10.30	78.23
72	Upper Bhavani	Nilgiri	Tamil Nadu	11.22	76.52

Analysis: Data collected from field and literature surveys were used to estimate habitat distribution and percentage endemism of odonates.

RESULTS

Diversity

Indian odonate fauna comprise 499 species representing 17 families and 3 suborders (Prasad, 1999). Current study shows that the Western Ghats odonate fauna comprise 176 species, representing two sub-orders, 13 families, 84 genera with 68 (38.63%) endemics (Tables 2, 3).

Table 2. Generic diversity and endemism of the Western Ghats odonates

Suborder	Family	No. Genera	No. Species	No. Endemics	Percent Endemism
Zygoptera					
	Coenagrionidae	10	25	4	16
	Platycnemidae	1	2	0	0
	Platystictidae	2	8	7	87.5
	Protoneuridae	7	15	11	73.3
	Lestidae	2	8	3	37.5
	Calopterygidae	2	3	0	0
	Chlorocyphidae	3	3	1	33.3
	Euphaeidae	2	4	3	75
Zygoptera Diversity		29	68	29	42.64
Anisoptera					
	Gomphidae	15	27	18	66.66
	Aeshnidae	4	8	1	12.5
	Cordulegasteridae	1	2	2	100
	Corduliidae	5	21	16	76.19
	Libellulidae	28	50	2	4
Anisoptera Diversity		53	108	39	36.11
Total Odonata Diversiy		84	176	68	38.63

Table 3. Endemic odonates of the Western Ghats and their habitat distribution (RI=Riverine, LA= Lacustrine, PL = Palustrine)

S NO:	Suborder	Family	Species	Habitats		
				RI	LA	PL
	Zygoptera					
I		Coenagrionidae				
1			*Agriocnemis keralensis*	-	-	-
2			*Agriocnemis pieris*	-	+	+
3			*Archibasis oscillans*	+	-	-
4			*Pseudagrion indicum*	+	-	-
II		Platystictidae				
5			*Platysticta maculata*	+	-	-
6			*Protosticta antelopoides*	+	-	-
7			*Protosticta gravelyi*	+	-	-
8			*Protosticta hearseyi*	+	-	-
9			*Protosticta mortoni*	+	-	-
10			*Protosticta rufostigma*	-	-	-
11			*Protosticta sanguinostigma*	+	-	-
III		Protoneuridae				
12			*Caconeura ramburi*	+	-	-
13			*Caconeura risi*	+	-	-
14			*Disparoneura apicalis*	+	-	-
15			*Disparoneura canningi*	-	-	-
16			*Elattoneura souteri*	+	-	-
17			*Elattoneura tetrica*	+	-	-
18			*Esme cyaneovittata*	+	-	-
19			*Esme longistyla*	+	-	-
20			*Esme mudiensis*	+	-	-
21			*Melanoneura bilineata*	+	-	-
22			*Phylloneura westermanni*	+	-	-
IV		Lestidae				
23			*Ceylonolestes davenporti*	+	-	-
24			*Indolestes pulcherrimus*	-	+	+
25			*Lestes patricia*	-	+	+
V		Chlorocyphidae				
26			*Calocypha laidlawi*	+	-	-
VI		Euphaeidae				
27			*Euphaea cardinalis*	+	-	-
28			*Euphaea dispar*	+	-	-
29			*Dysphaea ethela*	+	-	-
	Anisoptera					
VII		Gomphidae				
30			*Acrogomphus fraseri*	+	-	-

31		*Asiagomphus nilgiricus*	+	-	-
32		*Burmagomphus cauvericus*	+	-	-
33		*Burmagomphus laidlawi*	+	-	-
34		*Davidioides martini*	-	-	-
35		*Gomphidia fletcheri*	+	-	-
36		*Gomphidia kodaguensis*	+	-	-
37		*Heliogomphus kalarensis*	+	-	-
38		*Heliogomphus promelas*	+	-	-
39		*Macrogomphus wynaadicus*	+	-	-
40		*Megalogomphus hannyngtoni*	+	-	-
41		*Megalogomphus superbus*	+	-	-
42		*Merogomphus longistigma*	+	-	-
43		*Microgomphus souteri*	+	-	-
44		*Onychogomphus acinaces*	+	-	-
45		*Onychogomphus malabarensis*	-	-	-
46		*Onychogomphus nilgiriensis*	+	-	-
47		*Onychogomphus striatus*	+	-	-
VIII	Aeshnidae				
48		*Anaciaeschna donaldi*	-	+	-
IX	Cordulegasteridae				
49		*Chlorogomphus campioni*	+	-	-
50		*Chlorogomphus xanthoptera*	+	-	-
X	Corduliidae				
51		*Idionyx corona*	-	-	-
52		*Idionyx galeata*	+	-	-
53		*Idionyx minima*	-	-	-
54		*Idionyx nadganiensis*	-	-	-
55		*Idionyx nilgiriensis*	+	-	-
56		*Idionyx periyashola*	+	-	-
57		*Idionyx rhinoceroides*	-	-	-
58		*Idionyx saffronata*	-	-	-
59		*Idionyx travancorensis*	-	-	-
60		*Macromia annaimallaiensis*	-	-	-
61		*Macromia bellicosa*	-	-	-
62		*Macromia ellisoni*	-	-	-
63		*Macromia flavoclorata*	-	-	-
64		*Macromia ida*	+	-	-
65		*Macromia irata*	+	-	-
66		*Macromidia donaldi*	+	-	-
XI	Libellulidae				
67		*Epithemis mariae*	+	-	-
68		*Zygonyx metallicus*	+	-	-
		Total	49	4	3

The suborders Zygoptera (damselflies) and Anisoptera (dragonflies) are represented respectively by 8 and 5 families in Western Ghats. Zygopteran fauna comprises 29 genera and 68 species, of which 29 species are endemic. On the other hand, Anisopteran fauna comprises 53 genera, 108 species with 39 endemics. Percentage endemism is higher in Zygoptera (42.64%) than in Anisoptera (36.11%).

Families Libellulidae, Gomphidae and Coenagrionidae are most species-rich in the Western Ghats. These families constitute 58 percent of the odonate fauna and 36 percent of endemic species. Families with high percentage of endemism (>30%) are Platystictidae, Protoneuridae, Lestidae, Chlorocyphidae, Gomphidae, Cordulegasterdiae and Corduliidae. Families like Platycnemidae and Calopterygidae have no endemic species in the Western Ghats.

Four monotypic endemic odonates are known from the Western Ghats. Previously, all the four monotypic species, *viz.*, *Phylloneura westermanni, Melanoneura bilineata* (Protoneuridae), *Calocypha laidlawi* (Chlorocyphidae) and *Davidioides martini* (Gomphidae) were reported from Nilgiri-Wayanad-Coorg region, north of the Palaghat Gap. The current study discovered populations of *Phylloneura westermanni, Calocypha laidlawi* and *Davidioides martini* from Kathlekan reserve forest (Uttara Kannada dist.), Shenduranay Wildlife Sancturay (Trivandrum dist.) and Silent Valley National Park, respectively (Plate 1).

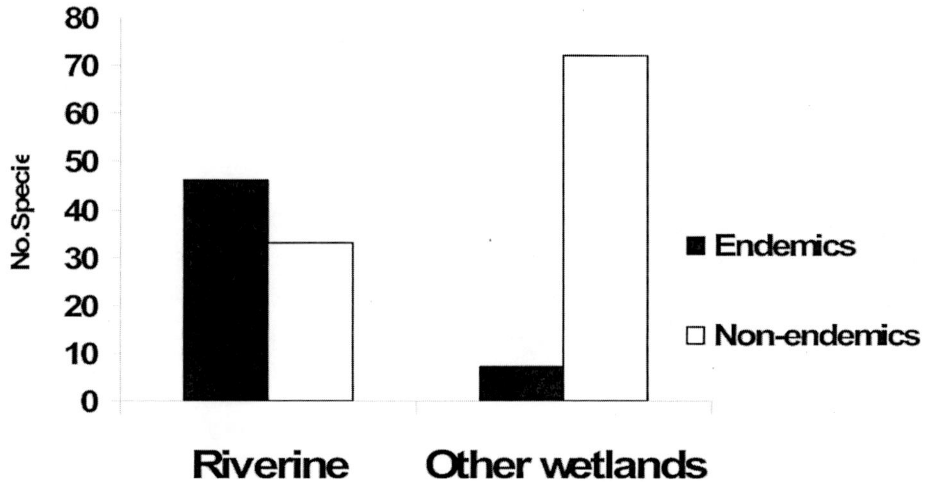

Figure 1. Habitat distribution of endemic and non-endemic odonates.

Habitat Preferences

Habitat preferences of 53 endemics and 105 non-endemic species are known in the Western Ghats. Endemic species, in both suborders have narrow habitat preferences than the non-endemics. Most of the endemic species (0.87, N=46) uses exclusively riverine and rest (0.13, N=7) uses lacustrine and palustrine habitats (Fig. 1). Frequency distribution of habitat preference shows that endemic species use less than three habitats compared to non-endemics (Figs. 2, 3).

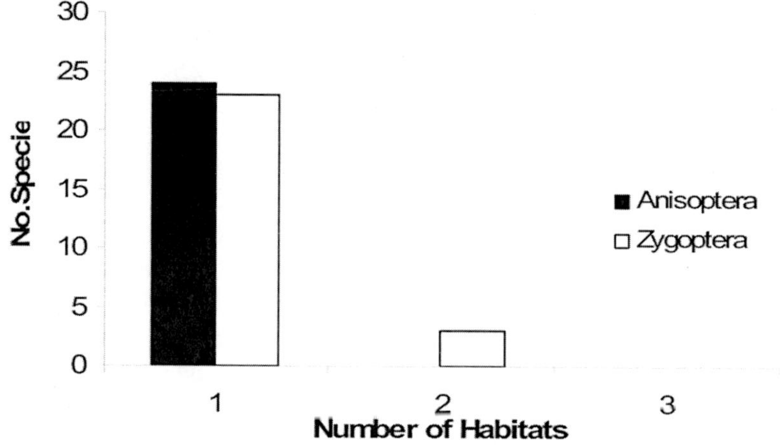

Figure 2. Frequency distribution of habitat preference of endemic odonates.

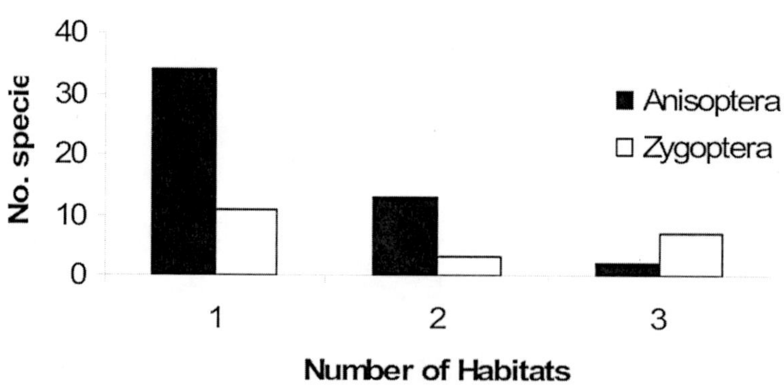

Figure 3. Frequency distribution of habitat preference of non-endemic odonates.

Threats to Endemic Odonates

Habitats of endemic odonates are exposed to a range of anthropogenic threats. Myristica swamps are rare formations in evergreen forests and important habitats for endemic odonates, especially the monotypic species such as *Phylloneura westermanni* and *Calocypha laidlawi.* Encroachment of swamps for areca nut and other plantations are fast denuding these important odonate habitats. Encroachments and draining of swamps were found throughout the Western Ghats.

Streams with good riparian forest cover are other important habitats for endemic odonates. Diversion of streams and indiscriminate construction of check dams drastically alter the flow dynamics of the streams and fundamentally changes the larval habitats. In addition to this, mining in river catchments, riparian deforestation, soil erosion and dumping of solid waste are other major threats to the habitats of endemic odonates (Plate 2).

Impact of non-point source chemical pesticides such as organochlorines, organophosphates and synthetic pyrethroides on odonate fauna is not known. However, total absence of endemic fauna in streams running through various commercial plantations in the current study indicates that chemical pesticides may indeed be causing serious damage to the odonate fauna.

DISCUSSION

High percentage of endemism (38.69%) characterizes the odonate fauna of the Western Ghats. This is comparable with other taxa such as angiosperms, freshwater fishes, amphibians and reptiles of the Western Ghats (Daniels, 2002). This high degree of endemism in Western Ghats odonates could be because of their unique life history. Larval stage is aquatic and requires specific breeding habitats for completing the life cycle. Adults depend upon forested riparian corridors for foraging and dispersal. Dispersal success depends upon availability of undisturbed riparian forests and suitable larval breeding habitats. Isolation of the Western Ghats from other odonate rich areas such as Indo-Burmese region could have promoted the origin of high endemicity.

Endemicity of the Western Ghats odonates is very high compared to of other known biodiversity hotspots of the world such as Cape Floristic Province of South Africa, where only 9% of odonates are endemic (Samways, 1992). On the other hand the odonate endemism of the Western Ghats is comparable to Sri Lanka, which has 117 species with 49 endemics (42% endemism) (Fonseka, 2000). Families such as Platystctidae and Gomphidae are particularly diversified in that island. On the other hand, in the Western Ghats families such as Protoneuridae, Gomphidae and Cordulidae are highly diversified.

Endemic odonates of Western Ghats particularly prefer riverine habitats such as montane streams and rivers. A large proportion (0.87, N=46) of the endemics use exclusively riverine habitats. Non-endemics on the other hand have wider habitat preference than endemics. Non-endemics use both natural and man-made lentic habitats such as pools, marshes and swamps. Nearly half of (0.47, N=33) of non-endemic species use riverine habitats, they are not exclusive to the habitat. In other terms they are generalists adapted to use a spectrum of lotic to lentic habitats.

The family Libellulidae with 51 species including 2 endemics is most diverse in the Western Ghats. Habitat preference for thirty-six species is known from this family. All the species uses lentic habitats. Only ten species uses riverine habitats in addition to lentic habitats. High species richness and wide geographic spread of this family is probably related to its lentic habitat, which is widely available. This may be generalized to infer that lentic habitats, though may increase species richness of an area, can potentially encourage colonization of widespread generalists species such as libellulids.

Long term conservation of endemic odonates is directly linked to the conservation of riverine ecosystem of the region. Recent studies have shown that inland wetlands of the region are facing serious threats (Vijayan *et al.,* 2005). Regional scale destruction of riverine habitats by hydro-electric and irrigation projects threatens the survival of endemic odonates, which depend on fast flowing torrential streams. Many of the endemic odonates such as *Disparoneura apicalis* (Protoneuridae) or *Idionyx* sps (Corduliidae) are very narrowly distributed within the Western Ghats and destruction of alteration of a small catchment essentially means extinction of these species. Local scale impacts on riverine habitats such as check dams, draining of swamps, canalization, dumping of wastes and pesticide pollution alters the quality of habitats and promotes the colonization of generalist species such as libellulids. These small local scale habitat alterations are very wide spread throughout the Western Ghats and are very difficult to assess its impact on the endemic odonates.

Though the odonate fauna of the region is fairly well documented taxonomically, the ecology of most of the species is poorly known. Larval stages of only 76 Indian species are known and full life history

Plate - 1. Monotypic species of the Western Ghats: (1) *Phylloneura westermanni* (Protoneuridae) (2) *Davidioides martini* (Gomphidae) (3) *Calocypha laidlawi* (Chlorocyphidae)

Plate 2. Threats to odonate habitats: (1) Riparian deforestation (2) Road development (3) Check dam (4) Solid waste disposal (5) Agriculture expansion (6) Draining of myristica swamps (7) Reservoirs.

is worked out only for 15 species (Prasad & Varshney, 1995). Detailed ecological information, especially of the endemic species is very essential for long term conservation of the odonates. Current conservation management measures of the region need to consider the importance of riverine ecosystems. A conservation management programme focusing on threatened odonate habitats such as Myristica swamps and forest streams are very essential for long term conservation. Current odonate conservation assessment lists only five Indian species (IUCN, 2006) and Indian Wildlife Protection Act, 1972 (Schedule I, Part IV) provide legal protection to only one species, *Epiophlebia laidlawi* Tillyard (Anisozygoptera). Comprehensive conservation assessment of Indian odonates and identification of "odonate hotspots" for long term conservation is immediately required to develop conservation and management strategies.

ACKNOWLEDGEMENTS

Department of Science and Technology, Government of India is sincerely acknowledged for supporting the Western Ghats odonate project of the author through DST Young Scientist Programme.

REFERENCES

Bhat, A. (2002). A study of the diversity and ecology of the freshwater fishes of four river systems of Uttara Kannada District, Karnataka, India. *Ph. D. Thesis submitted to Indian Institute of Science*, Bangalore. 178pp.

Chandran, M.D.S. (1997). On the ecological history of the Western Ghats. Curr. Sci. Vol.73 (2): 146-155.

Daniels, R.J.R. (2001). Endemic fishes of the Western Ghats and the Satpura Hypothesis. Curr. Sci. Vol. 81(3):240-244.

Daniels, R.J.R. (2002). A Report on the National Biodiversity Strategy and Action Plan-the Western Ghats Ecoregion: Submitted to the Ministry of Environment and Forests, India. URL. *http://sdnp.delhi. nic.in/nbsap/ index1.html*

Davis, D.Allen L. & Pamela Tobin (1984). The Dragonflies of the World: A Systematic List of Extant Species of Odonata. Vol I & II. Societas Internationalis Odontologica Rapid Communications (Supplements). No. 3.

Dugan, P.J. (Editor) (1990). Wetland Conservation: A Review of Current Issues And Required Action. Iucn, The World Conservation Union, Gland, Switzerland. 96 Pp.

Emiliyamma, K.G. and Radhakrishnan, C (2000). Odonata (Insecta) of Parambikulam Wildlife Sanctuary, Kerala, India. Rec. Zool. Surv. India. 98(1): 157-167.

Emiliyamma, K.G. and Radhakrishnan, C. (2002). Additions to the Odonata of (Insecta) of Thiruvananthapuram district, Kerala. Zoo's print Journal. 17(10): 914-917.

Emiliyamma, K.G., Radhakrishnan, C. and Jafer Palot, M (2005). Pictorial Handbook on Common Dragonflies and Damselflies of Kerala. Zoological Survey of India. 67pp.

Fonseka, Terence de (2000). The Dragonflies of Sri Lanka. WHT Publications (Private) Limited, Colombo, Sri Lanka. 303pp.

Fraser, F.C. (1924). A Survey of the Odonate (Dragonfly) Fauna of Western India with Special Remarks on the Genera Macromia and Idionyx and Descriptions of Thirty New Species with Appendices I, II. Records of the Indian Museum (A Journal of Indian Zoology). Vol. 26, Part V.423-522.

Fraser, F.C (1932). Additions to the survey of odonate fauna of Western India, with descriptions of nine new species. Records of the Indian Museum. 32:443-474.

Fraser, F.C. (1933-36). The Fauna of British India, including Ceylon and Burma, Odonata. Vols. 1-3. Taylor & Francis Ltd., London.

IUCN Red list of Threatened Species (2006). URL <www.iucnredlist.org>. Date accessed 05th December, 2006.

Jafer Palot, M., Dinesan Cheruvat, Emiliyamma, K.G. and C. Radhakrishnan (2002). Dragonfly Mencae at the National Fish Seed Farm, Malampuzha, Kerala. Fishing Chimes, 22 (5): 56-60.

Mathavan, S and P.L. Miller (1989). A collection of dragonflies (Odonata) made in the Periyar National Park, Kerala, South India, in January 1988. Soc. Int. Odonatol. Rapid. Coom. (Suppl.),

Menon, S. and Bawa, S.K. (1997). Applications of geographic information systems, remote sensing, and a landscape ecology approach to biodiversity conservation in the Western Ghats. *Curr. Sci.* 73 (2): 134 –145.

Mitra, T.R. (1994). Observations on the Habits and Habitats of adult dragonflies of Eastern India with special reference to the Fauna of West Bengal. Records of Zoological Survey of India. Occasional Papers No. 166.

Mitra, T.R. (1999). Geographical Distribution and Zoogeography of Odonata (Insecta) of Meghalaya, India. Occasional Papers No.170.

Mitra, T.R. (2002). Geographical Distribution of Odonata (Insecta) of Eastern India. Memoirs of the Zoological Survey of India. Volume 19 (1):1-208.

Mitra, T.R (2003). Ecology and biogeography of Odonata with special reference to Indian fauna. *Rec. Zool. Surv. India.* Occ. Paper No. 202:1-41.

Myers, Norman, Russell A. Mittermeier, Cristina G. Mittermeier, Gustavo A.B. Da Fonseca and Jennifer Kent (2000). Biodiversity Hotspots for Conservation Priorities. Nature. Vol. 403, 853-858.

Nagendra, H. and Gadgil, M. (1998). Linking regional and landscape scales for assessing biodiversity: A case study from Western Ghats. *Curr. Sci.* 75 (3): 264 –271.

Peters, G (1981). Trockenzeit-Libellen aus dem indischen Tiefland. Dtsch. Ento. Z., N.F. 28 Heft I-III, seite 93-108.

Prasad M. (1999). Faunal diversity in India: Odonata. Zoological Survey Of India, Calcutta. 171-178.

Prasad, M (1987). A note on the Odonata from South India. *Fraseria*, No. 12:50

Prasad, M. and Varshney, R.K. (1995). A checklist of the Odonata of India including data on larval studies. *Oriental Insects*, 29:385-428.

Radhakrishnan, C. (1997). Ecology and conservation status of Entomofauna of Malabar. *Zoos Print*, 11:2-5.

Radhakrishnan, C. and Emiliyamma, K.G. (2003). Odonata (Insecta) of Kerala: A Systematic database. In: Advancements in Insect Biodiversity. Ed. Rajiv K. Gupta. Agrobios, Jodhpur.

Rao R and Lahiri A.R. (1982). First records of Odonates (Arthropoda:Insecta) from the Silent Valley and New Amarambalam Reserved Forests. JBNHS 79(3): 557-562.

Rao, K.L. (1975). India's Water Wealth. *Orient Longman*, Hyderbad, India. 267pp.

Samways, M.J. (1992). Dragonfly conservation in South Africa: A Biogeographic Perspective. Odonatologica 21(2): 165-180.

Subramanian, K.A. and Sivaramakrishnan, K.G. (2002). Conservation Of Odonate Fauna in Western Ghats- A biogeographic perspective. Pages 11-22. In: Vistas of Entomological Research for the New Millenium. (Eds.: K.P. Sanjayan, V. Mahalingam and M.C. Muralirangan). G.S. Gill Research Institute, Chennai.

Subramanian. K.A. (2005). Damselflies and dragonflies of peninsular India-A field Guide. E-book of the Project Lifescape. Indian Academy of Sciences and Centre for Ecological Sciences, Indian Institute of Science, Bangalore, India. 118 pp. Available online in PDF format: http://ias.ac.in/ initiat/sci_ed/lifescape/odonates.html

Vijayan, V.S., Prasad, S.N., Vijayan, L. and Muralidharan, S (2005). Inland Wetlands of India-Conservation Priorities. Salim Ali Centre for Ornithology and Natural History, Coimbatore. pp XXiV+532.

The Wildlife (Protection) Act, 1972 (Ammended, 2002). Professional Book Publishers. New Delhi.

World Conservation Monitoring Centre (WCMC) (2000). Global Biodiversity: Earth's living resources in the 21st century. By: Groombridge, B. and Jenkins, M.D. World Conservation Press, Cambridge,UK.

Odonata : Biology of Dragonflies
Ed.: B.K. Tyagi
Scientific Publishers (India), 2007, pp. 273-322

18

DRAGONFLIES OF THE MADURAI KAMARAJ UNIVERSITY CAMPUS (TAMIL NADU, INDIA)

P.L. MILLER†

Department of Zoology, South Parks Road, OXFORD OX1 3PS, United Kingdom

ABSTRACT

Twenty five odonate species were observed on the Madurai Kamaraj University campus during the periods September 1987 and February 1988. Of these, proof of breeding on the campus was obtained for sixteen species and there was strong, circumstantial evidence of breeding in at least a further two species. Five bred only in permanent habitats and seven only in temporary habitats, the remainder probably doing so in both types. Sixteen species were sexually active at the largest habitat, a seasonal lake. Five further species, although commonly present, showed no sexual activity at campus sites.

INTRODUCTION

While there have been many reports on the odonate faunas of various parts of northern India and adjacent countries (*e.g.* Bhargava, 1989; Biswas *et al.*, 1990; Kumar, 1972, 1985; Ram *et al.*, 1983; Pajni & Tewari, 1988; Mahato, 1989), only a few deal with the southern part of the sub-continent (Larsen, 1987; Mathavan & Miller, 1989; Thomas & Gladstone, 1989) and none seems previously to have considered the dragonflies of the Madurai district. A brief report is presented here of the species of Odonata observed as adults at various habitats on the campus of Madurai Kamaraj University in 1987 and 1988.

The aim of this paper is to allow readers to identify all the common adult dragonflies they are likely to find on and in the vicinity of the campus of Madurai Kamaraj University and to indicate a few interesting aspects of their behaviour. The identification of larvae still presents many problems, and even for some common species they remain undescribed. Description in this paper is based on observations which were made at Madurai between October, 1987, and the beginning of February, 1988, - the best time of the year for dragonflies. There is no attempt to cover all the species which might be found and only a few of the possible 'vagrants', which do not breed on the campus, are included. I hope to draw attention to a fascinating order of insects on which many original behavioural observations can be made by the patient naturalist. Most dragonflies can easily be identified without capture, and they are large enough and often sufficiently localized for much of their behaviour to be followed for long periods.

All over the world the numbers of dragonflies have been much reduced and many species are threatened with extinction. In the main this is due to the destruction or pollution of their larval habitats by agriculture, commercial development and industry. Cutting down forests may be directly detrimental to the adults of some species as well as destroying larval habitats. This matters not only because dragonflies are directly beneficial to us but also because our survival is directly threatened by the degradation of our environment and by the destruction of the organisms with which we share it.

ODONATA

There are about 5000 species of dragonflies worldwide with the majority occurring in the tropics. India has an exceptionally rich and interesting odonata fauna with over 10% (about 600 species) of the world's species (Fraser, 1933, 1934, 1936; Tyagi, 1997). There are many endemic species, particularly in the montane regions of the south, as well as representatives of the African, Palaearctic and east Asian faunas. India's dragonfly fauna is still poorly known and there is a great need for basic information on the occurrence, distribution and seasonality of all species. This is so not least because dragonflies are active predators in the larval as well as adult stage but because their disappearance can give early indications of the degradation of aquatic habitats.

In much of India, mosquitoes transmit malaria, various types of filariasis and several viral diseases, including encephalitis (brain fever). In many habitats mosquito larvae are poorly a major source of food for the aquatic larvae of dragonflies, while adult mosquitoes, together with other medical and agricultural pests, are preyed on by winged adult dragonflies. However because they are opportunistic feeders they are unlikely to eradicate a pest completely although they may reduce its numbers. It is vital to know more about the biology of dragonflies, particularly those species which are reasonably abundant, and to study their life histories so that we can ensure that suitable and unpolluted habitats remain for their larvae and that the environment remains attractive to the adults. Untold damage can be done to dragonfly populations by the ill-considered use of insecticides, by the drainage of swamps, or by the removal of trees and bushes which are essential for adults. Such practices may greatly increase in the numbers of mosquitoes and other pests.

For identification of Indian species we still depend largely on the pioneering works of Fraser (Fraser, 1933-36. Fauna of British India, Dragonflies, Vols 1-3. Taylor & Francis, London) although these are somewhat out-of-date. Today there are many active biologists carrying out research on dragonflies in Indian universities and certain other specific institutions both governmental and otherwise.

CLASSIFICATION

Dragonflies (*i.e.*, all members of the order Odonata) are divided into three sub-orders. The Zygoptera (the damselflies) are easily distinguished by the similar shapes of their fore and hind wings (Fig. 1), by the side separation of their eyes, and by their small slender shape and relatively week flight. In contrast the Anisoptera, (sometimes thought of as the true dragonflies) have fore and hind wings of different shapes, eyes either touching each other or at least not well separated, and normally a much larger and more robust form capable of powerful flight. The Anisozygoptera with only 2 living representatives (*Epiophlebia* spp.) form an evolutionarily fascinating intermediate between the first two sub-orders. They combine the wings of a zygopteran with the body plan of a gomphid anisopteran. One of the two species is to be found in the Himalayas, the other in Japan.

STRUCTURE

Larvae

Almost all dragonflies have aquatic larvae. They are active predators and inhabit a great variety of aquatic habitats. One of their most striking features is the extensible labium, the mask, which can be shot out to grasp prey. Prey is detected visually in some species and by tactile means in others.

Dragonflies larvae depend on dissolved oxygen and not need to visit the surface for respiration. Zygopteran larvae respire by means of 3 external gills at the end of the abdomen. Anisopteran larvae do so by gills which lie inside the rectum, water being pumped in and out to irrigate them. They can also use the rectal pump as a means of jet propulsion and this enables them to swim very rapidly.

Adults

Adult dragonflies are highly adapted for an active aerial life and most of their behaviour takes place on the wing. They have very large compound eyes which may each contain 30,000 ommatidia (separate visual unite) in some large aeshnids. There are also 3 prominent simple eyes on top of the head, each with a large single lens which are probably important for stabilizing flight. The head bears a powerful set of mouthparts, capable of giving quite a painful bite to the handler in larger species. Dragonflies possess no sting, nor so they use poisons in their bites although they are capable of chewing up large and powerful prey.

Dragonfly wings have a prominent pterostigma (Fig. 1), usually seen as a small dark area towards the wing tip, which probably has an aerodynamic function. There is also a nodus or break in the leading with margin which may give increased flexibility. In flight the fore and hind wings usually beat more or less alternately, but when extra power is needed in some species they can beat in synchrony. In a large aeshnid they may beat 30-40 times a second, but in small zygopterans the rare in slower.

They thorax slopes backwards and the legs are therefore placed anteriorly, a position which helps prey capture on the wing. Most Anisoptera perch with their wings more or less parallel, but many libellulids flex their wings more or less parallel, but many libellulids flex their wings forwards and downwards when perched during the day. Zygoptera, however, usually perch with the wings folded along the abdomen, except for lestids which commonly leave their wings open.

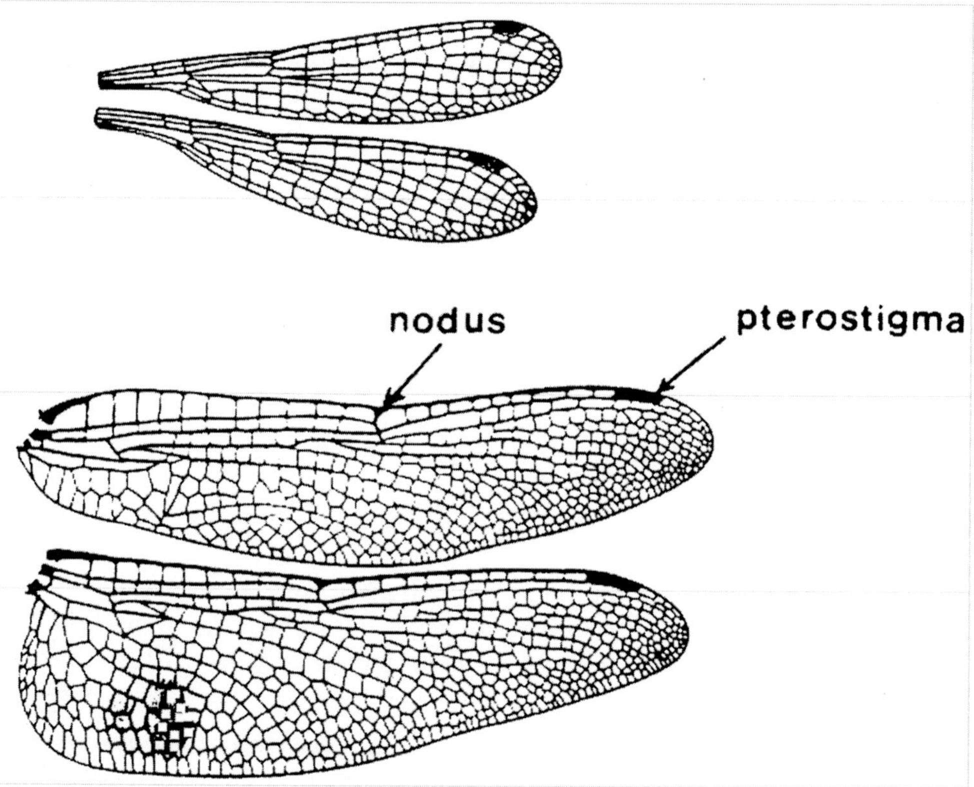

Fig. 1. Fore and hind wings of a zygopteran damselfly (above) and of a anisopteran dragonfly (below).

In females the genital opening is on the 8th segment. In Zygoptera and Aeshnidae there is a large obvipositor which is used for inserting eggs into plant tissue. Males are unusual in possessing a second set of external genitalia: the primary genital opening is on the 9th segment and sperm are transferred to the temporary sperm store on the third abdominal segment by bending up the abdomen shortly before copulation. The second segment of Zygoptera bears the penis, but in Anisoptera the penis is formed by an extension of the store on the 3rd segment. Why male dragonflies should have a secondary set of genitalia is not understood although many hpotheses have been proposed. One suggestion is that it allows the male to hold the female in tandem with the terminal claspers during copulation.

EMERGENCE FROM THE WATER

When mature the aquatic larva seeks a suitable support on which to climb out of the water. Anisopteran larvae may leave the water shortly before dawn but zygopteran larvae usually do so during the day. The larva sheds its skin (exuviae) and within 1-2 hours the adult dragonfly has expanded its body and wings, hardened its cuticle and is ready for flight. The teneral adult, recognizable by its pale colours, shiny wings and weak flight, then flies off, usually deprting from the water. By collecting exuviae useful information about those species which breed in a particular habitat can be obtained, provided they can be identified. Keeping larvae in an aquarium so as to obtain the emerging adults is the best way to identify difficult larvae.

After emergence most dragonflies remain sexually immature for a week or more. During this period they usually leave the vicinity of the water thereby preventing interactions with mature individuals. They feed voraciously and as they do so the cuticle thickens, the flight muscles develop and the gonads become mature. In many species the males become brightly coloured while the females become darker.

REPRODUCTINVE BEHAVIOUR

The reproductive behaviour of dragonflies has attracted the attention of biologists for many years. It is surprisingly flexible and the same species may behave differently at different times or in different habitats.

Once mature, males return to the water where they may adopt territories, areas of the habitat usually containing suitable oviposition sites which they defend against conspecific males and where they expect to meet and mate with females (e.g. *Brachythemis contaminata)*. However this behaviour may be dependent on the population density and many species are not normally territorial although mature males may interact aggressively at the breeding sites (e.g. *Anax guttatus; Tholymis tillarga*).

When females approach the water they are likely to be seized by males either in flight or when settled; Courtship is known in some Odonata, but was not seen in nay campus species. Once seized, the male forms a tandem with the female, gripping her head (in most Anisoptera) or her prothorax (in Zygoptera) with his terminal claspers. The male then transfers sperm from his primary genitalia to the secondary genitalia. Copulation starts by adopting the wheel position in which the female swings her abdomen up so that her genital opening can meet the male's secondary genitalia. Copulation may take place entirely on the wing and last for just a few seconds (*e.g.* in *Tholymis tillarga*), or it may continue after perching and last for up to an hour or even more (e.g. in some aeshinds).

Although copulation usually occurs near water where the female may oviposit, sometimes males attempt to capture females while they are feeding elsewhere or while they are flying towards the water. Such males may have been unsuccessful at the water and their approaches are often resisted by females.

Males produce many millions of sperm and they try to ensure that their sperm are used to fertilize the eggs of as many females as possible. Females, however, produce fewer eggs which they attempt to lay in carefully chosen sites after they have been fertilized by the best males available. There may therefore be a difference of interest between males and females, males being 'in a hurry', but females being more 'cautious and selective'. This hypothesis may help to explain some features of dragonfly behaviour.

Males compete with one another to mate with females and also subsequently to fertilize their eggs. How does the latter come about? When females store sperm for long periods and accept several mates, as is known in many animals, 'sperm competition' can occur. For example, the male in some dragonfly species attempts to place his sperm in the most advantageous position in the female so that it is the first to be use for fertilization. In other species, however, the male first scoops out some or all of the sperm of rivals from the female's storage organs, the bursa and spermathecae, before introducing his own sperm. He does this by using specialized hooks and bristles on the penis, and he can thus ensure that a female uses only his sperm until she mates again. After copulation a male commonly guards the female while she oviposits thereby reducing the likelihood that another male will displace his sperm. This may be done by remaining in tandem, as for example in *Ceriagrion coromandelianum* (Fig. 3), or by hovering above the ovipositing female and driving off intruding males, as in *Orthetrum pruinosum*.

DRAGONFLY HABITATS ON THE CAMPUS

Madurai is a city of more than 600,000 people situated in Tamil Nadu towards the southern tip of India (9°55'N and 78°07'E). The sea lies about 90 km to the east, and 210 km to the south is Cape Comorin, the southernmost tip of India. This foothills of the Cardamon hills rise about 40 km to the west and the Palani hills are only a little further to the north-west, both being extensions of the Western Ghats range. In the vicinity of Madurai many isolated rocky granitic hills rise about 100 m above the rich flat plain which is itself about 100 m above sea level.

The N.E. monsoon season usually provides central Tamil Nadu with plenty of rain between September and December, but the region is in the rain shadow of the western Ghats, and so receives little rain from the S.W. monsoon. The mean annual rainfall is only 30 cm. The mean mid-day temperature in December is about 30°C, but it rises to about 40°C in June and July.

The University lies about 14 km to the west of the city at the foot of a long granite ridge, the Nagamalai Hills ('Hills of Snakes') which run along its northern margin and separate the campus from the Vaigai river. It is situated in a highly cultivated region with numerous paddy fields, irrigation canals and reservoirs (tanks), some of which provide good habitats for dragonflies. The campus extends for about 1 km from east to west and it covers an area of 750 acres although not all this has yet been developed. It was founded in 1972, and many trees of various species were planted at that time so that today we can enjoy a green and shady environment which provides a refuge for much wildlife.

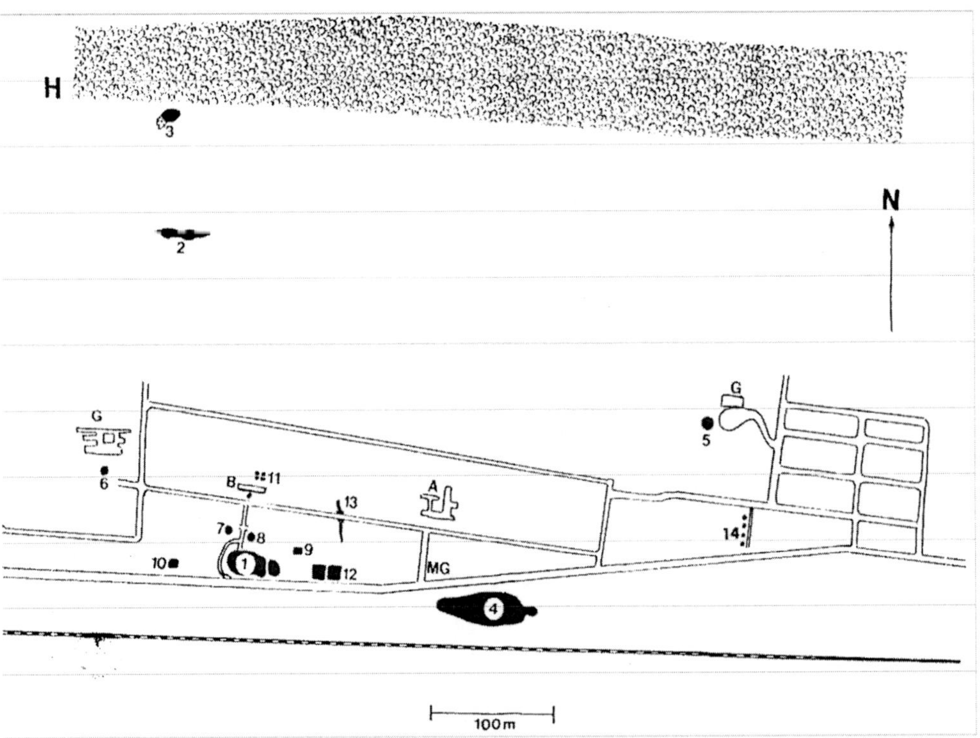

Fig. 2. Sketch map of the campus of Madurai Kamaraj University showing the location of some of the aquatic habitats where dragonflies occurred. (See Table I). A. Administration Buildings; B. Biology Department; G. Guest House; H. Hill of Snakes; MG. Main University Gate.

Fig. 2 shows a sketch map of the campus with a few of the dragonfly breeding sites indicated. The sites are listed in Table 1.

Table 1. Dragonfly breeding sites in the Madurai Kamaraj University campus.

S.No.	Location	Period when Water-Filled	Comments
1	Botanic garden lake	Sept – Jan	The most productive site. It contains no aquatic vegetation but has many small saplings and other plants which grow in the water.
2	Goat pond	Oct – Feb	Used by goats for drinking
3	Tree pond	Oct – Jan	Foot of hills, under a large tree
4	Village pond	Oct – March	A large pond, which remains water-filled until March because of irrigation. Much used for washing and by cattle, and very eutrophic.
5, 6	Guest House ponds	Permanent	One at Main and one at Faculty Guest House. Contain fish and aquatic vegetation, water lilies etc.
7, 8	Two Botanic garden ornamental ponds	Permanent	Both contain water lilies and other plants, and many fish. One is very shaded, the other is in the open.
9, 10	Two deep tanks: one in and one near Botanic garden	Permanent	Each is about 4 x 4 m, and the water level is ca. 3 m below ground level. Contain fish.
11	Biology School ponds. One is ornamental	Permanent	Several are used by the Dept. of Environmental Physiology for rearing animals.
12	Two reed-choked ponds in Botanic garden	Permanent	Fed by streams. Polluted
13	Sewage stream	Intermittent	Frequented by dragonflies but probably none breeds in it.
14	Four very small ornamental ponds in Horticultural garden	Permanent	Contain much aquatic vegetation
15	Artificial streams near buildings	Permanent	Overflows from the piped water system, *e.g.* by Biology labs: at least one species breeds in them.

There are probably other dragonfly-breeding sites on the campus, and altogether 250 mosquito-breeding sites have been identified.

Many of the dragonflies which appear on the campus, perhaps attracted there by its numerous trees, may breed elsewhere, for example in the Vaigai river, in nearby reservoirs or in paddy fields;

the presence of adults by water does not necessarily indicate that the species breeds there. Table 2 shows the estimated numbers of each species observed with evidence of breeding. Behavioural notes are given here for these species and in some cases comparisons are made with Kumar's (1972, 1985) results from Dehra Dun.

Table 2. The Occurrence and estimated abundance of Odonata on the campus of Madurai Kamaraj University. [Six categories of abundance were estimated during each visit to a habitat; 0 = 0; A = 1; B = 2-5; C = 5 – 10; D = 10 – 50; E = >50 Species seen to be reproductively active and ovipositing at a campus site......* Species for which proof of breeding was obtained (observation of larvae, exuviae or tenerals) ..... **]

Species	Abundance Oct - Feb	Evidence of reproductive activity	Site of activity (p = permanent) (t = temporary)
ZYGOPTERA			
Lestidae			
Lestes elatus Hagen	E	*	t
Coenagrionidae			
Ceriagrion coromandelianum (Fabricius)	D	**	p
Ischnura aurora aurora Brauer	C		
ANISOPTERA			
Gomphidae			
Ictinogomphus rapax (Rambur)	B	**	p (t?)
Paragomphus lineatus (Selys)	B		
Aeshnidae			
Anax guttatus (Burmeister)	C	**	t
Gynacantha dravida Lieftinck	B		
Hemianax ephippiger (Burmeister)	C	**	t
Libellvlidae			
Brachythemis contaminata (Fabricius)	D	**	t (p)
Bradinopyga geminata (Rambur)	E	**	p
Crocothemis s, servilia (Drury)	D	*	p, t
Diplacodes trivialis (Rambur)	E	**	t
Indithemis carnatica Fabricius	C	**	t
Orthetrum chrysis (Selys)	A		
Orthetrum pruinosum neglectum (Rambur)	A		
Orthetrum s. sabina (Drury)	E	**	p, t

Orthetrum taeniolatum (Schneider)	B	**	t (p?)
Pantala flavescens (Fabricius)	E		
Potamarcha congener (Rambur)	E	**	t
Rhyothemis v, variegata (Linnaeus)	A		
Tholymis tillarga (Fabricius)	E	**	t
Tramea basilaris burmeisteri Kirby	D	**	p, t
Tramea l, limbata (Desjardins)	C	*	p, t
Trithemis aurora (Burmeister)	D	**	p
Trithemis pallidinervis (Kirby)	C		
Zyxomma petiolatum (Rambur)	D	**	p

Irrigation comes to an end in March in most years, and the paddy fields are then dry until the following October. A prolonged dry season poses problems for all aquatic animals. Dragonflies may survive the dry season in one of several ways: they may lay drought-resistant eggs; their larvae may burrow into damp mud where they can aestivate (*i.e.* remain dormant) until the rains return. They may survive they dry season as long-lived adults, or finally the adults may migrate to wetter regions. On the campus the adults of some species appear all the year round either because they may utilize the permanent artificial habitats or because they have very long-lived adults which remain in a state of diapause (arrested development) until the rains return. Different species may use more than one strategy but there is an urgent need for more information about this in many Indian dragonflies.

In January 1988, as the Botanic garden lake dried out, its fauna became concentrated into the few remaining puddles. There was a remarkable concentration of several species of dragonfly larvae (Anax, Indothemis, Orthetrum, Ceriagrion *etc.*) as well as of aquatic water bugs (Heteroptera) such as Anisops, Ranatra, and Laccotrephes. However by the time the puddles had completely dried out all had vanished as the numerous paddy birds, kingfishers, mynahs, crows and other animals made the most of the banquet.

THE CAMPUS DRAGONFLY FAUNA

The dragonflies occurring on the campus belong to 3 anisopteran families and to two zygopteran families. The adults of twenty five species are described here most of which are abundant on the campus, many breeding there too.

Most of the common campus species are widely distributed throughout the warmer parts of the old World. Some occur throughout Africa and others in most of southern and eastern Asia even including Australia and New Zealand. One atleast is know also in the New World. The campus, representing a small and unnatural 'oasis' of trees in a heavily agricultural 'desert' attracts species which can disperse themselves effectively, some of them being regular migrants. In spite of their wide distribution, however, details of the biology of many of these common species are lacking. Not surprisingly, there is no localized endemic species with specialized habitat requirements on the campus.

MEASUREMENTS

Measurements of body length (which includes the head and the cerci), and of the hindwing are given for each species as a guide to identification. However not that all species show a range of size, some more than others, and the values are only an indication of approximate size. Where a greater range is given for hindwing than for body-length measurements, more data were available for the former.

ZYGOPTERA

The Zygoptera comprise a great variety of damselflies able to exploit numerous types of habitat. They were poorly represented on the campus only 3 species normally being present, although several vagrants were also seen.

LESTIDAE

Lestids are believed to be close to the ancestors of the Anisoptera on the basis of their wing venation. Many have shiny green metallic colouring and they commonly perch with open wings.

Lestes elatus Hagen

This species is common at many sites in southern India and it has also been found in Sri Lanka and in Thailand. On the campus it was seen at all times between October and January in the vicinity of the lake perching among grass or on bushes and it was also common on the hill side.

Body length: 42 mm.

Hindwing length: 21-24 mm.

The thorax has 2 prominent green metallic dorsal stripes, each of which becomes broader posteriorly, and a pair of brown lateral stripes. Abdominal segments are metallic green dorsally and pale brown below. The terminal abdominal segments are black. Males have some grey-blue pruinescence of the sides of the thorax and a small amount on the terminal segment of the abdomen. The strongly hooked claspers of the male are a good species-specific character.

BEHAVIOUR

Individuals were to be seen perched among grass or bushes, often in the shade and with their wings open. During December, pairs were sometimes seen in tandem ovipositing on twigs or grass stems usually about 10-20 cm above the water level, the eggs being inserted deep into the plant tissues. Some lestids lay eggs which have a long diapause (arrested development) during an unfavourable season and the eggs of *L. elatus* probably do not hatch until the following monsoon season.

Larvas and emerging adults were found abundantly in the lake during November and December, and Fraser has described the adults as being found in all months of the year.

COENAGRIONIDAE

This is a large, widespread and ancient family, which arose in the Cretaceous about 100 million years ago. It has representative species in almost every type of adonate habitat.

Ceriagrion coromandelianum (Fabricius)

Found in most parts of India and as far east as South Chine and Malaysia, *C. coromandelianum* was present on the campus throughout the period October – January being most abundant at the small well-shaded permanent Botanic garden pond among water lilies, but also usually present at the lake and some other ponds. The bright yellow colour of the male makes the species easy to identify.

Body length: 34 mm

Hindwing length: 18-20 mm.

The male has an alive green thorax and an entirely lemon-yellow abdomen. The female is duller and more robust, having a brownish thorax and green-to-gold abdomen.

BEHAVIOUR

Individuals perched low on vegetation making periodic swift flights in pursuit of prey. They were also sometimes seen to fly at immobile prey on vegetation, or sometimes at dark marks presumably mistaken for prey. They were reproductively active throughout the observation period. Copulation lasted about 20 min and pairs oviposited in tandem among water lily leaves with the female partially submerged (Fig. 3). It was noted by Kumar (1972) as a pond and stream breader and its reproductive behaviour has been discussed by Srivastava & Suri Babu (1985) and Suri Babu & Srivastava (1990) who observed four generations per year in Madhya Pradesh in central India.

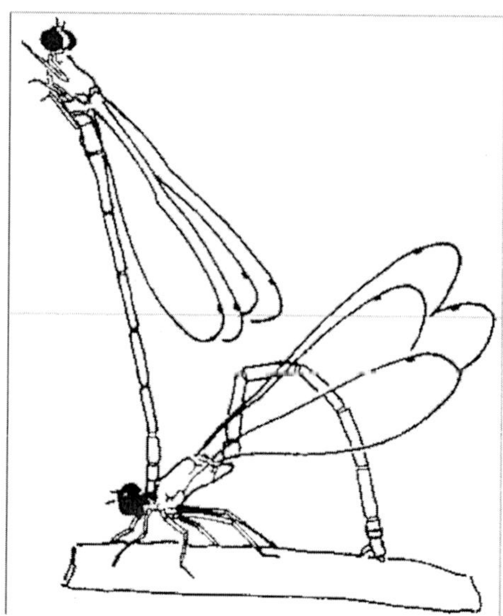

Fig. 5. Ceriagrion coromandelianum ovipositing. The male clasps the female by the prothorax

An unidentified zygopteran, perhaps this species, was seen in Meenakshi temple in Madurai city where, well after dark, adults were flying around fluorescent light strips, feeding on the small flies attracted to the lights. Perhaps they bred in the sacred lake.

Ischnura aurora Brauer

This tiny delicate species has a remarkably wide distribution being found in Japan, Australia, New Zealand, New Guinea, Fiji, French Polynesia and West Samoa as well as in mainland China and other parts of east Asia. In every region it occurs in almost the same form. In India it is widespread, being found in the plains and up to altitudes of more than 2000 m. On the campus it was present intermittently between October and January, but was never abundant. However it occurred commonly at the Vaigai river from where campus individuals may have come. Its yellow blue-tipped abdomen make it easy to identify.

Body length: 13 mm.

Hindwing leghth: 10-15 mm.

The mature male has a bronze-black thorax and lemon-yellow abdomen with segments 8-10 being light blue. The female is stouter and less brightly coloured with a green thorax and a large black dorsal band running the whole length of the abdomen above yellow sides.

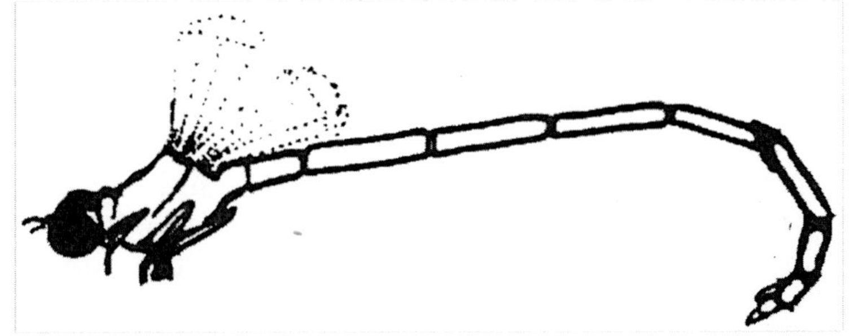

Fig. 4. Ischnura aurora female giving refusal signal to a male

BEHAVIOUR

Reproductive behaviour was not observed on the campus, but at Periyar lake (in the Cardamom Hills) I saw males congregated in large numbers at small ponds and in the surrounding vegetation with many tandem and copulating pairs. A very interesting feature of this species, first reported by Rowe in New Zealand, is that males form tandems and mate with teneral females shortly after the females have emerged. Many instances of this were witnessed at Periyar. Teneral females could be identified from their soft pale brown abdomens and their shiny, sometimes crumpled wings. Males were seen *in tandem* with females periodically inviting copulation by flexing the abdomen and opening the wings, but teneral females were sometimes reluctant perhaps because their soft cuticle was not yet sufficiently hard. During copulation, which lasted about 30 min, movements associated with sperm removal in other Ischnura spp were seen. Competition between males may have caused them to mate with teneral females even though those contained no rips egg. Mature females were seen to refuse to mate by signaling with a downturned abdomen (Fig. 4). It would be of great interest to know if teneral females accept more than one mate.

In New Zealand, *I. aurora* is known for possessing monogamous females which accept mating while teneral (Rowe, 1978, 1987), behaviour which has also been observed in the Periyar National Park, Kerala (Mathavan & Miller, 1989).

ANISOPTERA

The Anisoptera comprise a greater variety of dragonflies breeching in diverse habitats. They are richly represented on the coupons with close to two dozen species.

GOMPHIDAE

Gomphids belong to an ancient family represented in the Jurassic (140-170 millions years ago). India has over 70 species of gomphid, but several of these are now uncommon or rare. Many gomphids are stream and river-breeding species but a few breed in lakes and ponds. Gomphids have well separated eyes, short legs and narrow wings, usually shorter than the abdomen. They are commonly marked black and yellow and the sexes are similar, but in the males the posterior basal region of the hindwings is strongly angled, while the first abdominal segment bears a pair of lateral ear-like auricles, features similar to those in Gynacantha. The function of both is unknown. The hindwings of the female are rounded and there is no visible ovipositor.

Only two species of gomphid were seen on the campus and neither was common.

Ictinogomphus rapax (Rambur)

This large and handsome species is described by Fraser as occurring widely in southern India in a variety of habitats. It is also found in many parts of south-east Asia as far as China. It is a large 'ferocious' looking gomphid which cannot be mistaken for any other species on the campus. It occurred in October and November, occasionally in December and not at all in January, although mature larvae were found in the lake in January.

Body length: 70 mm

Hindwing length: 40 mm.

The large thorax has broad yellow stripes on a black background and the slender abdomen is banded with black and yellow. The anterior 2 segments of the abdomen are considerably swollen: segments 7-9 are also slightly swollen, segment 8 bearing foliations (downwardly extended flaps) (Fig. 5B). The wings are clear but may become slightly clouded in older individuals.

The larva is very flattened and rounded (characteristic of the genus) and after being handled it shows 'thanatotic' behaviour for several minutes remaining motionless as though dead even when returned to the water.

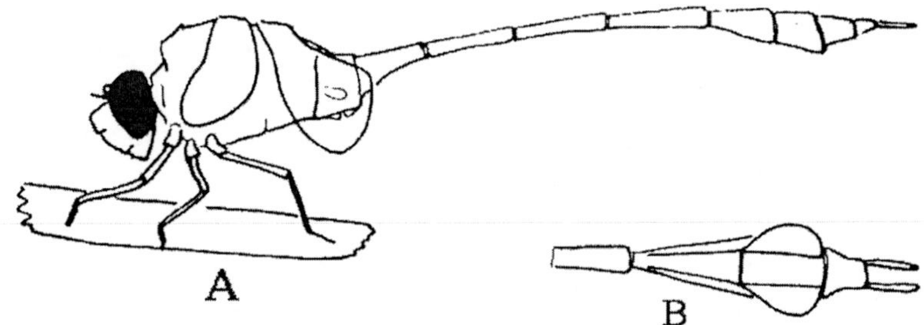

Fig. 5. Ictinogompus napax: A, typical perching positing with abdomen slightly raised. B. the expanded foliations of segment 8

BEHAVIOUR

Males perched on tall reeds or sticks near or over the water for long periods without movement. They have a hunched look when perched and the abdomen is slightly elevated (Fig. 5A). They normally left their waterside perches about 2 hours before sunset. Intruders were challenged and, if male, a fast chase often ensued. Copulation has not been seen but in a similar African species (*I. ferox*) it lasts for only about 9 seconds and takes place in the air, both of which are unusual features for gomphids. Afterwards the female departs to oviposit alone as in most other gomphids.

Paragomphus lineatus Selys (Mesogomphus lineatus)

This is a common species in southern India and its range extends to Burma and westwards to Turkey. It was seen intermittently at the lake on the campus most commonly in October and also in January, but there is no evidence that it bred there. It is common in the Vaigai river from which it may periodically wander, and it is said to breed in both running and still waters. Many of the smaller gomphid species are difficult to distinguish, but in *P. lineatus* the posterior swollen orange segments of the male's abdomen make identification easier.

Body length: 49 mm.

Hindwing length: 24-27 mm.

The well separated eyes are grey above and creamy on the sides. The thorax bears strongly marked 'tiger' stripes of yellow and black and the slender abdomen is also black and yellow. The male abdomen has prominent auricles on the first segment, segments 8-10 are swollen and the 8th tergite is extended ventrally into foliations. At the end of the abdomen the dorsal claspers are held together and curve strongly downwards whereas the ventral claspers are short and curve upwards. The ventral surface of the last 3 abdominal segments in both sexes is bright orange while other segments are grey-brown ventrally. The female is a bulkier and larger insect with identical patterning on thorax and abdomen but no posterior foliations.

BEHAVIOUR

Individuals were observed perched low on vegetation from which they made periodic short flights. When held in the hand the male made ventrally and dorsally directed arching flexions of the abdomen, which caused the expanded terminal segments to give it a cobra-like appearance (Fig. 6). Its reproductive behaviour has not been observed.

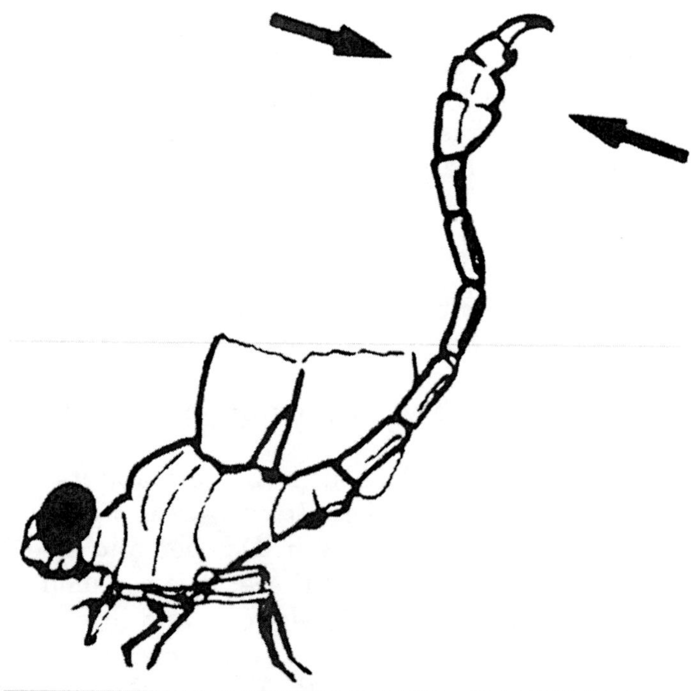

Fig. 6. Paragomphus lineatus: male (photo by Jill Silsby).

AESHNIDAE

Aeshnids belong to an ancient family going back to the Upper Jurassic (about 140 million years ago). Most are large hawker dragonflies which fly continuously for long periods. Females have large ovipositors and insert eggs into plant tissues or sometimes into mud. Males in some genera are more brightly coloured than females. The eyes have a large area of convergence and the abdomen is long and narrow.

Anax guttatus Burmeister

A. guttatus has a wide distribution occurring throughout south and east Asia and as far as Australia. It is also found in the Seychelles. It is the largest dragonfly likely to be seen on the campus where it was common throughout October to January. It was seen flying in all parts of the campus but most frequently in the vicinity of the lake, even after it had dried out.

Length: 80 – 85 mm

Hindwing length: 50-55 mm.

The hindwings of tenerals and mature insects have a large pale amber patch in the proximal half, a feature which varies in specimens from different parts of India. The wing shape of males and females is the same (*cf*. Gynacantha).

The thorax is slight green and the abdomen has a prominent blue mark on the dorsal side of segment 2. Segment 3 is white ventrally and other segments have orange and black markings.

The eyes which are very large and have a long confluent region are dark green above and yellow on the sides. The sexes are coloured similarly but females can be recognized from the prominent ovipositor on segment 8 and 9 while the secondary genitalia of the male are easily seen.

BEHAVIOUR

When not sexually active, adults spent much time resting high among tree branches with intermittent periods of patrolling and feeding. They often perched suspended under branches, sometimes upside down or aligned with the sun's rays during the hot part of the day.

At sunset there was usually an intense period of feeding activity with many individuals flying at tree-top height, sometimes interacting momentarily. No comparable period of activity at dawn was observed.

Individuals were found to be sexually active at intervals throughout the day in October and November but not sexual interactions were observed after mid-December although the species remained abundant in the area. Males searched intensively among vegetation in the lake for females. They did not establish territories but frequently interacted aggressively with other males. Aeshinds are believed to have excellent thermoregulatory powers and this may allow males to remain very active in the middle of the day without getting too hot.

Males seized females close to the water, formed a tandem and then flew swiftly towards the trees where copulation took place. Males commonly attacked pairs in tandem before or after copulation, occasionally forming a triple tandem (male-male-female) (Fig. 7A) or knocking a pair into the water. Oviposition took place into vegetation (mainly into the trunks of small saplings) above the water surface in October, but more often at or below the water surface in November, the eggs being thrust deep into the plant tissue. It would be interesting to know if the choice of oviposition site always changed consistently in this way during the wet season.

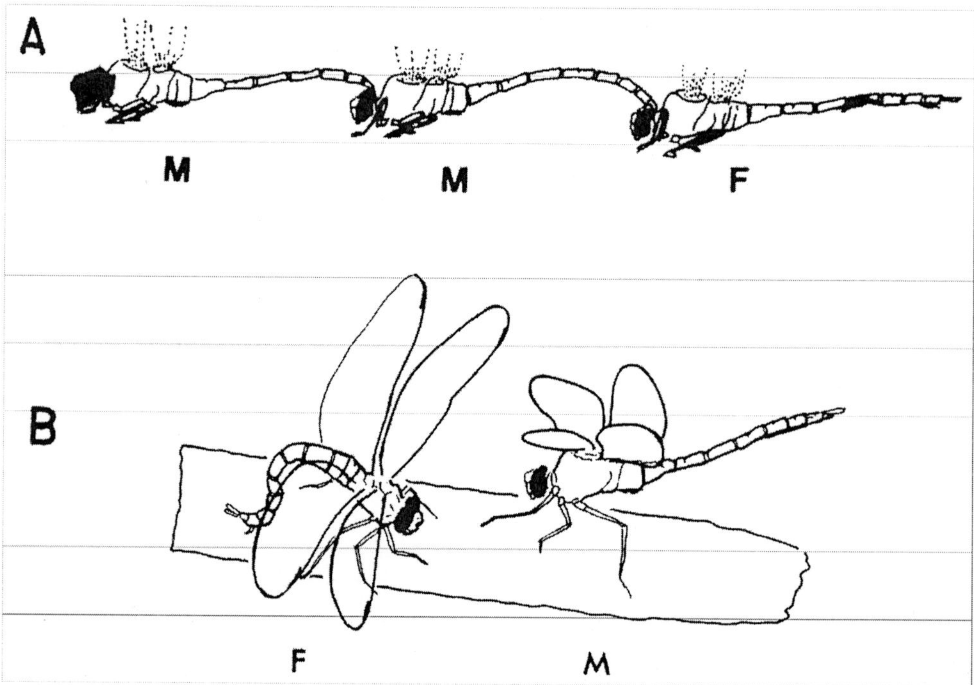

Fig. 7. Anax guttatus: A. in triple connection with 2 males (M) leading one female (F). B. a male guarding a female during oviposition

Ovipositing females were commonly guarded by males which remained in tandem with them, or less frequently by the male perching close to the ovipositing female, following her from one site to another and driving off other males which approached" on one occasion such 'non-contact guarding', which in unusual in the Aeshnidae, persisted for 18 minutes (Fig. 7B). Late in the evening, after sunset, when most males had departed, many females came to the water and oviposited without male guarding. This enables them to avoid harassment by males, and it may also have reduced predation.

A captive female, imprisoned in a tank with some water, oviposited into a block of polystyrene foam in which the eggs were easily located and the larvas hatched 8-10 days later. The first-instar larvae are strikingly banded as in other Anax spp. Larvae of all stages were found abundantly in the lake during the 4 months until it dried out. Mature larvae were sometimes preyed on by small kingfishers. Many exuviae were found in January.

Hemianax ephippiger **(Burmeister)**

This species which occurs throughout much of Africa and southern Europe is widespread in Indian and it is a well-known migrant. It is similar in colour to *A. guttatus*, but considerable smaller. It was abundant and reproductively active at the lake during October. Adults started to emerge from the lake in December and many were then found on the nearby hill. A few reproductively active pairs reappeared at the lake early in January before it dried out.

Body length: 62-65 mm

Hindwing length: 44-47 mm.

The hind wings have a pale amber patch on the proximal half which varies in intensity and tends to disappear in mature individuals. The thorax is browner than in *A. guttatus* and the abdominal markings including the blue spot at the base of the abdomen are less prominent.

BEHAVIOUR

Reproductive behaviour was similar to that of *A. guttatus* with males searching for females over the lake. Copulation took place perched on trees and lasted in one case for 31 min. Females oviposited into the trunks of small saplings at and below water level with the pair remaining in tandem. In Africa this species has been found to complete larval development in 100 days.

Gynacantha dravida **Lieftinck (=** ***hyalina*****)**

G. dravida is a large dull brownish aeshnid which occurs throughout much of India and is also found in many parts of east Asis.

Its large wings and thin delicate appearance make in unmistakable among campus species.

Length: 68-72 mm

Hind wing: 46-50 mm

The head bears large dark brown eyes and a black T-shaped mark on the frons. The thorax is light brown with dark brown stripes and is much smaller than in Anax while the large wings are much broader and more rounded at the tips (Fig. 8). They are transparent in young specimens but may become brownish in older individuals. The pterostigma is pale. The hind wings of the male are angled in the proximal posterior region, but they are smoothly rounded in the female. The abdomen is long and slender, and has dark and light brown markings. In both sexes there is a marked narrowing of the abdomen in segment 3. The male has a prominent pair of auricles on the 1st segment. The cerci of the female are long and delicate and are usually broken off in mature specimens, probably during oviposition or mating.

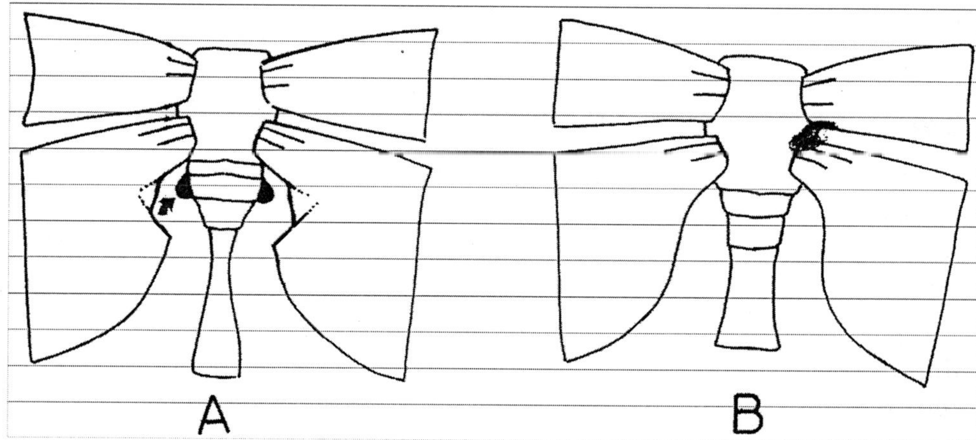

Fig. 8. Gynacantha dravida: A. male, showing the angled hind wings and the auricles (arrow) on first abdominal segment. B. female, showing round hind wings and no auricles.

BEHAVIOUR

Gynacantha spp. are well know to be active only around dawn and dusk when the light intensity is low, and occasionally in shaded areas of forests at other times of the day. The were commonly caught at lights and occasionally seen hawking low in the botanic garden after dusk, at the same time at which *A. guttatus* was patrolling at tree-top height.

No reproductive activity was observed. Elsewhere it has been suggested that Gynacantha spp. remain in forested regions throughout the dry season and become sexually active at the onset of the rains breeding in forest streams or ponds. Fraser describes some Indian species turning green and blue with sexual maturation.

There is little known about the behaviour of this species and observations on its reproductive behaviour would be valuable.

LIBELLULIDAE

This is the most recent dragonfly family appearing only in Miocene deposits some 15-25 million years ago. Libellulids are very varied in habits and habitats. They are typically smaller than aeshnids with short, stout abdomens and no visible ovipositor. Mature males are commonly brightly coloured and may develop a pruinescence (a waxy bloom) on the body. Many of them make short flights from perches and hence are known as darters or chasers. Some, however, such as Pantala, Tramea, Zyxomma and Tholymis fly continuously when active.

Most campus dragonflies belong to this family.

Brachythemis contaminata (Fabricius)

This species is common throughout India and in many parts of eastern Asia. Its bright orange colouration makes the male easily recognized.

Males and females were found throughout the period from October to January on the campus, but they were never very abundant. A few were normally to be seen at the Botanic garden lake, and rather more at the village pond.

Body length: 30-32 mm

Hindwing length: 22-28 mm.

The thorax and abdomen are thickest and in mature males they are uniformly orange. Immature individuals are yellow to brown with small black markings, and they have a yellow pterostigma and a pale orange patch on the wings. In mature males the wing patch becomes deeper orange and extends nearly as far as the pterostigma which turns bright red. Mature females become a dun-brown colour.

BEHAVIOUR

During the day both sexes perched low on vegetation or on the ground in the vicinity of water, sometimes over water, making occasional feeding flights. After disturbance they took off and sometimes turned to fact the intruder while hovering, a behaviour pattern resembling that of *B. leucosticta* in Africa where several may follow a mammal as it walks through grass or swamps, catching some of the small insects which are disturbed. It would be interesting to know to what extent this behaviour occurs in *B. contaminata.*

Sexual activity was not seen until 2-2.5 hours before sunset. Males perched on vegetation at the side of the lake or pond and defended territories against other males, returning repeatedly to the same perch. Two males were sometimes seen to interact by hovering in parallel, facing the same way: they gradually approached each other until their wing tips touched where upon a high-speed chase ensued. Copulation took place on the wing and usually lasted less than 30 s. Females oviposited by making a sequence of small dipping movements at one place, gluing their eggs to floating or upright vegetation at the water surface while being guarded by a hovering male. Larvae were found abundantly in pools in the river Vaigai. They have the habit of burrowing rapidly into sand when disturbed, quickly disappearing from view.

Bradinopyga geminata (Rambur)

B. geminata is found throughout lowland India and it has also been described from Thailand. Its grey-black colouring, bicoloured pterostigma and habit of perching on walls or rocks make it unmistakable. It was common on the campus, probably breeding in many of the permanent aquatic sites.

Body length: 41 mm

Hind wing length: 33 mm

Males and females are very similar in appearance and have the same colouring. The eyes are dark grey on top and pale grey on the sides. The thorax has a white ground colour thickly sprinkled with fine black spots making it look grey. The parallel-sided abdomen has a mottled black and white pattern which again gives it a grey-black look. The wings are clear and the distinctive pterostigma has a black central patch between two white bands.

BEHAVIOUR

Throughout the day both sexes could be found perched on rocks, the walls of buildings and on tarred roads where their cryptic colouring made them difficult to see. The never perched on vegetation. They were very alert and flew rapidly and low when alarmed. They were commonly seen on the walls of the deep tanks, usually choosing shaded areas in the middle of the day and sunlit patches early or late. At Dehra Dun, Kumar (1972) describes it as breeding specifically in cemented permanent tanks and as being on the wing from February to December.

Sexually active males were normally to be seen on the low stone walls of ornamental ponds where they would frequently engage in chases with other males. Near mid-day they would elevate their abdomen pointing it at the sun (the 'obelisk' position), probably as a means of preventing overheating (Fig. 9).

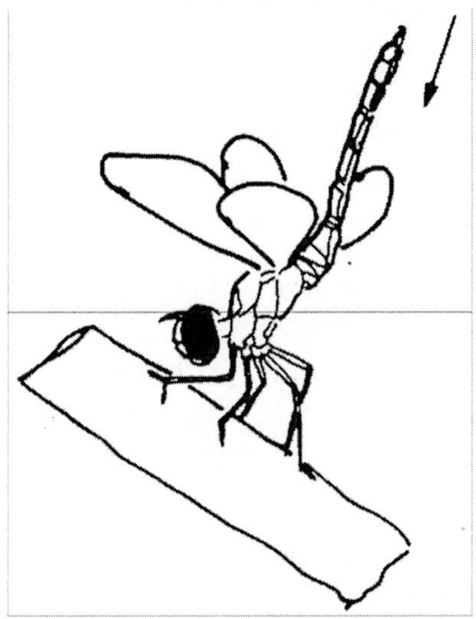

Fig. 9. Bradinopyga geminata in the 'obelisk' position in which it is parallel to the sun's radiation (arrow).

Copulation lasted only 5-10 s and was completed with the pair perched beside the water. The ovipositing female sometimes made a series of up to 20 tentative dips about 10 cm above the water before she descended and dipped into the water releasing a large batch of eggs usually in the vicinity of water lilies or other aquatic vegetation.

Crocothemis servilia (Drury)

C. servilia occurs as a number of subspecies in different parts of its range which extends from Egypt and Madagascar (Malagasy) through much of south and eastern Asia to Japan, the Philippines and Taiwan: it has recently been introduced into the United States of America. The form which occurs in India, *C. servilia servilia*, is the most widespread. The bright red thorax and abdomen of the male make it easily recognized on the campus where it was abundant from October to January, occurring both at the temporary lake and at most of the permanent ponds including the deep tanks. Many emerging adults were found in December. In Dehra Dun it is a multivoltine species breeding in temporary and permanent habitats (Kumar, 1972, 1985).

Body length: 41 mm

Hind wing length : 31 mm.

Tenerals of both sexes have a yellow thorax and abdomen with a creamy strips between the wings. Pale narrow stripes also run from the anterior base of the forewings towards the neck and there is a black line along the top of the abdomen. The thorax and abdomen of mature males are bright red. The wings are clear except for a saffron patch at the base of the hindwing, somewhat variable in size, and a small patch at the forewing base. The pterostigma is yellow.

BEHAVIOUR

Males and females were seen to perch low on vegetation or on the ground in the vicinity of the ponds throughout much of the day. Teneral males were also commonly able to perch close to or over the water without interacting with the sexually mature males. They provide an exception to the rule that immature forms normally avoid water.

Mature males commonly perched on the perimeter stones of the ornamental ponds, 'obelisking' (*cf.* Fig. 9) in hot sunlight and making frequent sallies at other males. Copulation was brief, lasting only a few seconds and was usually completed after perching. At the end of copulation the female sometimes continued to perch and she might then be quickly abandoned by the male. During oviposition the female dipped repeatedly in the vicinity of water plants while the male normally hovered above on guard. Occasionally females were seen to make dipping movements over dew-laden short grass apparently mistaking it for water.

Some mature males perched 10-15 m from the pond and pursued females when they approached the water, although they normally did not succeed in catching them. These males may have been unable to compete successfully with others which had already obtained a territory at the water or they may have waited until a resident male departed. In a related species *C. erythraea*, mature males commonly spend only 1-2 hours at the water although this depends on the density. By marking male *C. servilia*, it would be possible to seen if the same was true for this species.

Diplacodes trivialis **Rambur**

This is the smallest libellulid on the campus and it is not likely to be mistaken for any other species. It was described by Fraser as one of the commonest species in India. It also occurs throughout east Asia and in Australia. It was abundant from October to January at many sites on the campus and on the slopes of the hill. Few were sexually active and most individuals were seen away from water.

Body length: 27-33 mm

Hindwing length: 20-24 mm

The immatures and females are coloured yellow and black and the cerci are white. The eyes are brown above and yellow-green at the sides. The thorax is yellow with darker markings and a scattering of small black spots. Mature males develop a light blue pruinescence over thorax and abdomen. The abdomen is thin and segments 2 and 3 are only slightly swollen. The wings are clear except for a tiny saffron patch at the hindwing base and the pterostigma is brown or black. Immature *D. trivialis* and *Indothemis carnatica* might be confused, but apart from its larger size the latter may be distinguished by its purple-black frons (region in front of the eyes).

BEHAVIOUR

Most individuals seen had the cryptic immature or female colouration. The perched on very low vegetation or on the ground and flew only a few cm above the ground, making swift darting flights. This made them easily overlooked.

Mature males sat on low vegetation over the lake where they waited for females. Copulation started in the air and was sometimes completed after perching. It lasted from a few seconds to 1 minute. Males also encountered females elsewhere and copulation was observed up to 50m from water.

Females oviposited by flying low and fast over the water making quick dipping movements, followed 10 cm behind by a guarding male. At the lake, ovipositing females were not seen to select specific sites for oviposition. Larvae and exuviae, distinguishable by their small size, were found abundantly at the lake. Pandian & Mathavan have found that larval development takes 135 days at 32 C. It would be of interest to know more about the occurrence and activity of this species during the dry season.

Indothemis carnatica (= *caesia*) Fabricius

I. carnatica was present at the Botanic garden lake during October, became much more abundant during November and December but was then only rarely seen in January. It occurs in many parts of India and Sri Lanka, and has been recorded in Thailand.

The genus is close to Trithemis but separated by details of wing venation. On the campus it might be confused with *Trithemis festiva* (although this species was rare on the campus), but can be distinguished by its larger size and prominent white cerci, but cerci of *T. festiva* being black.

Body length: 37-39 mm

Hindwing length: 28-31 mm

Immature individuals are coloured bright yellow-green with black markings. Mature females are also bright green and black but become brownish when older. Mature males develop a dark grey, almost black pruinescence first over the thorax and then over the abdomen, through which the deeper black markings can just be distinguished. They appear uniformly black from a distance, except for the white cerci. The eyes are dark purple-brown above and grey on the sides. The frons is a shining metallic black. The wings are clear.

BEHAVIOUR

Near the Botanic garden lake they perched on low vegetation or on the ground while making intermittent flights in pursuit of prey. Sexually mature males also perched on low vegetation, seldom more than 10 cm above the ground or water surface. The were not seen at other aquatic sites on the campus. The black colour of the males coupled with their habit of perching near the ground probably helps them to absorb radiant heat efficiently.

The species was not territorial and few interactions between males were seen. Copulation took place after settling and lasted about 1 min. After copulation the pair flew in tandem round the lake several times before commencing oviposition, apparently searching for a suitable region (reminiscent of the behaviour of Sympetrum and Tramea spp). When ovipositing, they made a series of up to 20 small incipent dipping movements from a height of about 10 cm while hovering or moving forwards very slowly: they then descended and the female made a single contact with the water surface, releasing a large batch of eggs (Fig. 10). Egg batches which were released only 3 – 5 times per minute were deposited in the vicinity of emergent vegetation. Oviposited eggs were surrounded with a large mass of sticky jelly which caused the egg batch to remain united. Females were also sometimes observed ovipositing alone.

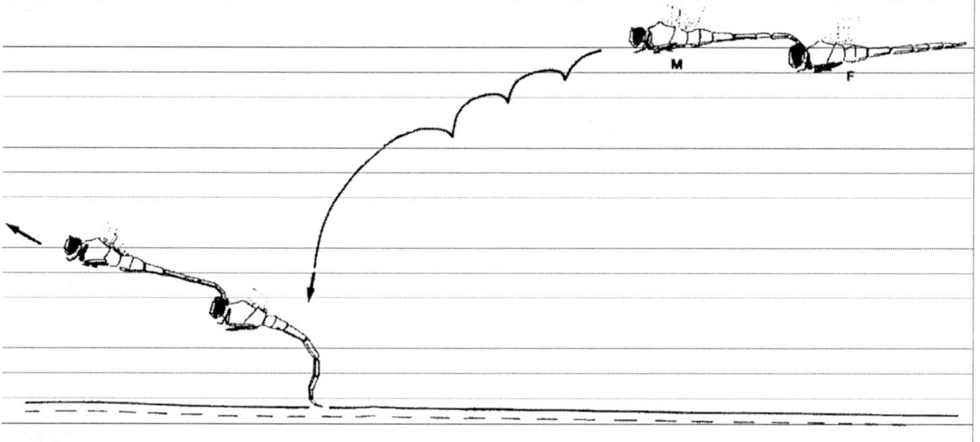

Fig. 10. Indothemis carnatica: pair in tandem oviposition, making a sequence of small dipping movements before the female touches the water surface and releases eggs.

The larvae were very numerous in the lake, many emerging in December and January. A large number of the larvae found stranded in puddles as the lake dried out proved to be of this species.

Orthetrum pruinosum Burmeister

O. pruinosum was seen only rarely on the campus but since it was seen elsewhere in the area between October and January it is included here. It has a widespread distribution in India and throughout south east Asia, occurring as several subspecies. It is found at up to 3000 m as well as in the plains.

Body length: 44-48 mm

Hindwing length: 33-37 mm.

The features which distinguish all Orthetrum spp are the dorsally enlarged and hairy pronotum, appearing like a furry collar, which is easily visible when viewed from the side, and the specialized secondary genitalia of the male in which the penis is equipped with a long bristle, the flagellum (not always easily seen in dry specimens).

Mature male *O. pruinosum* can be distinguished from other Orthetrum spp by their magenta (purplish red) abdomen and dark brown thorax with a grayish pruinescence. Male *Trithemis aurora* have an abdomen of the same colour but they are much smaller. The wings of both sexes are just tipped with brown at their extremities and there is a dark brown patch at the base of the male hind wings. The wings of old females become brownish.

BEHAVIOUR

Mature males, observed elsewhere, were highly territorial and aggressive towards each other, perching at the water's edge usually on the ground. Copulation took place on the ground or on low vegetation and lasted for several minutes. After copulation the female often remained perched while the male then hovered over her sometimes butting or ramming her and even landing on her abdomen a number of times (Fig. 11). Such aggressive behaviour sometimes persisted for up to 2 minutes before the female finally took off and commenced oviposition, with the male pushing her towards the water.

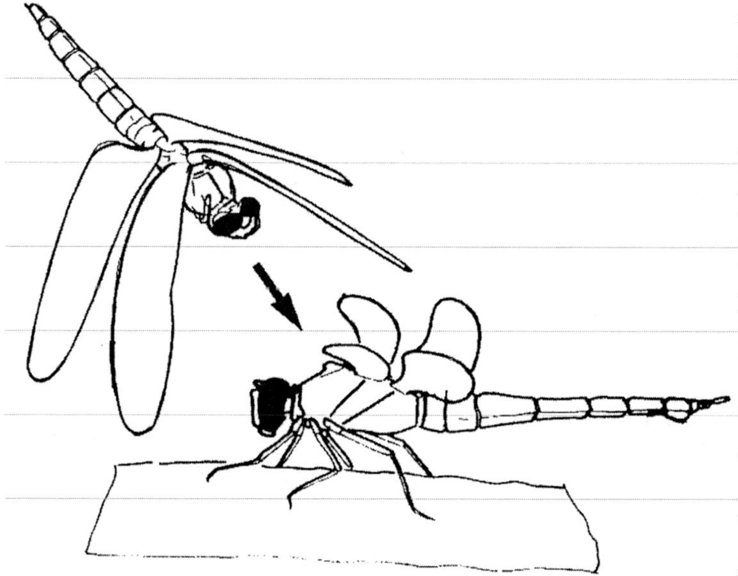

Fig. 11. Orthetrum pruinosum: After copulation the female commonly perches and the male in flight then rams her several times as though urging her to oviposit.

How is this behaviour to be explained? After copulation the male seems to urge the female to start ovipositing as soon as possible, but the female delays. By hastening oviposition, the male may increase the chances of obtaining a copulation with another female. The female however postpones oviposition possibly while she prepares eggs, or so as to test the male's willingness to remain with her and be an effective guard. In some other species, females sometimes delay oviposition until others are seen to oviposit, their presence indicating safety from predators. Ovipositing females sometimes suffer severe mortality from fish, frogs, spiders and water bugs as well as from terrestrial predators such as birds, and they are also often interrupted by intruding males. It would be interesting to know more about the significance of this behaviour.

During oviposition female *O. pruinosum* flicked droplets of water, probably containing eggs, towards or onto the bank, a type of oviposition also observed in *Potamarcha congener*. The tergite of the 8th segment in the female is extended ventrally on each side into two flaps (foliations) which may assist this activity. The male guarded very closely, sometimes with his wings hitting those of the female.

At Dehra Dun this species breeds in ponds and streams and adults appear in April-May and July-October (Kumar, 1972, 1985; *cf.* Mathavan, 1985).

Orthetrum sabina Drury

O. Sabina has a very wide distribution occurring throughout much of Africa and Asia and as far as Japan and Australia. It is an opportunistic species with larvae able to thrive in variety of habitats including both running and still waters. It was found to be present at many sites on and in the neighbourhood of the campus from October to the end of January. At the Botanic garden lake its numbers increased greatly in November and December but then declined in January as the lake dried up. However oviposition was seen to continue into the lake until 3 or 4 days before it was completely dry.

The very thin abdomen coloured with black and cream bands in both sexes make it easily recognized.

Body length: 45-53 mm

Hindwing length: 31-35 mm max. breadth: 10 mm.

The head has relatively small eyes with very little convergence in the mid-dorsal region and no marked dorsal "eyeshade". The Orthetrum character of the hairy 'collar' is easily seen. The thorax has black stripes on a yellow ground, while the abdomen is alternately black and creamy-white, darkening to pale yellow in older individuals. The 1st to 3rd abdominal segments in both sexes are much swollen whereas the remaining segments are very thin with a slight expansion of segments 7-10. *Zyxomma petiolatum* has an abdomen of similar shape but different in colour. In both species this may bring the center of gravity forwards and is perhaps an adaptation for increasing manoeuvrability.

The sexes are very similar in appearance, the mature male having no pruinescence, an unusual feature for an Orthetrum. The cerci are white.

The unmarked wings are long and narrow compared with species such as *Tholymis tillarga*. Such a wing shape may be associated with high manoeuvrability but poor gliding ability.

BEHAVIOUR

Many individuals were seen in all parts of the campus and on the hillside perching on low vegetation or on the ground. Similarly at the water, low perches were preferred, often being half way along a twig or stem. In the evenings they ceased activity about one hour before sunset and they were then sometimes found in trees 2-3 m above ground. They were seen to feed on smaller dragonflies several times.

Sexually mature males at the water did not defend localized sites although they sometimes interacted aggressively with each other. Copulation was always completed after settling, usually on vegetation or on the ground near water. Copulations normally lasted between 2 and 10 minutes, but occasionally they continued for up to 50 minutes. During copulation small rapid movements of the male's anterior segments could be seen.

What causes copulations to be so variable in duration? In some other Orthetrum species, long copulations have been correlated with the more complete removal rivals' sperm from the storage organs of females, while only a small amount is removed if copulation is short. Dissection of females of *O. sabina* caught during copulations suggests that the same is true in this species.

After copulation at the water the female normally remained perched, but the male then took off and flew in a circle round the female making small dipping movements towards the water as he did so (Fig. 12). He might then perch beside here, only to repeat the manoeuvre several more times. The dipping movements resembled those of an ovipositing female and might perhaps be used to encourage the female to commence oviposition. The similarity of male and female colouring may enhance this type of intersexual behavioural mimicry.

Females oviposited by dipping towards emergent or floating vegetation or towards the bank, rotating slowly about a point as they did so. Sometimes a droplet of water was seen to be flicked a short distance. Males normally guarded ovipositing females, flying above them and driving off intruding males, but many females oviposited alone sometimes doing so at a slower dipping rate and with more changes of site that when guarded. Confirmation of this observation would be valuable.

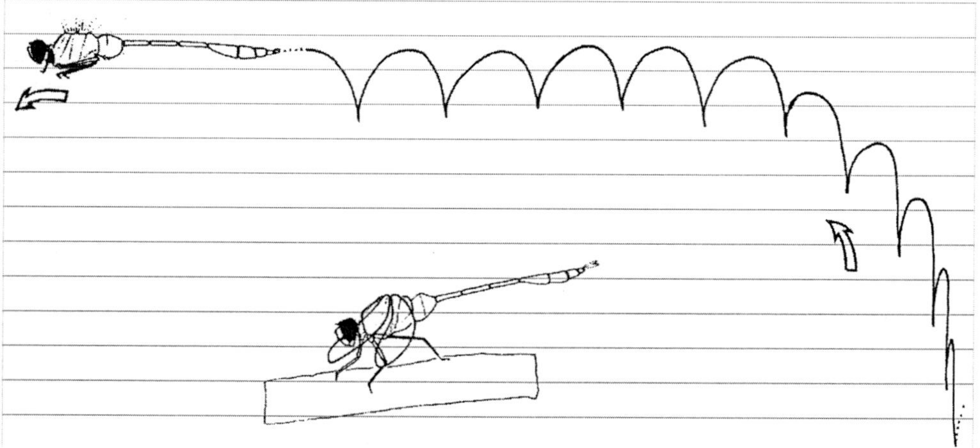

Fig. 12. Orthetrum Sabina: After copulation the female commonly perches and the male flies around her making a series of small dipping movements

Orthetrum taeniolatum Scheider

This species is widely distributed in India and westwards into North Africa and Southern Europe. It is the smallest Orthetrum sp to be seen on the campus and the blue colouration of mature males together with the hairy collar make them easily recognized, although some other Orthetrum spp, not seen at Madurai, are similar. It was not very abundant on the campus but it was probably breeding in a permanently running drain outside the Biology Department (dragonflies will suffer if plumbing maintenance improves). Adults were seen occasionally in November, December and January at several pond sites. Elsewhere it breeds in streams and ponds and in the deep pools of rivers, and adults have been found in most months of the year. In Dehra Dun it breeds in streams and adults are active from May to September (Kumar, 1972).

Body length: 39-41 mm

Hindwing length: 25-28 mm

The immature colouration is yellow-brown with black and cream stripes on the thorax and dark abdominal markings. In maturing males the abdomen first develops a light blue pruinescence which then spreads to the thorax. One male, observed in the same place for 10 successive days, was teneral on the first day, began to show some abdominal pruinescence on the 6th day and was fully pruinosed by the 10th day. The blue pruinescence of several libellulids has been shown to reflect ultraviolet radiation. Since dragonflies are able to detect ultra-violet, this may make them more visible to each other.

Mature females gradually turn a dull brown. The wings are clear but in old females they become darkened. The abdomen is gently tapering and segments 2 and 3 are only slightly broader than the remainder.

BEHAVIOUR

Both sexes were often found perching on walls or on the ground. Sexually active males were seen in the vicinity of the drain, where some had probably emerged, near the laboratory ponds, and at goat, tree and village ponds. As in other Orthetrum spp, copulation occurred on the ground lasting several minutes and subsequently males guarded during oviposition. There is a paucity of information about the detailed behaviour of this species, and further observation would be rewarding.

Pantala flavescens (Fabricius)

P. flavescens is one of the best known and most widely distributed libellulids, occurring in all the tropical and some temperate regions of the world. It is a regular migrant and it appears in very large numbers after heavy rain, drifting and sailing 3-15 m in the air. It was abundant in all 4 months, October – January.

Body length: 46 mm

Hindwing length: 39-41 mm.

The large eyes have red dorsal "eyeshades" and yellow-grey sides. Immature insects have a yellowish abdomen with black dorsal markings. In mature males the abdomen becomes red while in mature females it turns brown. The hindwings of the male have a brown tip whereas the female's hindwings have a saffron patch at the base. The hindwings in both sexes are broad at the base, a feature which reduces the wing loading and is probably an adaptation for gliding.

BEHAVIOUR

Large flocks of Pantala are one of the most familiar sights throughout the rice-growing areas of Tamilnadu although single individuals are also commonly seen. They tend to be concentrated where food is abundant, for example circling on the leeward side of trees when the wind is strong, scouting along city streets or at rubbish dumps where flies accumulate, wheeling in the air above fields during ploughing, flying low over water or bushes where midges swarm, circling round termite mounds when alates (winged forms) are emerging, diving among feeding elephants as they toss grass in the air, or sailing high above hills where thermals may lift small insects into the air.

In Africa they migrate twice a year across the equator following the wet weather associated with the Inter-Tropical Convergence zone. In America they migrate north into temperate regions during summer and the next generation returns in the autumn. They have been found flying at 6000 m over the Himalayas, and they have been known to reach New Zealand involving a non-stop flight of 2000 km. In India Pantala is said to migrate westwards during the north-west monsoon, but much uncertainty surrounds its behaviour and further study is much needed.

On the campus they were common every-where. About one hour after sunrises on several days in October many were seen flying steadily in a westward direction. In other months, however, no consistent movement was observed although they remained very abundant.

Throughout the period from October to January flocks contained males and females as well as mature and immature forms including tenerals with soft cuticle and shiny wings. Thus there seemed to be a continuous recruitment of young dragonflies.

At hot times of the day their flight was slow and leisurely, and they made long glides with only occasional wing flaps. This may have enable them to remain airborne without the danger of overheating. Late in the evening, in contrast, their flight was fast and erratic with many sharp turns, which may have helped them to avoid predators. They sometimes continued to feed until well after sunset and individuals commonly came to lights. They sometimes perched during the day, particularly during storms, always doing so vertically in bushes and trees.

Reproductive activity was not commonly seen after the end of November although their numbers remained high, but gravid females were found in all 4 months. Males sometimes became temporarily localized and defended small boides of water, even road puddles, where females were obtained. Copulation took place at such sites and also sometimes within the flock, although most dragonflies in the flock were not sexually active. It lasted up to 5 minutes and took place on the wing with the male actively flying and the female held above the male so that her stationary wings were horizontal (Fig. 13). The pair then either separated or, if near water, commenced oviposition in tandem. Oviposition occurred in a wide variety of small temporary bodies of water but not in permanent ponds possibly because of their vegetation.

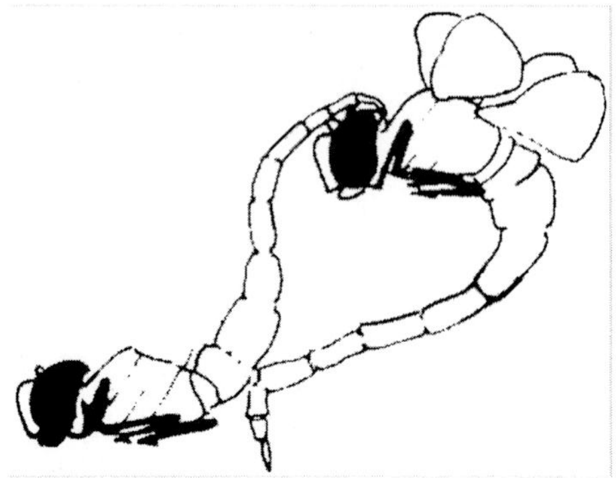

Fig. 13. A pair of Pantala flavescens in copulation in the air. The male flies actively while the female glides.

The many Pantala seen in the regional probably breed in the numerous paddy fields and their associated irrigation systems. At 35°C Pantala eggs hatch in about 3 days and can develop into adults in as little as 35-40 days, giving a minimal generation time of under 2 months. The long dry season is presumably overcome by emigration to wetter regions. Eggs can easily be obtained from a ripe female by repeatedly dipping her abdomen into water.

The contribution of Pantala towards the control of pests is probably considerable, and a knowledge of the biology of this species is particularly important. The excessive use of chemical pesticides in rice fields in some other parts of the world has greatly reduced the numbers of dragonflies which breed in them. It is essential to avoid similar mistakes in Tamil Nadu.

Potamarcha congener Rambur

P. congener occurs commonly in wooded regions throughout south and south-east Asia and as far as Australia. It was very abundant in many parts of the campus throughout the 4 months of observations. The blue pruinescence on the thorax and anterior segments of the abdomen make the male distinctive and easy to identify.

Body length: 43 mm

Hindwing length: 33-35 mm.

The relatively small eyes are dark brown dorsally and grey at the sides. The thorax and abdomen in immature individuals are yellow with prominent black markings. Mature males develop a grayish-blue pruinescence over the thorax which gradually spreads down to the 4th

or 5th segment. The posterior part of the abdomen remains black and yellow. In populations of the same species in the Pacific (e.g. Sulawesi), the whole abdomen may become pruinosed. In several libellulid species, old females come to resemble the males in colour and it has been proposed that this reduces their attractiveness to males and allows them to oviposit with less interruption. This may be so, for example in *Tholymis tillarga* (q.v.), but in p. congener, old females develop pruinescence over the whole abdomen and thus do not resemble males. In both sexes the yellow areas on thorax and abdomen become progressively obscured by dark pigmentation.

The 8th tergite of the female bears a pair of prominent 'foliations' or ventral extensions which probably assist oviposition. The abdomen of gravid females is noticeably plumper than that of immature females.

The wings are narrow like those of Orthetrum spp. and in both sexes they may be tipped with dark brown although this is somewhat variable. Immature individuals have a light brown colouration along the leading edges.

BEHAVIOUR

Throughout the day individuals perched on the ends of twigs, sticks, branches, or fences, usually 1-2 m above the ground, from which they periodically took off in pursuit of prey. They perched with the wings flexed forwards and downwards, a characteristic position of many libellulids. Considerable competition occurred for perches with males or females sometimes chasing each other off repeatedly. In the middle parts of the day they commonly perched in the shade or in sun-dappled areas.

In December and January several roosting clusters were observed in the evenings on the dead twigs of several small trees (*Syzygium cumini*) in the Botanic garden. About 1 hour before sunset individuals congregated in the vicinity of the roosting sites and then by a series of short flights they grouped themselves close together on the chosen twigs. By marking males and females I found that the same individuals returned to the same twigs night after night. Females outnumbered males at some of the roosts by 2 or 3 to 1, and as many as 100 individuals, all sexually immature, might be clustered close together. They seemed to be in a state of adult diapause (arrested development) perhaps waiting until suitable breeding conditions returned with the next wet season: the clustering may have been a form of anti-predator behaviour whereby they could alert one another to the approach of the predator, perhaps ants or geckoes, on the tree.

During October and early November, males adopted and defended territories along the banks of the Botanic garden lake. Some individually marked males were found to return repeatedly to the same territories and to use the same perches over a period of 3 weeks. When females appeared at the water they were usually seized by a territorial male and copulation took place while airborne lasting only 5-10 seconds with both partners actively flying (*cf.* Pantala).

Females oviposited facing towards the bank and flicking drops of water, each with a small batch of eggs, onto the bank (Fig. 14). The 'foliations' probably assist in propelling the eggs, as suggested in *O. pruinosum*. The female usually rotated about one spot thereby throwing her eggs along the bank in an arc. Eggs survive probably only if they fall among damp vegetation or into humid cracks in the ground. When kept damp in the laboratory, the embryos developed in about 5 days at 35°C but did not hatch until they were submerged in water. In nature eggs would probably hatch only after a rise in water level.

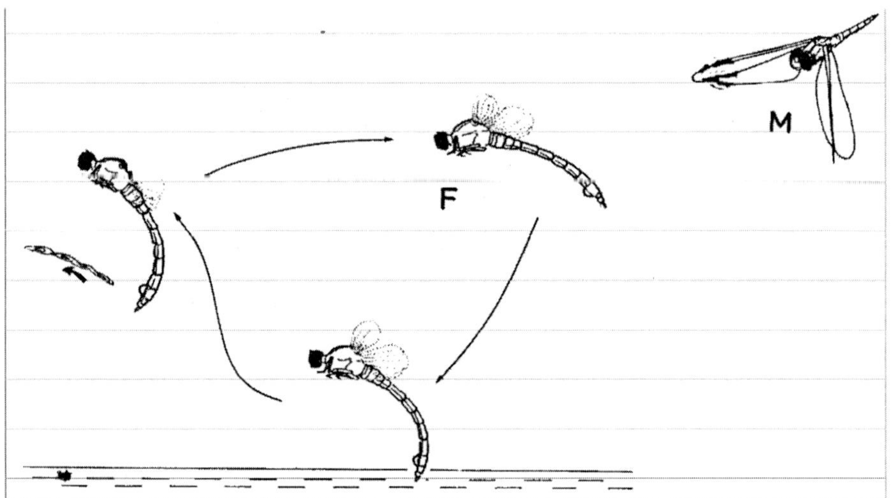

Fig. 14. A female Potomarcha congener (F) oviposits by scooping up droplets of water and flicking them onto the bank while a male (right, M) guards her.

During November the amount of sexual activity declined rapidly although the insects remained abundant with many newly emerged ones joining the roosts. Unlike *Orthetrum Sabina, Trithemis aurora* and *Crocothemis servilia, P. congener* seems to be incapable of breeding all they year round, and it may be finely adjusted to a cycle of intermittent breeding in temporary ponds which appear with the monsoon season. Gregarious roosting was observed during December and January with groups of 100 individuals assembling on the same branches each night (Joseph & Lahiri, 1989; Miller, 1989). Similar behaviour, but involving maximally only 7 individuals, was seen early in August. At Dehra Dun this species breeds in ponds and is on the wing from June to September (Kumar, 1972). However there is a need for more information on the dry-season behaviour of this species.

***Rhyothemis variegata* Linnaeus**

R. variegata occurs in much of India and also in parts of east Asia including Japan. Several subspecies are known. It was not common on the campus but occasionally was sighted in the vicinity of the Botanic garden lake where individuals resided for a few days. The wings with their unmistakable patterning make it easy to identify.

Body length: 32 mm

Hindwing length: 31 mm max. breadth: 15 mm.

The hindwings are very broad and their posterior borders reach 2/3rd of the way down the abdomen. The thorax and abdomen are relatively small and delicate. The frons and thorax have a greenish metallic shine, while the abdomen is more or less all matt black in mature individuals.

The wings of the male are pale yellow with isolated small brown patches in the basal, mid and tip regions. The wings of the female are more strongly marked and have much more extensive brown markings on a strong yellow background, except for the wing tips which are clear. It is unusual for females to have bolder wing markings than males.

The thorax is differently shaped from that in many other libellulids: the mesepisternum, the sloping region immediately in front of the wings is short, as it is also in Pantala, Tramea and Tholymis.

BEHAVIOUR

Members of this genus have a characteristically weak fluttery flight which may allow their strongly marked wings to be visible to conspecifics. Some species appear to be gregarious and flutter at tree top height in large groups, but this was not seen in R. variegata on the campus. Individuals perched during several days at the lake margins but no sexual behaviour was observed and the species apparently did not breed there.

***Tholymis tillarga* (Fabricius)**

T. tillarga has a very wide distribution which extends throughout much of Africa, south and east Asia and as far as Australia. It is found in many lowland and hilly regions of India. The dark and white wing spots of the male make identification easy. On the campus it was to be seen throughout October-January, being extremely abundant in November and early December but with declining numbers in January.

Body length: 43 mm

Hindwing length: 31-37 mm

In both sexes the eyes are a striking apple-green at the sides; the male has a reddish dorsal patch while in the female it is brown. Immature individuals are light brown with a prominent white stripe between the wings like that in some other immature libellulid species. Mature males have a red thorax and abdomen whereas mature females are brown.

The hind wings are broad and in both sexes there is a brown patch (stronger in males) which extends to the nodus. In males there is also a more distal white patch (due to a wax) which is invisible with transmitted light but shows up well with reflected light. Thus an ovipositing female might see only the brown patches of a male above her whereas a male looking down on another would see both the brown and white patches of his rival. Patrolling males with their white patches make a fine sight in the late evening. Old females come to resemble males in body colouring but they never develop white wing patches.

BEHAVIOUR

During the day individuals were commonly seen perched hanging vertically in shady parts of bushes in the vicinity of the lake. When disturbed they either immediately sought a similar place of refuge or they flew for several minutes in one area apparently feeding. Like *P. flavescens* they never perched horizontally. Many immature individuals usually started to fly shortly before sunset, doing so with a very jittery flight about 1 m above the ground or over the water. They were usually ignored or given only a cursory inspection by patrolling males which were apparently easily able to distinguish them from mature females. Mature and immature individuals were both occasionally found at lights.

The main periods of reproductive activity of this species are late in the day and at dawn. Mature males commenced to patrol at the Botanic garden lake, and also at ponds and the deep tanks, about 2-2 ½ h before sunset. When the population density was high in November, however, some males were active as much as 3 ½ h before sunset (at about 14.30 h). They continued to patrol without perching until 20-25 min after sunset, defending a smaller area as their numbers increased and perhaps becoming territorial under these conditions. Aggressive encounters were common, often ending in spiraling chases high in the air. Males appeared in the morning about 10 min before and continued to patrol actively until 20 min after sunrise.

Copulation lasted only 5-10 s and always took place on the wing. The pair often sought a sheltered place under a bush while copulating, thereby avoiding the attention of other males. Males guarded females during oviposition hovering close above them, but they were often unsuccessful and females were frequently seized by other males as soon as they started to oviposit. Thus a female might mate with several different males during one brief visit to the water.

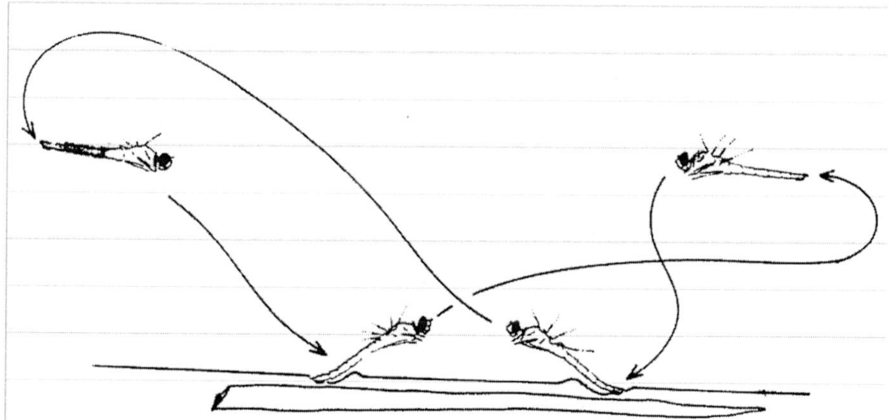

Fig. 15. Tholymis tillarga: The flight path of a female ovipositing onto a reed just below the water surface

Females arrived flying low over the water and quickly commenced to oviposit, but this usually attracted males and most bouts of unguarded oviposition were soon interrupted. Females attempted to avoid capture by rocketing several metres up into the air, after which they usually soon returned to the same oviposition site. Unreceptive females curled the abdomen downwards. During oviposition they repeatedly dipped onto pieces of floating reed, grass or twigs, releasing small batches of about 9 eggs each time which sometimes piled up on top of each other (Fig. 15). They preferred sites which were well concealed (to avoid male capture) but not too near the bank (where a drop of water level might strand their eggs). Many females were seen to arrive at the water and ovisposit about 30 min after sunset when most males had left and by when it was almost too dark to see them.

The eggs were coated in a substance which became sticky in water and then set into a hard porous mesh gluing the eggs firmly to vegetation. Eggs started to hatch after 5-6 days at 30°C. The first instar larvae have a black head and a black abdomen. This makes them resemble animated colons (:) as they swim about. They also have prominent horns on the head close to the eyes and above the antennae.

Tramea basilaris Palisot de Beauvois

T. basilaris has a very widespread distribution, occurring commonly in much of Africa, throughout India and eastwards to Vietnam, with two subspecies, one African and one Asian. The Indian subspecies is *T. basilaris burmeisteri.*

It was very abundant on the campus from October to January breeding extensively in the lake and remaining in the vicinity after the lake had dried out. It was also common on the hillside above the campus in November and December. Its distinctive hindwing marks make it easy to identify. Tramea spp occur world-wide and their form and habits are similar.

Body length: 52 mm

Hindwing length: 40-45 mm.

The eyes are very large and the frons is yellow. The thorax is greenish-yellow with black markings turning brown in mature individuals. The immature abdomen is yellow with black bands. In mature females it darkens and in males it turns orange-red. The hind wings are very broad, as in Pantala, and they bear two prominent dark red-brown patches near the base, somewhat variable in shape and surrounded and united by a yellow halo. The pterostigma in this and the next species is longer in the forewing than in the hindwing.

The male secondary genitalia are prominent when viewed from the side, and the female has an extended sub-genital plate below the 9th segment.

BEHAVIOUR

On the campus *T. basilaris* perched high on vegetation and were often to be seen at the tips of tall reeds, or high on bushes or trees. They commonly flew several metres above the ground, sailing on their broad wings during the heat of the day and sometimes among flocks of Pantala, perhaps migrating with them.

The larvae were abundant in the Botanic garden lake until it dried out, and many emerged during November and December. The adults became particularly abundant in late January and many females were seen perching in the Botanic Garden during the day in and near the dry lake bed. Gravid females ere found in all 4 month.

Between October and December, mature males patrolled or perched in the vicinity of the lake and several of the permanent ponds, but did not defend territories although they sometimes interacted aggressively with other males. Copulation took place with the pair perched in trees or bushes near or over the lake, and it lasted 5-10 min. The pair then flew in tandem circling the lake many times and giving chase vigorously to single males.

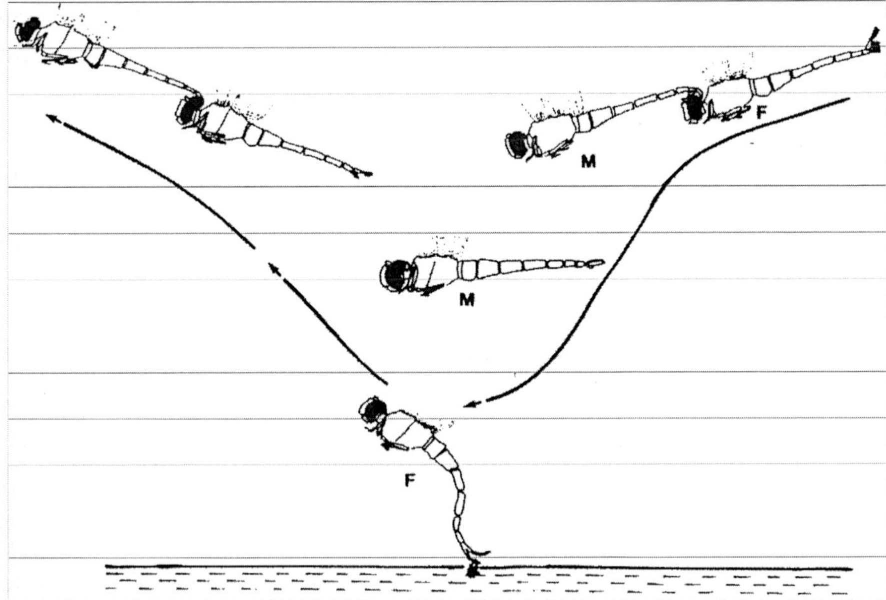

Fig. 16. Tramea basilaris: An ovipositing pair is shown with the female (F) momentarily released by the male (M) while she dips to the water surface

Oviposition took place in the vicinity of aquatic vegetation. The male released the female 10-20 cm above the water surface whereupon she made one dip to the water surface and was then reclasped in tandem by the male. The pair then sometimes made a further circuit of the lake before ovipositing again (Fig. 16). They remained in tandem for up to 25 min, spending much of that time circling the habitat and dipping sometimes less than once per minute (e.g. one pair made 6 dips in 10 minutes). While flying the female bent the 9th and 10th abdominal segments upwards and accumulated a large batch of eggs on her extended sub-genital plate. After their release the eggs were found to be surrounded by a mass of sticky jelly which probably caused them to adhere to aquatic vegetation. After oviposition the female rapidly flew upwards, sometimes with the male in pursuit, and she then usually escaped. Females also sometimes oviposited alone, doing so with a series of quick dipping movements in one region.

Tramea limbata Desjardins

As in the previous species, *T. limbata* has a very widespread distribution in Africa and in much of India. It has also been reported from islands such as the Seychelles, Madagascar (Malgasy), the Maldives and Diego Garcia which suggests that it is a long-range migrant.

On the campus it was common from October to December, but in January it was not as abundant as *T. basilaris.* It is easily distinguished from *T. basilaris* by the black hindwing patch.

Body length: 51 mm.

Hindwing length: 40-46 mm.

In structure it resembles the foregoing species closely, but the colouring is much darker. The frons in front of the large eyes is a metallic purple. The thorax and abdomen are orange-brown in immature insects but become black in mature ones, except for the dark red abdomen of the male. The base of the hindwings bears a single narrow black patch extending anteriorly to the costa (leading edge) but stopping before the trailing edge. The male's secondary genitalia are even more prominent that in *T. basilaris* and the female's subgenital plate is much enlarged.

BEHAVIOUR

Their perching and patrolling habits were very similar to those of *T. basilaris.* The commonly sailed on motionless wings for long periods at treetop height in the middle of the day with the abdomen drooping and the mid and hindlegs trailing (Fig. 17). This position might increase the convective heat loss from the body surface and help them to remain on the wing in strong sunlight. *T. basilaris* also showed this behaviour but less commonly, perhaps because its paler colouring allowed it to reflect more of the incident radiation. When perched in the hot sun both species were seen to bend the abdomen down to a position in which it caught the minimum of sunlight.

Mature males were seen to patrol over many of the permanent ponds and the Botanic garden lake for long periods, perching only occasionally. They sometimes flew low over the water making shallow dipping movements, an activity which resembled oviposition and perhaps induced females nearby to approach the water. On several occasions 2, 3 or even 4 tandem pairs appeared simultaneously after a long interval with no female. The site of one tandem pair may have caused other females to approach the water where they in turn encountered males.

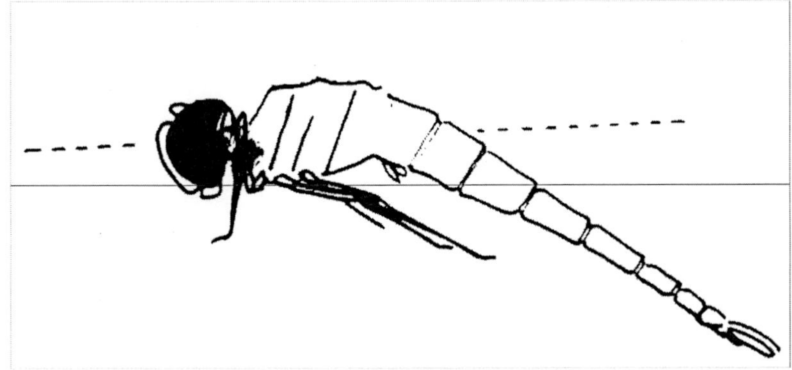

Fig. 17. Position adopted by Tramea limbata flying during hot time in the middle of the day. The legs trail and abdomen droops, perhaps enhancing cooling

Copulation took place on trees or bushes sometimes well away from water, and lasted 5-7 minutes. As in *T. basilaris* after copulation the pair flew in tandem for several minutes chasing single males away from the area before commencing to oviposit. The male normally released and then reclasped his female after one dip to the water surface, but on several occasions the male was seen to hover above the female while she made repeated dips to the surface in one area, this type of 'non-contact' guarding resembling that of many other libellulids. This illustrates again the variability of odonate reproductive behaviour.

Trithemis aurora Burmeister

T. aurora is widespread in India and throughout eastern Asia occurring as far east as Japan. It was seen on the campus from October to January, being abundant in October and November, declining in December, and being relatively uncommon in January. It tended to frequent the permanent ponds at all times, but was also sometimes seen at the Botanic garden lake. The magenta colour of the male and its small size make it unmistakable.

Body length: 33 mm.

Hindwing length: 24-30 mm.

The dorsal part of the eyes is magenta and the sides are purple grey. The frons is a metallic purple. The thorax and abdomen of immatures is yellow marked with black. Mature males have a uniformly magenta or purplish red thorax and abdomen, and the anterior wing veins are reddish. There is a large saffron patch at the base of the hindwings and usually a smaller one on the forewings, but these features are variable. Females have a paler hindwing patch, and in some the wings are broadly tipped with yellow as far as the pterostigma. Tipped and untipped females were both seen on the campus.

BEHAVIOUR

T. aurora typically perched very low on vegetation or on the ground. Mature males perched on stones round the permanent ponds frequently chasing each other and returning to the same perch again and again. Copulation took place in the air and lasted about 20 seconds. While the female hovered above, the male then sometimes flew down to the water surface where he in turn hovered as though to indicate the oviposition site. The female then usually started to oviposit flying low and rapidly along the water surface dipping her abdomen at a high rate as she went. The male followed close behind her guarding her from other males. Oviposition did not appear to be directed towards or onto vegetation.

In Dehra Dun, this species is reported breeding in streams and adults are active between September and December (Kumar, 1972).

***Trithemis pallidinervis* Kirby**

This species is known in east Asia as far as the Philippines and it is also widely distributed in India. It was not common on the campus, being seen only during December and January, mainly at the village and goat ponds.

Body length: 41 mm.

Hind wing length: 30-36 mm.

T. pallidinervis is a larger but thinner species than T. aurora and it has unusually long legs. The sexes are similar in colour and appearance and no pruinescence is developed. The frons is shiny metallic purple. The thorax and abdomen are patterned in yellow and black, the yellow darkening with maturity so that the whole insect looks somber. The wings bear a very small saffron patch at the base, considerably smaller than in *T. aurora*.

BEHAVIOUR

Males sat perched on prominent places of vegetation about 1 m above the ground facing into the wind and close to the water edge. Their reproductive behaviour was not observed.

***Zyxomma petiolatum* Rambur**

This species is common throughout south and east Asia being found as far away as in Australia: in India it is widespread in the lowlands and up to 1000 m. It is an unmistakable insect, crepuscular in habits and somber in colouring. On the campus it was commonly seen in the evening at small permanent ponds, such as the ready ponds in the botanic garden, the deep tanks and the permanent ponds in the Botanic garden and at the Main Guest House.

Body length: 46 mm.

Hind wing length: 30-35 mm.

The two sexes are similarly coloured. The eyes, which are relatively as large as those of any other libellulid, are uniformly green. The thorax and abdomen are dark brown to black and the wings are a smoky brown, with deeper tips. There are very small patches of dark brown at the wing bases. Its dark colouration probably allows it to roost cryptically and also may increase its absorption of radiation.

The abdomen is swollen in the 1st and 2nd segments, narrows to a waist in the 3rd and then remains slender in the more posterior segments, resembling O. Sabina in shape (q.v.).

BEHAVIOUR

The behaviour of *Z. petiolatum* resembles that of *T. tillarga*. During the day *Z. petiolatum* was found perching vertically low down among shady grass or bushes, but I was not been able to observe its feeding behaviour.

Towards dusk, and sometimes at other times of the day in densely shaded areas, males patrolled incessantly over water without perching. They did so over most of the permanent ponds and deep tanks on the campus, but not over the lake. They sometimes chose very small bodies of water, for example corners of the reed-filled ponds, or some small (1 x 1 m), water-filled concrete tanks, which they remained inside by making repeated tight turns. They were also seen over the Botanic garden pond where early in the evening they flew in broad arcs over the surrounding grass, but later as the light intensity diminished they remained over the water. Two males over the same pond chased one another repeatedly exhibiting their extreme manoeuvrability until eventually one left. Males sometimes hovered in front of their human observer seeming to stare at the intruding monster with their large green eyes. Copulation took place in the air and lasted about 10-20 seconds. Females oviposited by dipping regularly while being guarded by the male.

Kumar (1972, 1985) describes this species as replacing *Bradinopyga geminata* in pools which become stagnant (eutrophic) but at Madurai both were found in the deep tanks and in some of the ornamental ponds.

DISCUSSION

The campus odonate fauna of 25 species included no unusual or rare species. The list probably contains all the species (18) regularly breeding at sites on the campus during the observations months, but it is incomplete in that a few other species, including a cordulegasterid and an aeshnid, visited the campus briefly but were not identified. Eighteen of the species are abundant and widespread throughout India and have been reported from many sites in the north including Dehra Dun (Kumar, 1972). Likewise 18 of them have been reported from the Periyar National Park in Kerala (Mathavan & Miller, 1989).

Seven species, namely *Lestes elatus, Anax guttatus, Hemianax ephippiger, Diplacodes trivialis, Indothemis carnatica, Potamarcha congener* and *Tholymis tillarga*, were seen to breed only in temporary habitats on the campus. A further species, *Ceriagrion coromandelianum, Ictinogomphus rapax, Bradinopyga geminata, Trithemis aurora* and *Zyxomma petiolatum* were seen to breed only at permanent campus sites. The remaining six species probably bred in both types and such species are likely to be seen as adults throughout the year. However habitat permanence is not necessarily the only relevant criterion in habitat choice since the temporary lake was much larger than any other habitat on the campus and I have observed *Anax guttatus* and *Tholymis tillarga* breeding in large permanent lakes elsewhere. *T. tillarga* was seen to oviposit at small pools as soon as they had been formed at the start of the monsoon, whereas other species were more cautious.

Species, which breed only in temporary habitats, may be able to survive prolonged dry seasons in one of several ways. They may lay drought-resistant eggs; have drought-resistant larvae, which become buried in mud, migrate as adults to other regions, or remain, in the area as adults in reproductive diapauses. Examples of each of these methods are known from species in other parts of the tropics, but there is little information concerning this problem for Indian species. The distribution of several of the species, which depend on permanent aquatic habitats and may have relatively short life cycles, has probably been much extended by the construction of numerous man-made dams, tanks and reservoirs during the last few thousand yeas in southern India. Since these store water during prolonged dry seasons their formation may been considerably enhanced the distribution of certain species in such areas, some of which may in turn have become adapted to identify and exploit such habitats.

ACKNOWLEDGEMENTS

I am grateful to the Royal Society, the Indian National Science Academy, the Trustees of the Aneurin Bevan Foundation and the Indian Council for cultural Relations, which made my visits possible. I am deeply grateful to Professor T.J. Pandian and Dr. S. Mathavan for hospitality at Madurai.

REFERENCES

Bhargava, R.N. 1989, On a collection of Odonata from West Kameng district, Arunachal Pradesh. Indian Odonatol. 2, 45-46.

Biswas, V., M.A. Bashar & A. Begum 1990. On a collection of Odonata from Bagerhat District, Khulna, Banladesh. Indian Odonatol. 3: 65-66.

Fraser, F.C. 1933-1936. Fauna of British India including Ceylon and Burma: Odonata, Vols 1 – 3. Taylor & Francis, London.

Joseph, K.J. & A.R. Lahiri 1989. The diel patterns of communal roosting behaviour in *Potamarcha congener* (Rambur) (Anisoptera: Libellulidae). Adv. Odonatol. 4: 45-52.

Kumar, A. 1972. The phenology of dragonflies in the Dehra Dun Valley (India). Odonatologica 1: 199-207.

Kumar, A. 1985. A review on the bio-ecology of Indian dragonflies. Proc. 1st. Ind. Symp. Odonatol. (Ed. S. Mathavan): 73-90.

Larsen, T.B. 1987. A migration of *Pantala flavescens* in South India. Notul. Odonatol. 2 : 154.

Mahato, M. 1989. Dragoflies of Kathmandu valley, Nepal. Indian Odonatol. 1: 15 – 20.

Mathavan, S. 1985. Reproductive behaviour of *Brachythemis contaminata* and *Orthetrum Sabina* (Odonata: Libellulidae). Proc. 1st. Ind. Symp. Odonatol. (Ed. S. Mathavan): 93 – 106.

Mathavan, S. & P.L. Miller 1989. A collection of dragonflies (Odonata) made in the Periyar National Park, Kerala, South India. Soc. Int. Odonatol. Rapid Comm. No. 10: 1-10.

Miller, P.L. 1989. Communal roosting in *Potamarcha congener* (Rambur) and its possible functions (Anisoptera: Libellulidae). Odonatologica 18: 179 – 194.

Miller, P.L. 1991. The effect of oxygen lack on egg hatching in an Indian dragonfly. Physiol. Ent. (In Press).

Pajni, H.R. & P.K. Tewari 1988. Some notes on the Odonata of Chandigarh and other areas of north-western India. Indian Odonatol. 1: 139-141.

Ram, R., V.D. Srivastava & M. Prasad 1983. A note on a collection of Odonata from Eastern Uttar Pradesh, India, Notul. Odonatol. 2: 15-16.

Rowe, R. 1987. Ischnura aurora (Brauer) a dragonfly with unusual mating behaviour (Zygoptera: Coenagrionidae) Odonatologica, 9: 285 – 292.

Rowe, R. 1987. The dragonflies of New Zealand. Auckland University Press.

Srivastava, B.K. & B. Suri Babu 1985. Reproductive behaviour of *Ceriagrion coromandelianum* Ramb. (Zygoptera: Coenagriidae). Proc. 1st Ind. Symp. Odonatol. (Ed. S. Mathavan) : 209 – 216.

Suri Babu, B. & B.K. Srivastava 1990. Breeding biology of *Ceriagrion coromandelianum* (Fabricius) with special reference to seasonal regulation (Zygoptera: Coenagrionidae), Indian Odonatol. 3: 33-43.

Thomas, M. & M. Gladstone 1989. Odonata occurring at the Madras Christian College Campus, South India. Ind. Odonatol. 2: 49-50.

Tyagi, B.K. 1997. Dragonfly Wealth of India: Extant species – a preliminary systematic list. Zoo's Print 12: 3-10.

---

† *This original contribution by the late Prof. P.L. Miller, a close associate of Prof. B. Kiauta, both academically and odonatologically, but also because of long years of membership on the Editorial Board of ODONATOLOGICA, was earlier submitted to the INDIAN ODONATOLOGY, edited by Dr. B.K. Tyagi. In 1990, Prof. Miller visited India, on a fellowship facilitated jointly by the Royal Society of the Royal Society, the Indian National Science Academy, The Trustees of the Aneurin Bevan Foundation and the Indian Council for Cultural Relations, and worked both at the Madurai KamrajUniversity, Madurai (TN) and at Jodhpur(Rajasthan) where I was posted and also headed the SIO National Office in India. During a week long sojourn of Dr. Miller in Rajasthan, we together made odonatological expeditions in both the deep forests of Mount Abu and in the deserts dotted with a chain of man-made lakes (cf.,* Tyagi B.K. & P.L. Miller, 1991. A note on the odonata collected in south western Rajasthan, India. Notulae odonatologicae 3:134-135*). I hope my resolution in thus publishing his article, with commensurate editorial adjustments, on such a noble occasion will be viewed with appreciation and the readers will find it most enjoying and enriching!*

Odonata : Biology of Dragonflies
Ed.: B.K. Tyagi
Scientific Publishers (India), 2007, pp. 323-341

19

LARVAL AND ADULT BEHAVIOURAL PATTERNS OF SOME ODONATE SPECIES FROM DEHRADUN VALLEY

AMIT MITRA

Senior Lecturer, Sherubtse College, Kanglung, Bhutan
e.mail: amitodonata@yahoo.com

ABSTRACT

Different behavioural patterns of five odonata species have been studied in and around Asan reservoir, Dehradun, India from May 1995 to April 1997. The odonate imago feeds only on living prey and predominantly on flying insects. *Orthetrum s. sabina* has been seen to detect the perching prey more accurately. The mature males of all the five species arrive at the rendezvous earlier in the season during the day than females and form the territory. Only except *Brachythemis contaminata,* the males of other four species either show aggressive behaviour and chase display (in Anisoptera) or show threat display by 'abdomen raising' or 'wing opening' (in Zygoptera). Before wheel tandem lasts for a few seconds in *Crocothemis s. servilia* and *Brachythemis contaminata* while it lasts from 2-5 minutes in *Orthetrum s. sabina.* Comparatively, in zygoptera, before wheel tandem lasts longer and intramale sperm translocation occurs at that time. The copulatory wheel generally lasts for 6-22 seconds in *Crocothemis s. servilia*; 4-15 seconds in *Brachythemis contaminata;*

15-25 minutes in *Orthetrum s. sabina*; 10-15 minutes in *Ischnura a. aurora* and 35-45 minutes in *Ceriagrion coromandelianum*. All the three anisoptera oviposit by frequent dipping of their abdomen under water surface, whereas, the two zygopterans oviposit endophytically. Among the five species, only *Ceriagrion coromandelianum* oviposits in tandem. The duration of oviposition varies from 20-30 seconds in *Crocothemis s. servilia;* 3-6 minutes in *Orthetrum s. sabina;* 4-6 minutes in *Brachythemis contaminata;* 20-25 minutes in *Ischnura a. aurora* and 20-30 minutes in *Ceriagrion coromandelianum.* Odonata larvae are generalized predators and early instars prefer *Paramoecium* spp., *Daphnia* spp., diatoms etc., while later instars prey on chironomid larvae and pupae, mosquito larvae, ephemeropteran nymphs, *Branchiura* spp., *Limnodrilus* spp. and some nematodes. *Orthetrum s. sabina* larvae shows cannibalism. The last instar of all the five species stop feeding 2-3 days before emergence. Emergence occurs in *Crocothemis s. servilia* and *Orthetrum s. sabina* during 5.00 to 7.00 hours at a height of 6-25 cm and 10-30 cm from the water level, respectively. In *Brachythemis contaminata* emergence occurs during 12.30 to 2.00 hours at a height of 2.5-10 cm. Emergence occurs in *Ischnura a. aurora* during 12.00-2.00 hours and in *Ceriagrion coromandelianum* during 3.00-5.00 hours at a height of 8-12 cm and 10-15 cm from the water level, respectively.

INTRODUCTION

Life cycle of Odonata is well demarcated into different behavioural patterns. The adult life is dominated by two main behaviour, *i.e.*, feeding and reproduction. The total life span is well marked into two periods, viz., pre-reproductive and reproductive. Pre-reproductive period is the period between emergence and reproduction. The main behaviours during this period are dispersal after emergence and feeding to help the teneral adults becoming fully mature adults within a few days. The adults are opportunistic feeders, rather hunters, and start feeding soon after emergence. They are generalized, obligate carnivores (Corbet, 1962) and predate on most of the cacheable and sizeable small animals, predominantly during flight. Hence, the pre-reproductive period is also called maturation period.

It does not mean that in the reproductive period odonates remain on fast. Together with feeding behaviour the most spectacular phenomenon of this period comprise the different types of reproductive behaviour, including territoriality, pre-copulatory tandem formation, copulatory wheel formation, post-copulatory flight and oviposition. The reproductive behaviour has been studied in details in different parts of the world. In Indian subcontinent, Kumar & Prasad (1977) have studied different aspects of reproductive behaviour of *Neurobasis c. chinensis* while Prasad & Ghosh (1982) have recorded the reproductive behaviour of *Urothemis signata signata*. Begum *et al.*, (1982) have commented on the sexual behaviour and oviposition of *Zyxomma petiolatum*. Studies on the reproductive behaviour of *Ischnura a. aurora, Ceriagrion coromandelianum* and *Chloroneura quadrimaculata* have been carried out by Srivastava & Babu successively in the years 1984, 1985a and 1985b. Prasad (1986) has studied the reproductive behaviour of *Tholymis tillarga* where as Mitra (1987) made some comments on its breeding habitats. Begum *et al.*, (1990) have documented the reproductive behaviour of *Rhodothemis rufa*. Prasad (1990, 1991) again has made some important observations on the reproductive behaviour of *Ceriagrion coromandelianum*; *Pseudagrion rubriceps* and *Brachythemis contaminata*. Chowdhury & Karim (1994) have observed the reproductive behaviour of *Copera annulata,* while Srivastava *et al.* (1994) has made a detailed study on the behaviour of reproduction and oviposition in *Pseudagrion decorum*. Sangal *et al.*, (1994) have described the reproductive ethobiology of *Orthetrum s. sabina* and *Neurothemis i. intermedia*. I have studied the reproductive behaviour of *Pseudagrion rubriceps* at Asan Reservoir (Mitra, 1996), which is the same habitat where the present study was carried out.

The dragonfly larvae on the other hand show the characteristic feeding behaviour and the crucial emergence behaviour. The larval feeding behaviour of dragonflies has been studied worldwide and it is well known that they are carnivorous, preying on almost any kind of ingestible animals they can perceive. In Indian subcontinent, however, little work has been done so far in this context. Mathavan (1976) has analysed the satiation time and predatory behaviour of *Paragomphus (Mesogomphus) lineatus* larvae in south India. Begum *et al.* (1985) have studied the food preference of final instar of six species under the laboratory condition, viz. *Ictinogomphus angulosus, I. rapax, Brachythemis contaminata, Tholymis tillarga, Crocothemis s. servilia* and *Diplacodes trivialis*. Thomas (1987) has studied the comparative predatory efficiency of three species of dragonfly naiads near Chennai (Madras), while Rose & Thomas (1994) studied the feeding behaviour of *Anax guttatus* in details.

Like many other aquatic insects emergence is one of the critical steps in the life cycle of dragonflies, both physiologically and ecologically. This is the turning point from aquatic to aerial life (Ubukata, 1973). The emergence pattern, sex-ratio and post emergence dispersal has been well documented in different parts of the world. However, as far as the Indian subcontinent is concerned, only a few species have been studied on account of their emergence. Kumar (1989) has studied the emergence pattern of *Pantala flavescens* in relation to the monsoon in India. Kumar & Negi (1989) have reported the emergence and sex-ratio of *Bradinopyga geminata* and *Crocothemis s. servilia* at Dehra Dun valley, India. Begum *et al.* (1992) have described the emergence patterns of *Tholymis tillarga, Neurothemis t. tullia, Diplacodes nebulosa, Crocothemis s. servilia* and *Ictinus* sp. in Ramna lake of Dhaka city in Bangladesh. Biswas *et al.* (1994) have also examined the emergence pattern of *Brachythemis contaminata, Tholymis tillarga, Crocothemis s. servilia, Diplacodes trivialis, Orthetrum pruinosum neglectum* and *Ictinus angulosus* under the laboratory conditions in Bangladesh. I have studied the emergence behaviour of *Trithemis festiva* under laboratory conditions at Dehra Dun, India (Mitra, 1998).

In the present study the detailed ethobiology in the life span of five odonata species, viz., *Crocothemis servilia, Brachythemis contaminata, Orthetrum sabina, Ischnura aurora* and *Ceriagrion coromandelianum,* have been documented from Asan Reservoir, Dehradun, India. Some part of their behaviour must have already been known but, the present study reveals their complete behavioural patterns with comparative notes on earlier studies.

METHODOLOGY

Five odonata species, viz., *Crocothemis servilia, Brachythemis contaminata, Orthetrum sabina, Ischnura aurora* and *Ceriagrion coromandelianum* were selected for the present study due to their well contrasted colouration. Moreover, their conspicuousness in the field and lack of visual similarity with any other species in the Asan reservoir area (Mitra, 1996) helped in their easy field observation. Field observations, collection of exuviae after measuring the height from the water surface and simultaneous laboratory rearing and analysis were made during May 1995 to April 1997.

The time span for different behavioural patterns has been recorded with the help of a stop-watch. The data range for each behavioural pattern along with the total number of observations (N) and the mean value (x) have been reflected in the text. The records for initiation of adult feeding have again been put forward in tabular form (Table 1). Tables for other behavioural patterns are skipped due to their wide range for the number of observations. A field binocular was used for careful observation.

OBSERVATION

Different behavioural patterns of five odonata species that have been observed during the present study are discussed under four broad categories, *viz.,* feeding behaviour of imago (irrespective of sex), reproductive behaviour, larval feeding behaviour (irrespective of sex) and emergence behaviour.

Table 1. Period between completion of emergence and first feeding (initiation of feeding)

Species name	No. of observations								
	1	2	3	4	5	6	7	8	Mean
Crocothemis s. servilia	45 min	33 min	20 min	55 min	40 min	1.12 hrs	---	---	44.16 min
Orthetrum s. sabina	55 min	43 min	56 min	1.35 hrs	---	---	---	---	1.02 hrs
Brachythemis contaminata	Not in the same day								
Ischnura a. aurora	3 hrs	3 hrs	5 hrs	2 hrs	4 hrs	---	---	---	3.40 hrs
Ceriagrion coromandelianum	36 min	32 min	39 min	47 min	41 min	51 min	1.15 hrs	50 min	46.38 min

CROCOTHEMIS S. SERVILIA (DRURY)

Feeding behaviour of imago: Adults start feeding soon after emergence (N = 6, x = 44.16 minutes: Table 1) and continue to do so throughout life. These perchers feed only on living prey and predominantly on flying insects. They usually perceive their prey when it is above them and therefore approach it from below. They generally prey upon adults of diptera, ephemeroptera, neuroptera, smaller lepidoptera and even upon smaller damselfly trespassers into their territory. On wing *Crocothemis servilia* has never been observed to catch a perching prey, though they may attack the resting members of other dragonflies, not for preying but for capturing the perch.

Reproductive behaviour: *Territoriality*: The males generally have shown a strong territorial behaviour, although some feral males also exist in the population. The appearance of males and females at the reservoir site depends chiefly on the sunlight. Generally the males arrive at the rendezvous (breeding ground) during 7.30 to 8.30 hours. The females appear from the surrounding vegetation little late at about 8.15 to 9.30 hours. On a cloudy day they frequent at the rendezvous well after the said period. Males, after arrival find out a base perch on the stems of *Ipomoea fistulosa, Memosa pudica, Cyprus* spp., *Polygonum* spp. or on some dry plants, stones etc. No particular preference for perch has been noticed. The perch forms the centre of a circular territory with a radius of about 1 meter, which has been defended by the resident male from the intra - and interspecies invasions. The resident male used to show chase display against the conspecific male intruders within the territory, whereas they generally do not respond aggressively against the heterospecific males, though may do so to defend the perch from the intruding male. Sometimes the males choose two to three perches within the territory and remain there alternately. However, it has been observed that territoriality is a density dependent factor and occurs more at low density, while the number of wandering or non-territorial mature males are found more with an increase in population density.

Courtship and wheel formation: When the female enters into the territory of a resident male, the later tries to catch on her prothorax by its anal appendages. Once the male succeeds the pair gets air-borne *in tandem* and forms the wheel position in which the vulvar region of the female gets interlocked with the secondary copulatory apparatus of the male, present beneath the second abdominal segment. The duration of the wheel position varies from 6-22 seconds (N = 30, x = 10.11), which is the shortest among the five selected species. Generally the wheel formation occurs in flight and the pair in courtship remains in the air without any support of perch till end. However, in some cases supports have been used as a hiding place to avoid interference by other males. Intra-male sperm translocation has not been witnessed and hence can well be predicted to occur before the males arrive at the rendezvous.

Oviposition: In most of the cases the female starts laying eggs soon after the breaking of the copulatory wheel. They do oviposit moving repeatedly forward and backward along a straight line in less dense areas without interference. Although it is a multivoltine species and occurs throughout the year, the density of imago increases rapidly during May to mid July. During this period the effect of interference by conspecific males during wheel formation and oviposition, increases. Though males generally respond aggressively in guarding their females during oviposition against molestation, still in most of the cases the females have been captured again by another male before completing their oviposition, sometimes even before starting the oviposition. The females that are "wheeled" simultaneously two to three times or more, takes some rest hiding at the surrounding vegetation and then proceeds towards the water surface for oviposition. Females generally takes 20-30 seconds for completing the oviposition without interference (N=35, x = 28). In some rare occasions, oviposition lasted 1 minute or more.

Larval feeding behaviour: The hatching starts 7-10 days after oviposition (N=33, x = 8.5). The early instars (2nd to 5th) preferably feed on *Paramoecium* sp., diatoms, *Daphnia* and some other unicellular organisms. The later instars gradually change their feeding habits towards larger prey like chironomid larvae (blood worm), mosquito larvae, ephemeroptera nymphs, some annelids and nematodes etc. They used to catch their prey by sudden extension of the labium or merely by movement of the labial palpi, when the prey is sufficiently small. It has been recorded that 9th instar onward a larva was able to take as many as 28 mosquito larvae per day or 31 blood worms per day but not more than 10 at a time. The ultimate instar usually stops feeding just 2-3 days before emergence.

Emergence behaviour: The fully mature larva crawls out of water generally during early morning hours (5.00 to 7.00 hours), climbing on some emergent aquatic or semi-aquatic plants in natural environment. Some bamboo sticks have been provided in laboratory rearing as emergence support. In laboratory conditions, three out of 17 larvae were found to emerge during day time (11.00 to 13.00 hours). After leaving the water, final instar larvae generally prefer to get slightly inclined emergence support, like that of *Trithemis festiva* (Mitra, 1998), so that their dorsal surface face the water with the head directing up. In search of suitable emergence site the larvae may cover a height of 6-25 cm (N=17, x = 14.5) from the water surface. After getting the suitable site for exuviation, the last instar becomes dry in air. Rupturing starts from the prothorax region at about 15-38 minutes (N=5, x=19.3) after leaving the water. The total process starting from the rupturing of prothorax of the larva to the sloughing off the larval skin takes 18-35 minutes (N=12, x=29.5 minutes). The length of rupturing area varies from 4-5 mm whereas the length of the exuviae varies from 16.5-18 mm.

ORTHETRUM S. SABINA (DRURY)

Feeding behaviour of imago: The adults start feeding around an hour after emergence (N = 4, x = 1.02 hours; Table 1). They capture their prey with the help of relatively long, articulated and spiny legs. They feed in flight when the prey is small (mainly diptera) using the forelegs to capture the prey and transfer it to the jaws; but settles to a perch to devour when the prey is large enough, such as some zygoptera or lepidoptera. They prefer flying insects, but are also able to detect the perching prey. I have observed *Orthetrum sabina* catching a perching male of *Rhodischnura nursei* in one instance and preying upon a settled pair of *Ischnura aurora* in courtship wheel in the other.

Reproductive behaviour: *Territoriality*: The males demonstrate strong territoriality and the number of wandering males are very few in the population. The males arrive at the rendezvous during 8.00 to 8.30 hours. The females appear little late from the surrounding vegetation at about 8.15 to 9.00 hours. Males, just after arrival, choose a base perch on the stems of *Lantana camara, Ipomoea fistulosa, Polygonum* sp., *Typha elephantina, Eupatorium adenophorum* etc. or on the nearby stones. The perch forms the centre of a circular territory with a radius of about 1/2 metre, however, it has been observed that the shape and size of territories occupied by the males within their chosen microhabitat are totally depended on the population density and the physical features of the habitat. The resident male shows aggressive behaviour and chase display against the intruding male of same or other species. Sometimes they feed on intruding males of *Ischnura, Ceriagrion, Pseudagrion* etc. Occasionally they change their base perch.

Courtship and wheel formation: The female upon encroachment into a male's territory has been chased by the later to catch on her prothorax with its anal appendages. The tandem pair settles on some vegetation for a while. Within 2-5 minutes, the courtship wheel is formed by interlocking of vulvar region of the female with the secondary copulatory apparatus of the male present beneath the second abdominal segment (Fig. 1). The duration of the wheel position generally varies from 15-25 minutes (N = 23, x=18.5), but during May - June when the population density reaches at its peak, the wheel may last much shorter (2-3 minutes only) due to interference.

Oviposition: After breaking of the wheel, the female generally takes a rest of 5-15 seconds and then starts oviposition without being assisted by the males. They dip their abdomen about 30-32 times and lay the eggs in water among the aquatic weeds. Occasionally, the female before completing the oviposition has been bound by another male in tandem and is forced to make the wheel again. During oviposition the female after releasing her first batch of eggs may take rest for a while (5-6 seconds) and then again heads to oviposition site for laying the next batch.

Larval Feeding Behaviour: The hatching starts after 9-11 days (N=15, x=9.5). The early instars (2nd to 5th) prefer *Paramoecium* sp., *Daphnia* sp. and some other zooplanktons. The succeeding instars enjoy larger prey like mosquito larvae, nymphs of ephemeroptera and plecoptera, blood worms and some other nematodes. The last two larval instars often show cannibalism and prey on the zygoptera larvae, smaller anisoptera larvae and even on conspecific early instars.

Emergence Behaviour: The emergence behaviour is quite same to that of *Crocothemis s. servilia* except that the emergence site from the water surface vary between 10-30 cm (N=11, x = 17.5). The length of rupturing area varies in between 5-6 mm where as the length of the exuviae varies from 18.2-20 mm.

BRACHYTHEMIS CONTAMINATA (FABRICIUS)

Feeding Behaviour of Imago: The adults unlike the above two species do not start feeding on the same day of emergence (Table 1). They are the typical perchers and spend most of their active time on a perch. From the perch they make short flights to capture their flying prey which they would perceive when it is above them. After catching the prey they again come back to the base perch to devour it. They have been seen to predate on different smaller flies like diptera, ephemeroptera, plecoptera etc., but never been noticed to prey on the smaller zygoptera.

Reproductive behaviour: *Territoriality:* The territorial behaviour is somewhat less displayed by the males of *Brachythemis contaminata* than the previous two species. The coexistence of more than five males along with the males of different species has been observed several times on a single stem of *Ipomoea*. The males arrive at the rendezvous at 9.00 to 10.00 hours and the females frequent during 9.30 to 10.30 hours. When the females arrive within the site of the males, a mild competition was noticed between the males. Changing of base perch within a limited area has also been noticed. The mature males remain in their territory till late evening.

Courtship and wheel formation: During fore noon the males are generally less responsive to the females. Their reproductive activity starts soon after the noon and continue till afternoon. The pre-wheel tandem lasts for a few seconds and then the courtship wheel forms in the air. The mating pair seldom settles on nearby vegetation. The wheel position lasts for 4-15 seconds (N=15, x=10). Intramale sperm translocation has not been observed and can well be assumed to occur before the arrival of males at rendezvous.

Oviposition: Just after breaking of the copulatory wheel, the females do oviposit in the marshy area of the reservoir. The females alone, without any male guarding, slowly flutter over the water surface and lay the eggs one after another. Sometimes after laying the first batch of the eggs, the female takes some rest before going for the next batch. The oviposition lasts for 4-6 minutes (N=10, x = 4.5).

Larval feeding behaviour: Hatching starts within 6-9 days (N = 17, x=7). Feeding behaviour is quite same as that of earlier two species. 9th instar onward a larva is able to take as many as 30 mosquito larvae per day or 25 blood worms per day but not more than eight at a time. The last instar usually stops feeding 2 days before emergence.

Emergence behaviour: The fully mature last instar usually crawl out of water during late night (12.30 to 2.00 hours), climbing on some emergent aquatic or semiaquatic plants. The exuviae have been found mainly on *Typha elephantina, Polygonum barbatum, Ipomoea fistulosa, Cyperus digitatus* and *Scirpus mucronatus.* The height of the exuviae from the water level generally varies in between 25-30 cm (N=14, x=27), however, a few exuviate near the water level. After getting suitable site to exuviate, the final instar gets dried up in the air and the cuticle starts rupturing from just below the head up to the end of meta-thoracic segment. The length of the rupturing area varies from 4-4.5 mm where as the length of the exuviae measures from 13-14.5 mm. The total process starting from the rupturing of the prothorax to the sloughing off the larval skin takes about 20-30 minutes (N=10, x=23).

ISCHNURA A. AURORA (BRAUER)

Feeding Behaviour of Imago: The males begin to feed 2-5 hours (N=5, x=3.40 hours: Table 1) after emergence, while the females start feeding only on the next day. They are able to perceive both flying prey as well as the settled ones. These delicate insects generally prey upon smaller diptera and ephemeroptera like *Caenis* sp. etc.

Reproductive Behaviour: *Territoriality:* The males arrive at the rendezvous during 6.30 to 7.00 hours, while the females appear from the surrounding vegetation around an hour late. After arrival, the males perch on some nearby vegetation like *Polygonum lanigerum, P. plebejum* and on some other grasses (Fig. 3A). The perch forms the centre of a very small territory with a radius of about 20 cm, which is defended by the resident male from the intruding conspecific males. Usually they display threat by wing opening (Moore, 1960) or abdomen raising to show the blue colour of the last three abdominal segments.

Before wheel tandem: With the sight of a female within the territory, the male follows her. As soon as she alights on some

Fig 1: Courtship wheel of *Orthetrum s. sabina*

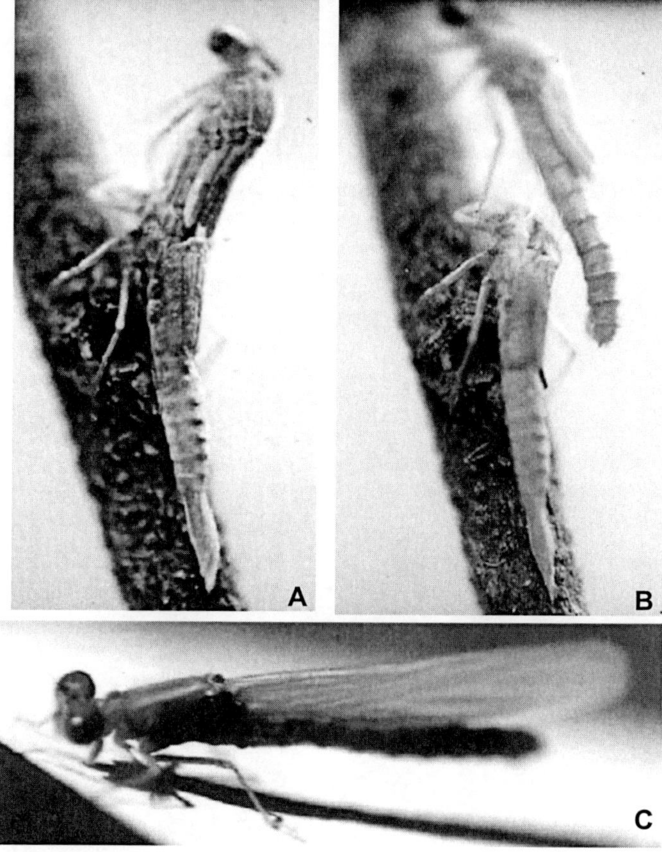

Fig 2: Emergence behaviour of *Ischnura a. aurora*

A. Imago is coming out of last instar larva. B. Imago is fully emerged.
C. Maturation is underway. Wing buds are developing into wings first.

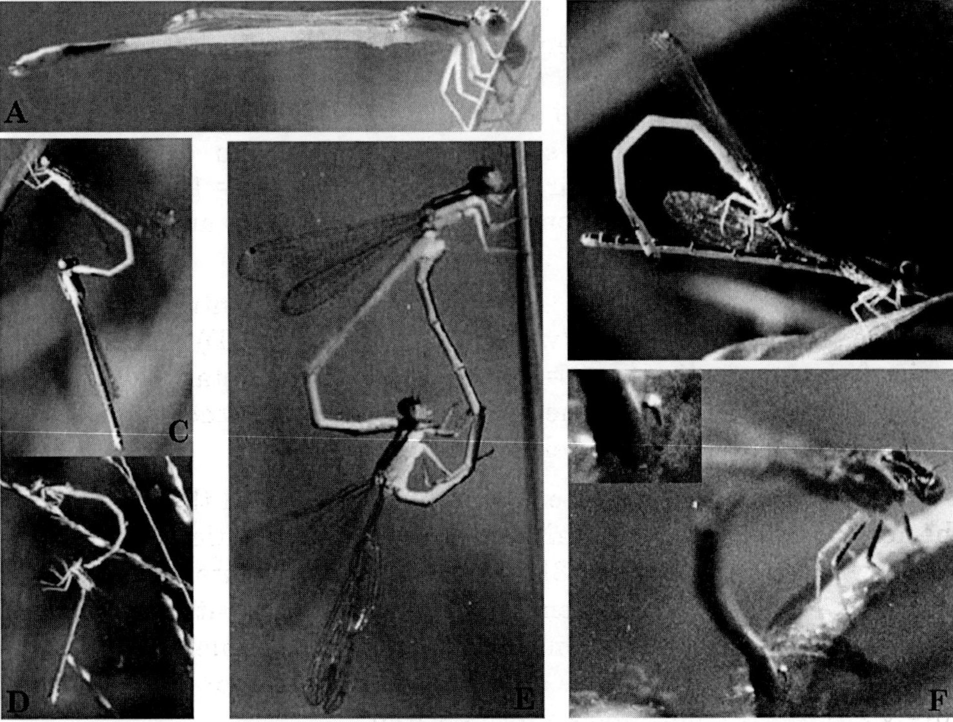

Fig 3: Reproductive behaviour of *Ischnura a. aurora*
A. Territorial male, B. Male catching a perched female, C. Before wheel tandem, D. Copulatory wheel position, E. After wheel tandem, F. Endophytic oviposition.

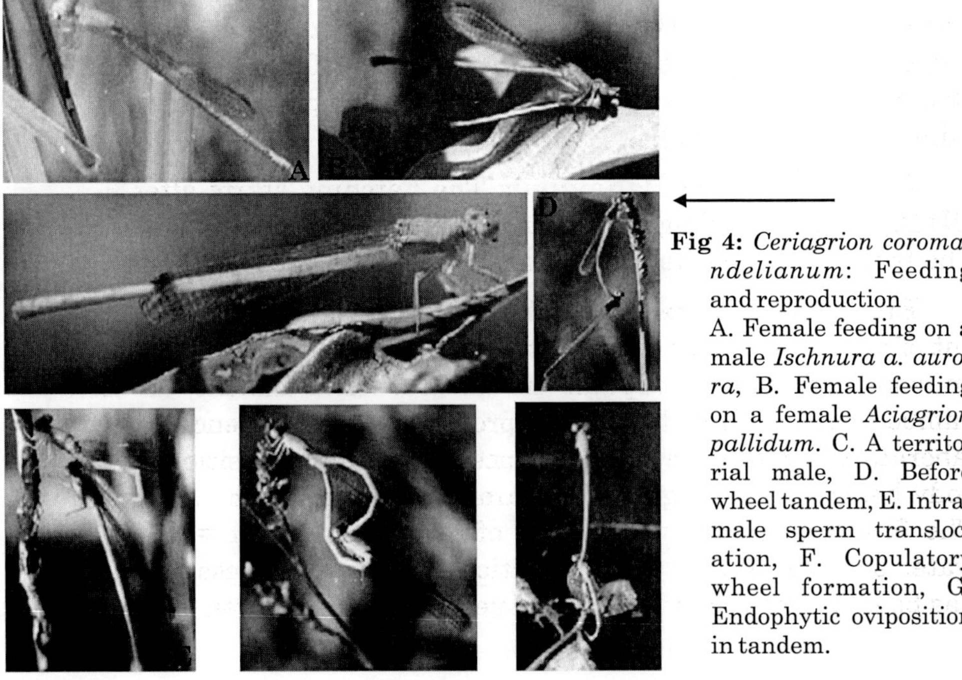

Fig 4: *Ceriagrion coromandelianum*: Feeding and reproduction
A. Female feeding on a male *Ischnura a. aurora*, B. Female feeding on a female *Aciagrion pallidum*. C. A territorial male, D. Before wheel tandem, E. Intramale sperm translocation, F. Copulatory wheel formation, G. Endophytic oviposition in tandem.

vegetation, the male hovers in the air remaining at the same place and observes her very carefully. Suddenly it pounces on her and catches her wings by its legs (Fig. 3B). The male then catches on the female's prothorax by its anal appendages and this is followed by an in tandem flight to perch on some nearby vegetation (Fig. 3C). The 'before wheel tandem' lasts around an hour (N=7; x = 1.10 hours) and intramale sperm translocation takes place 3-4 times, at an interval of about 10 minutes.

Wheel Position: After the completion of intramale sperm translocation, the courtship wheel is formed (Fig. 3D). The duration varies from 10-15 minutes (N=9; x=14), but in one instance it has been observed till 45 minutes. The mating pair changes the perch during disturbance keeping the wheel intact.

After wheel tandem: After breaking of the wheel the male lowers the female and the female also grasps some vegetation by her legs. 'After wheel tandem' (Fig. 3E) lasts for 5-10 minutes (N=9; x = 6 minutes). Finally the tandem breaks before oviposition. This is a species where there is no male guarding the female in her act of oviposition, and the presence of this long lasting 'after wheel tandem' surely arouses further interest for research.

Oviposition: After a rest of 10-15 minutes, the female starts oviposition unaccompanied by a male (Fig. 3F). She lays eggs in the emergent stems of aquatic vegetation. The female goes down underwater till their thorax region is above water and has never been found to submerge totally while laying eggs. The females frequently changes her places during oviposition. The oviposition lasts for 20-25 minutes (N=12, x = 22).

Larval Feeding Behaviour: The hatching starts after 6-8 days (N=15, x=6.5). Feeding behaviour is quite same as the earlier species. The last instar stops feeding 2 days before emergence.

Emergence Behaviour: The fully mature last instar larva crawl out of water generally during late night (12.00 to 2.00 hours), climbing on some semi-aquatic plant in natural environment or on some bamboo sticks that has been provided as emergence support in laboratory. Preference of emergence support and positioning of their body before the emergence is again quite same as the above species. The final instar climbs a height of 8-12 cm (N=7, x = 11) from the water surface in laboratory condition. The total process of emergence has been photographed in natural sequence and is given in the Fig. 2.

CERIAGRION COROMANDELIANUM (FABRICIUS)

Feeding Behaviour of Imago: The imago starts feeding within an hour of emergence (N=8, x=46.38 minutes, Table 1). They usually perceive the moving prey when it is above them and therefore approach it from below. Mostly they prey upon small diptera, plecoptera, ephemeroptera, some lepidoptera and even on smaller zygoptera like *Ischnura a. aurora* (Fig. 4A), *Rhodischnura nursei*, *Enallagma parvum* etc. They use to make short flights from their perch to capture the prey and again settle back to devour it.

Reproductive Behaviour: *Territoriality:* The males arrive at the rendezvous during 6.30 to 7.00 hours, while the females frequent little late (7.00 to 7.30 hours). The males perch on almost all types of aquatic grasses (Fig. 4C) and no specific preference has been observed. The perch forms the centre of a circular territory with a radius of about 50 cm and the resident male guards it from the intruding conspecific males.

Before wheel tandem: As soon as the female enters into the territory of a resident male, it chases her and generally after a short dual flight it succeeds to bind her in tandem link, catching hold of her prothorax with its anal appendages. In flight, the precopulatory tandem search for a suitable site for perching. This flight lasts for 40 seconds to 2 minutes (N=7, x=1.35 minutes). The pair in tandem then settles on some vegetation where the male anchors the plant and the female hangs vertically (Fig. 4D). The male then positions the abdomen in such a way that its gonopore on the 9th abdominal segment touches its secondary copulatory apparatus present ventrally on the second abdominal segment. At this time, intramale sperm translocation from the gonopore to the vesicula spermalis takes place and it lasts for 15-35 seconds (N=8, x=18.55) (Fig. 4E).

Wheel Position: After completion of the intramale sperm translocation, the male relaxes its abdomen. The female then tries to interlock her vulvar region with the secondary copulatory apparatus of male by curling the abdomen forward and upward to form the copulatory wheel position. After 2-3 attempts, the spectacular courtship wheel is formed (Fig. 4F). The duration of the wheel position varies from 35-45 minutes (N=12, x = 38.17). The transfer of sperms from male to female takes place during this "wheel position". During interference by the conspecific males, the male in courtship vibrates its wings vigorously to drive it away from the mating site. Sometimes the pair changes the perch while maintaining the wheel. If the wheel breaks in the process, the pair in tandem forms the wheel again.

After wheel tandem: After breaking of the wheel, the male lowers the female. The female also grasps some vegetation by her legs. After a rest of 5-8 minutes (N=5, x = 6.37), the tandem searches a suitable spot for oviposition in post-copulatory flight over the aquatic vegetation. The flight lasts for 6-10 minutes (N=4, x = 7).

Oviposition: In tandem, female oviposit endophytically among the aquatic plants, preferably *Potamogeton crispus* and *Ceratophyllum demersum* (Fig. G). The remaining behavioural aspects during oviposition are quite same as that of the previous zygoptera. The oviposition lasts 20-30 minutes (N = 9, x = 24.5 minutes).

Larval Feeding Behaviour: The hatching starts 7-8 days (N=8, x = 7.5) after oviposition. Larval feeding behaviour is quite same as that of earlier species. The last instar stops feeding 2-3 days before emergence.

Emergence Behaviour: The fully mature last instar larva crawls out of water generally during early morning (3.00 to 5.00 hours) for emergence. Behavioural aspects are quite same as all the previous species. The preference of emergence height is about 10-15 cm (N=5, x=12) from the water surface.

CONCLUSION

The adult odonata feed only on living prey and predominantly on flying insects whose movement, size and shape elicit the appropriate responses (*cf.* Corbet, 1980). The two Zygoptera often capture resting, immobile prey which coincides with the observations made by Goodyear (1970). Among the three Anisopterans, only *Orthetrum s. sabina* has been recorded to detect the perching prey but perhaps only if they first see the prey when it is moving or in flight. They have often been observed to predate on smaller Zygopterans where as *Crocothemis s. servilia* rarely do so, while *Brachythemis* never does it.

In all the five selected species the rendezvous is at or near the oviposition site. In every case mature males arrive at the rendezvous earlier in the season and in the day than females and form territory. The resident males of *Crocothemis s. servilia* show aggressive behaviour and chase display against the conspecific male intruders where as *Orthetrum s. sabina* males chase the intruding male of other species also. The males of *Brachythemis contaminata,* however, does not show any aggressive behaviour to intra-or interspecific males which contradicts the previous observation of Prasad (1991). Both the Zygoptera species i.e., *Ischnura a. aurora* and *Ceriagrion coromandelianum* defend their territory from the intruding conspecific males.

However, the 'abdomen raising' (Moore, 1960) has been observed only in the case of *Ischnura a. aurora.* The threat display that has been observed in *Ceriagrion coromandelianum* is only the 'wing opening' (Moore, 1960) not the 'abdomen raising' which again contradicts the previous observation of Prasad (1990).

The 'before wheel tandem' is very short lasting event in Anisoptera. In *Crocothemis s. servilia* and *Brachythemis contaminata* it lasts for a few seconds while in *Orthetrum s. sabina* it can last 2-5 minutes which is quite shorter than the observation (8-13 minutes) made by Sangal *et al.* (1994). Comparatively, in Zygoptera, before wheel tandem lasts longer and intramale sperm translocation occurs at that time. In *Ischnura a. aurora* it lasts for about one hour which contradicts the previous observation by Rowe (1978). In *Ceriagrion coromandelianum* the before wheel tandem lasts for 6-10 minutes, which is more or less similar with Prasad's (1990) observation.

In *Crocothemis s. servilia,* copulation wheel lasts for 6-22 seconds which extends its range of duration from the previous observation (10-15 seconds) by Prasad (1990). In *Orthetrum s. sabina,* duration of wheel position generally varies from 15-25 minutes, but during overcrowding it may also lasts for 2-3 minutes only. Sangal *et al.* (1994) have, although, mentioned the general range of duration of wheel position of *Orthetrum s. sabina* (12-18 minutes), they perhaps haven't observed the same during high population density. *Brachythemis contaminata* takes on an average 10 seconds to complete the copulatory wheel position which agree fairly well with the previous observation by Prasad (1991). However, as far as the wheel position of *Ischnura a. aurora* is concerned, the present observation contradicts greatly the previous observation by Rowe (1978) from New Zealand. According to Rowe (1978), the males of *Ischnura a. aurora* mates only with freshly emerged immature females. But according to the present observations, based on laboratory dissections and photographs of wheel position (Fig. 3) it is clear that in all the phases of reproduction the females are mature enough. However, the duration of wheel position (10-15 minutes) is somewhat same as Rowe's (1978) observation (20 minutes). Another interesting contradiction with Rowe's (1978) observation is that the females would start oviposition within 10-15 minutes after breaking of the 'wheel', while according to Rowe (1978) the females (still immature) may start oviposition after one to five days of the copulation. However, the duration of copulatory wheel of *Ceriagrion coromandelianum* (35-45 minutes) agree fairly well with the previous observation (an average of 33.6 minutes) from Kolkata (old name Calcutta) by Prasad (1990).

The females of *Ceriagrion coromandelianum* among the selected five have only been found to oviposit *in tandem*. In *Crocothemis s. servilia,* the duration of oviposition is same (20-30 seconds) as that of Prasad's (1990) observation. The observation of Sangal *et al.* (1994) on the duration of oviposition of *Orthetrum s. sabina* (3-6 minutes) also coincides with the present study. *Brachythemis contaminata,* however, takes more time (4-6 minutes) to complete the oviposition than the duration (50 seconds to 1.25 minutes) given by Prasad (1991). The remaining two Zygoptera oviposit endophytically. *Ceriagrion coromandelianum* has been found to oviposit preferably within the stems and leaves of *Potamogeton crispus* and *Ceratophyllum demersum,* whereas, no such preference for the aquatic vegetation has been observed in case of *Ischnura a. aurora.* Females of both the zygoptera, unlike *Pseudagrion rubriceps* have never been found to be fully submerged in water (Mitra, 1996) while laying their eggs. However, in Zygoptera there is another stage in between 'wheel position' and 'oviposition', *i.e.,* 'after wheel tandem'. In *Ischnura a. aurora* after wheel tandem lasts for 5-10 minutes where as in *Ceriagrion coromandelianum* this post copulatory flight in tandem to explore the appropriate place for oviposition (Mitra, 1996) continues till 6-10 minutes. Oviposition in *Ischnura a. aurora* lasts for 20-25 minutes. The duration of oviposition in *Ceriagrion coromande-lianum* has been found 20-30 minutes with an average of 24.5 minutes from nine observations which again contradicts the previous observation (2-3 minutes) of Prasad (1990).

Interestingly, the contradictions in the overall reproductive behaviour of *Ischnura a. aurora* between Rowe (1978) and the present observation suggest a DNA analysis to confirm the closeness of the Australian and Asian race of the species.

Odonata larvae are generalized predators that detect their prey by means of their compound eyes and/or mechanoreceptors (Pritchard, 1965, '66). It has been observed that the prey is usually caught by sudden extension of the labium or merely by movements of the labial palpi when the prey is sufficiently small as has been recorded by Lawton (1970) and Caillère (1972). The early instars of each of the five species usually prey on *Paramoecium* sp., *Daphnia* sp., diatoms and some other unicellular organisms. The later instars gradually prefer larger prey like chironomid larvae (blood-worms) and pupae, mosquito larvae, ephemeropteran nymphs, some annelids like *Branchiura* spp. and *Limnodrilus* spp., nematodes *etc.* The present study on the larval feeding habits agrees fairly well with the observation of Chowdhury and Rahman (1984), Begum *et al.* (1985). Among the five species only the *Orthetrum s. sabina* larvae shows cannibalism. The last instars of all the five species stop feeding 2-3 days before emergence as has been observed by Mathavan (1976), in case of *Paragomphus (Mesogomphus) lineatus* larvae.

Emergence in dragonflies is the transitional stage between aquatic and aerial mode of life. The fully mature larva crawls out of water during late night or early morning hours climbing on some emergent or semiaquatic plant. After getting dried it slowly casts off the larval skin. The teneral adult, within a course of time, becomes mature and flies away. In *Crocothemis s. servilia* and *Orthetrum s. sabina* emergence occurs during early morning (5.00 to 7.00 hours) which agrees fairly well with the observation of Begum *et al.* (1992) in Ramna Lake of Dhaka. However, their observation on the emergence time of *Brachythemis contaminata* (5.00 to 6.30 hours) differs from the present observation (12.30 to 2.00 hours). The emergence time of *Ischnura a. aurora* is generally the late night (12.00 to 2.00 hours) whereas in *Ceriagrion coromandelianum,* emergence occurs during early morning (3.00 to 5.00 hours). The larvae of all five species prefer slightly inclined emergence support as previously recorded by the author in case of *Trithemis festiva* (Mitra, 1998). In *Crocothemis s. servilia,* the height of the emergence place from the water level varies from 6-25 cm while Begum *et al.* (1992) and Biswas *et al.* (1994) have given it as 5-14 cm and 4.31-17.52 cm respectively. The larvae of *Orthetrum s. sabina* may cover a height of 10-30 cm from the water level to get a suitable site for emergence. Begum *et al.* (1992) have observed this emergence height in *Brachythemis contaminata* varying between 2.5-10 cm, where as Biswas *et al.* (1994) have recorded it as 25.40-27.94 cm. The present observation (25-30 cm) agrees well with the observation of Biswas *et al.* 1994. However in both the Zygoptera, the height of exuviae from the water level is more or less similar.

ACKNOWLEDGEMENTS

I am grateful to Dr. Arun Kumar, scientist SF and Officer-in-charge, Northern Regional Station, Zoological Survey of India, Dehradun and Prof. S. K. Kulshrestha, Head, Department of Zoology, D.A.V. PG College, Dehradun for their continuous guidance throughout the study period. Dr. J.R.B. Alfred, Director, Zoological Survey of India has kindly permitted the present study. Thanks are also due to Dr. B.P. Uniyal, Botanist, Botanical Survey of India for floral identification.

REFERENCES

Begum, A., Bashar, M.A. and Biswas, B.R. (1982). Studies on the life history of *Zyxomma petiolatum* Rambur (Odonata : Libellulidae). *Dhaka Univ. Studies* B 30(2): 131-138.

Begum, A., Bashar, M.A. and Biswas, V. (1990). The mating behaviour and development of *Rhodothemis rufa* (Rambur) (Anisoptera : Libellulidae) *Indian Odonatol.* 3: 1-9.

Begum, A., Bashar, M.A. and Biswas, V. (1992). Emergence pattern of some dragonfly nymphs (Odonata : Anisoptera) in Ramna lake of Dhaka city. *Dhaka Univ. J. Biol. Sci.* 1(1): 19-24.

Begum, A., Bashar, M.A. and Nasiruddin, M. (1985). Laboratory studies on the food preference of six species of final instar dragonfly larvae. *Bangladesh J. Zool.* 13(1): 31-34.

Biswas, V.; Begum, A.; Bashar, M.A. and Begum, S.A. (1994). Emergence Pattern of some Dragonfly (Odonata : Anisoptera) larvae under the laboratory condition. *Advances in Oriental Odonatology* (Ed. V.K. Srivastava) : 23-30.

Caillere, L. (1972). Dynamics of the strike in *Agrion* (syn. *Calopteryx*) *splendens* Harris 1782 larvae. *Odonatologica* 1: 11-19.

Chowdhury, S.H. and Karim, N. (1994). Observations on the reproductive behaviour of *Copera annulata* (Selys). *Advances in Oriental Odonatology* (Ed. V.K. Srivastava) : 69-76.

Chowdhury, S.H. and Rahman, E. (1984). Food preference and rate of feeding in some dragonfly larvae (Anisoptera : Odonata). *Ann. Entomol.* 2(1): 1-6.

Corbet, P.S. (1962). *A biology of dragonflies.* Witherby, London.

Corbet, P.S. (1980). Biology of Odonata. *Ann. Rev. Ent.* 25: 189-217.

Goodyear, K.G. (1970). *Lestes sponsa* (Hansemann) as a predator of *Tipula melanoceros* Schummel. *Entomologist London* 103: 215-16.

Kumar, A. (1989). Patterns of emergence of dragonflies (Insecta : Odonata) in relation to monsoon in India. *India J. Phy. Nat. Sci.,* 9(A): 10-15.

Kumar, A. and Negi, B.K. (1989). The emergence and Sex ratio of *Bradinopyga geminata* (Rambur) and *Crocothemis servilia servilia* (Drury) at Dehra Dun, India (Anisoptera : Libellulidae). *Fraseria,* 16: 70-71.

Kumar, A. and Prasad, M. (1977). Reproductive behaviour in *Neurobasis chinensis chinensis* (Linn.) (Zygoptera : Calopterygidae). *Odonatologica* 6: 163-171.

Lawton, J.H. (1970). Feeding and food energy assimilation in larvae of the damselfly *Pyrrhosoma nymphula* (Sulzer). *J. Anim. Ecol.* 39: 669-89.

Mathavan, S. (1976). Satiation time and predatory behaviour of the dragonfly nymph *Mesogomphus lineatus*. *Hydrobiologica* 50: 55-64.

Mitra, A. (1996). Reproductive ethobiology of *Pseudagrion rubriceps* Selys (Zygoptera : Pseudagriinae) at Asan Reservoir (Dehradun : India). *Ann. For.* 4(2): 139-144, 1 coloured plate.

Mitra, A. (1998). Notes on the emergence behaviour of *Trithemis festiva* (Rambur) (Odonata : Anisoptera) under laboratory condition. *Ann. For.* 6(1): 72-78.

Mitra, T.R. (1987). Note on *Tholymis tillarga* (Fabr.) (Odonata : Libellulidae) with special reference to its breeding habit. *J. Bengal. nat. Hist. Soc.,* 4(1985): 144-146.

Moore, N.W. (1960). The behaviour of the adult dragonfly. *In dragonflies (P.S. Corbet, C. Longifield and N.W. Moore). Collins, London; Chapter 9.*

Prasad, M. (1986). Reproductive behaviour in *Tholymis tillarga* (Fabricius) (Anisoptera : Libellulidae). *Fraseria,* 10: 43-44.

Prasad, M. (1990). Reproductive behaviour of *Ceriagrion coromandelianum* (Fabricius) and *Pseudagrion rubriceps* Selys (Zygoptera : Coenagrionidae). *Ann. Entomol.* 8(2): 35-38.

Prasad, M. (1991). On some aspects of reproductive behaviour in *Brachythemis contaminata* (Fabricius) (Anisoptera : Libellulidae). *Ann. Entomol.* 9(1): 1-3.

Prasad, M. and Ghosh, S.K. (1982). Studies on the estuarine odonata from 24 Parganas district of West Bengal with a note on the reproductive behaviour in *Urothemis signata signata* (Rambur) (Odonata : Insecta). *J. Bombay Nat. Hist. Soc.* 79(2): 290-295.

Pritchard, G. (1965). Prey capture by dragonfly larvae. *Can. J. Zool.* 43: 271-89.

Pritchard, G. (1966). On the morphology of the compound eyes of dragonflies, with special reference to their role in prey capture. *Proc. R. Entomol. Soc.* 41: 1-8.

Rose, S. & Thomas, M. (1994). Feeding Behaviour of *Anax guttatus* (Burmeister) (Odonata : Anisoptera). *Advances in Oriental Odonatology (Ed. V.K. Srivastava):* 19-22.

Rowe, R.J. (1978). *Ischnura aurora* (Brauer) a dragonfly with unusual mating behaviour (Zygoptera : Coenagrionidae). *Odonatologica,* 7(4): 375-383.

Sangal, S.K.; Bhandari, P. and Saxena, A. (1994). Reproductive behaviour of *Orthetrum sabina sabina* (Drury) and *Neurothemis intermedia intermedia* (Rambur) (Anisoptera : Libellulidae). *Advances in Oriental Odonatology (Ed. V.K. Srivastava)* : 63-68.

Srivastava, B.K. and Babu, B.S. (1984). Some observations on oviposition of *Ischnura aurora* (Brauer) in Indian Biotopes (Zygoptera : Coenagrionidae). *Fraseria* 6: 24.

Srivastava, B.K. & Babu, B.S. (1985a). Reproductive behaviour of *Ceriagrion coromandelianum* (Zygoptera : Coenagriidae). *Proc. First Indian Symp. Odonatol* : 209-216.

Srivastava, B.K. and Babu, B.S. (1985b). On some aspects of reproductive behaviour in *Chloroneura quadrimaculata* (Rambur) (Zygoptera : Protoneuridae). *Odonatologica* 14(3): 219-226.

Srivastava, V.K.; Srivastava, B.K. and Babu, B.S. (1994). The behaviour of reproduction and oviposition in *Pseudagrion decorum* (Rambur) (Zygoptera : Pseudagriinae) in central India. *Advances in Oriental Odonatology (Ed. V.K. Srivastava):* 77-84.

Thomas, M. (1987). Comparative predatory efficiency of three species of dragonfly naiads. *M. Phil. thesis, Univ. Madras.*

Ubukata, H. (1973). Life history and behaviour of a corduliid Dragonfly, *Cordulia aenea amurensis* Selys. I. Emergence and Pre-reproductive periods. *Journal of the Faculty of Science, Hokkaido University, Series VI, Zoology,* 19(1): 251-269.

Odonata : Biology of Dragonflies
Ed.: B.K. Tyagi
Scientific Publishers (India), 2007, pp. 343-348

20

COELICCIA HOANGLIENENSIS SPEC. NOV., A NEW PLATYNEMID DAMSELFLY FROM HOANG LIEN MOUNTAINS IN THE NORTH OF VIETNAM (ZYGOPTERA: PLASTINEMIDIDAE)

DO MANH CUONG

Hom thu so 16, Buu Dien 10210 – 35 Thai Thinh, Ha Noi, Vietnam
e.mail: cuongdm@hotmail.com; docuong@gmail.com

ABSTRACT

The new species (♂ holotype: Vienam, Lao Cai Prov., Sa Pa, Cong Troi, Hoang Lien National Park, 1945 m alt., 15.VII.2005; deposited in Do M. Cuong Collection) of genus *Coeliccia* is described with illustrations and a photo in nature.

INTRODUCTION

Coeliccia is a beautiful and diversified genus. Habitats of its species are forests on highland areas. This genus was first described by Rambur in 1842 and it differs with others in having a strongly angled medio-anal link (Laidlaw, 1932). Ten species in this genus have been recorded in Vietnam and among them eight species are indigenous to Vietnam (Kenichi, 2000). The first record of *Coeliccia* was from Vietnam probably published by Laidlaw in 1932 in his review of the genus and all specimens were collected from Tonkin. The second most important paper on this genus is Asahina' publication in 1997 with description of several unknown species collected from Northern Vietnam. However, the fauna of *Coeliccia* of Vietnam is still enigmatic as is the case of two species *C. tomokunii* Asahina, 1997 and *C. scuteluum* Laidlaw, 1932.

In the trip to Sa Pa, Lao Cai Province in 2005, I went to Cong Troi to collect dragonflies in the Hoang Lien National Park. It takes about twenty minutes to go by motorcycle from the town center to the park, which is the first base for climbing to the top of the Fansipan Mountain. It was a cold and rainy day as we were breaking into the forest near the trail which directed up towards "the Roof of Indochina". Following the other track, we got down about 100m and stopped by a running stream, where by some time later we collected a number of *Megalestes mican*, a shining green chlorolestid damselfly. We then turned left and walked along the stream, on a dark and humus path. As it stopped raining, under the slight sun light through the leaf canopy of the forest, I suddenly observed a slender damselfly resting on a fern leaf. It was a *Coeliccia* damselfly with two bright greenish stripes on the mesepisternum but it was quite different from any known species. Afterwards, I collected one more specimen; this one lacks a hind wing, however. After examining the two male damselflies, I believed they belonged to a new species and thus treated them into a new taxon, *Coeliccia hoanglienesis.* I will describe the details in the section below.

COELICCIA HOANGLIENENSIS SPEC. NOV. (FIG. 1 – 6)

Material. – Holotype ♂: Vienam, Lao Cai Prov., Sa Pa, Cong Troi, Hoang Lien National Park, 1945m alt. 15.VII.2005 - Paratype ♂: same collection date and locality as holotype. Both of holotype and paratype are deposited in Do Manh Cuongs' Collection.

Etymology. – The specific name is a reference to the locality where the species was discovered, Hoang Lien National Park.

MALE. – A beautiful and slim damselfly, characterized by large white blue marking on the pterothorax and a remarkably light green stripe on the mesepisternum (Fig. 1).

Head. – Black with bluish marking, ratio of length and width about ½, labium white yellow, labrum black. Parietal area includes four bulging parts with the front ocellus in the center which is contiguous point of four bulging parts. Two lateral ocelluses are on posterior bulging parts each having an anterior white marking in front. The posterior edge of the occipital is marked by two bluish spots on the dorsal view (Fig. 2).

Thorax. - Prothorax - whole black except 1/3 lower parts.

Synthorax (pterothorax)- Mesostigmal lamella is black. Mesepisternum is black with a light green stripe on each side. Mesepimeron and metepisternum connect together making the mesometapleural suture nearly disappeared. There is a large bluish area on the center of mesepimeron and meterpisternum. The black hem, which encompasses the center bluish area, is nearly enclosed except two discontinuous: a part of metapleural suture near the spiracle and the lower edge of bluish part between mid and hind coxa. Metepimeron is quite bluish. Metapostterpimeron is bluish with a small black spot at the upper angle (Fig. 1 and 3).

Wings. - Hyaline. In forewings 20 postnodals and 18 in hind wings at the first row: Anal bridge and anal cross meet at edge of wing. Discoidal cell of the hindwing is little longer than that of the forewing and the anal side of discoidal cell longer than the costal side in both wings. Pterostigma brown, parallelogram shape with costa and subcosta being longer sides; and cover 1 1/3 cell.

Abdomen. – Long and slender, the 1st segment is black and marked by a large bluish anterior and lower square, lateral view. Lower half of the 2nd segment of abdominal is bluish but the upper part is inky-black. The 3rd segment is black with a narrow bluish stripe and its width approximately ¼ the width of the whole segment. The segments from 4th to 8th are black with narrow bluish stripes on the lower parts; the 9th and the 10th are quite black; the width of the light colour stripes reduces gradually toward the end of the abdominal (Fig. 1 and 3).

Anal appendages. – Black and slightly hairy; the superior appendages are shorter than the inferior appendages with the length ratio about 4/5. In lateral view, the superior is narrow at the base, enlarges at the middle, then become narrower again near the tip. There is a small spine on the lower anterior edge of the extended structure of the superior appendage (Fig. 4 and 5). In dorsal view, the superior appendage is wide at base and gradually narrows down at 2/3 of superior from the base, then widen again at near the tip, which is a obtuse tooth developing from outer edge of superior (Fig. 5). In lateral view, base of inferior is board and it is gradually slender and slightly curved upward at approximately 1/3 of inferior from the base, remaining part is nearly constant in width and ended with a round curve downward tip (Fig. 4 and 5). In dorsal lateral view, the inferior appendages curve inwardly making plier-shaped touching together at tips. Inner side, the base of the inferior appendage is olive-yellow, remainder part is dark, except an inner area from the curve point to the tip of inferior appendage, marked with an inner side scar (Fig. 5).

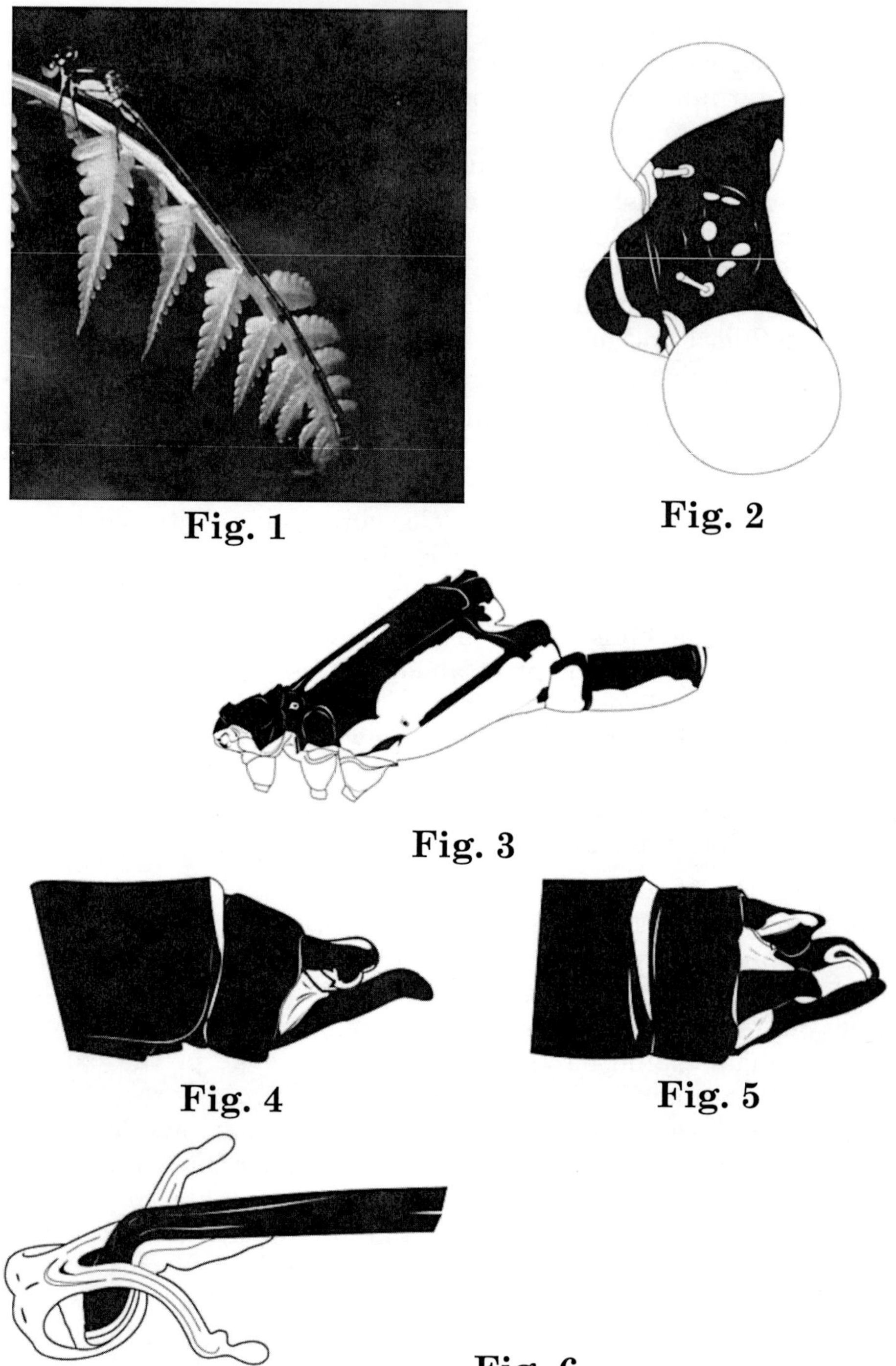

Fig. 1-6: *Coeliccia hoanglienensis*, male holotype: 1: photo in the nature; 2: Head; 3: Thorax and 1st, 2nd segments of abdomen – lateral view; 4: anal appendages – lateral view; 5: anal appendages – dorsal lateral view; 6: penis.

Genitalia. - Penis - Shaft of penis is colophony-brown and devoid of spines with a ventral hyaline membrane tape, it curves downward at tip and becomes flattened making spoon-shape. The terminal lobe curves upward then backward with the width growing narrower, about ½ of width flattened object, then deeply cleft and ending in a pair of long whip-like objects (slender bands). Two slender bands, whose length is about 2 time of the flattened object, grow backwards. The colophony-brown colour of shaft spreads to the inner surface of the flattened object then it is prevented by a ridge line, remainder parts of spoon-shape object and two slender bands are hyaline (Fig. 6).

Measurement (in mm). – Hind wing 28, abdomen (including anal appendage) 44.

FEMALE: - Unknown

DISCUSSION

This description intensifies the number of *Coeliccia* of Vietnam to eleven species. *C. hoanglienensis* spec. nov. owns all typical characters of the genus and differs from others in having a remarkably light green stripe on the mesepisternum of pterothorax. Unfortunately in this research, I could not check type of *C. pyriformis* described without illustration by Laidlaw in 1932. However, the characters of pterothorax, abdomen in and anal appendages of *C. pyriformis* are not in agreement with *C. hoanglienenis* spec. nov. and *C. hoanglienensis* spec. nov. is a distinct species.

ACKNOWLEDGMENT

This publication is dedicated to Professor Emeritus Dr. B Kiauta, Executive Editor of Odonatologica, in celebration of his 70th birthday. On this occasion, I am grateful to him for his continuous support for my dragonfly research. I am also thankful to Dr. M. Hämäläinen, Dr. M. Schoor, Dr. H. Karube and Mr. A. Sasamoto for providing literature and to Miss Dang Thanh Hoa for assisting me in the field trip.

REFERENCES

Asahina, S. 1993. A list of the Odonata recorded from Thailand. Part VI. Platycnemididae – Genus Coeliccia: 2-20

Asahina, S. 1997. Records of the Northern Vietnamese Odonata taken by the expedition members from the National Science Museum, Tokyo. 5.

Coenagrionidae, Protoneuridae and Platycnemidae. Bulletin of the National Science Museum Series A (Zoology) 23(1): 17-34

Kenichi W. 2000. Coeliccia of Thailand and Malaysia. The Nature & Insect, New information on Odonata Vol 35, No 11: 2-5

Laidlaw F.F. 1932. A revision of the genus Coeliccia (Order Odonata). Records of the Indian Museum Vol. XXXIV: 7-43

Hämäläinen, M. 1986. Description of *Coeliccia furcata* spec. nov. from upper Burma (Zygoptera: Platycnemididae). Odonatologica 15 (4): 445-447.

Odonata : Biology of Dragonflies
Ed.: B.K. Tyagi
Scientific Publishers (India), 2007, pp. 349-355

21

OBSERVATIONS ON MATING AND OVIPOSITION BEHAVIOUR OF *TETRATHEMIS PLATYPTERA* SELYS, 1878

C.G. KIRAN[1] and F.K. KAKKASSERY[2]

[1]*Mayooram, Pulari Nagar, Kodunganoor,*
P.O, Thiruvananthapuram, Kerala, India 695 013
e.mail : cgkiran@gmail.com
[2]*Reader, Research and Postgraduate Department of Zoology,*
St.Thomas' College, Trichur, - 680 001, Kerala
e.mail: kakkassery@yahoo.com

ABSTRACT

A detailed study was carried out of the mating and unusual oviposition behaviour of *Tetrathemis platyptera* during October 2005. It was revealed that unlike other dragonflies, the female *Tertrathemis platyptera* lays eggs on the twigs or leaves of plants hanging on to the water bodies, a special strategy for their larval development, and hatched nymphs were dropped into water for their further development. The total time of courtship, tandem position and wheel position are also discussed in this paper.

INTRODUCTION

The reproductive behaviour of different families of Dragonflies has been studied extensively by scientists around the world. But, many of them concentrated in the suborder Zygoptera, comprising damselflies (Bick & Bick, 1972; Heymer, 1972, 1973; Corbet, 1962;

Kumar, 1972, 1976, 1980) whose courtship, mating and ovipositon are lengthy processes and easily observable, whereas the Anisopterans' (dragonflies) reproductive pattern is so fast, and in most cases the sperm transfer occurs while they are on the wings. Elsewhere in the world the reproductive behaviour of *Nannophya pygmaea* Rambur (Tsubaki & Ono, 1986) and *Eleuthemis buettikoferi* Ris (Lempert, 1988) has been studied well. As far as Indian subcontinent is concerned, the comparative reproductive study between two species in the family Libellulidae, *Ortherum sabina sabina* and *Neurothemis intermedia intermedia* was also thoroughly observed (Sangal *et al.*, 1994)

Females of most of the members in the family Libellulidae oviposit by washing their eggs off in flight, by tapping the water surface or on floating vegetation (Corbet, 1962). However, in genus *Tetrathemis,* it has been observed that the females lay eggs on the objects, which overhang on water, and from there newly hatched larvae drop in to their future habitat (Fraser, 1933). More or less same type of observation was made in the Dragonfly, *Micrathyria dictynna* (Förster, 1998). However, after the ecological note on the reprodu-ctive behaviour of *Tetrathemis* (Fraser, 1933) no serious attempt has been made to study and describe the details of the mating of this group. This paper is an effort to describe the detailed account of the mating and unusual oviposition behaviour of *Tetrathemis platyptera* Selys.

MATERIAL AND METHODS

The site selected for the study was a temporary cement pond, 3-metre in diameter and 75 cm in depth, in a house garden in Trivandrum, Kerala, South India (8°30'40"N and 77°0'8"E). Because of the artificial nature of the pond, the water is full throughout the year. The pond is inhabited by tadpoles, water skaters, a single large carp fish etc. The main flora found are night blooming red water lilies and *Salvinia* species, which cover the whole water surface of the pond, whereas the spirally coiled algae are in the bottom.

These have altogether made an ideal environment for laying and hatching of the eggs, hiding of the larvae, and for the successful completion of aquatic stages of odonates with low predatory pressure. After monsoon, besides *T. platyptera,* a good number of other odonates also visited the pond. Different types of garden plants also surround the pond. While one side of the pond is exposed to morning sunlight, the other side is covered with small trees having orchids, with their overhanging roots seen above the pond.

After the mating and oviposition, the male and female were captured and examined in detail and compared with the descriptions of Fraser (1933) in order to confirm the species, and released to nature.

RESULTS

In India, there is only one species seen under genus Tetrathemis, *T. platyptera* Selys, which is one of the smallest dragonflies found in this region. The closest relative of this species is *Tetrathemis yerburyi*, which is confined to the submontane areas of Srilanka. The characteristic features of *T. platyptera*, are: the abdomen is black in colour, marked with citron yellow and is shorter than the wings. The base of the wings are coloured partly with yellow. And, sometimes, at first look, it may be confused as the young ones of *Orthetrum sabina sabina* or gomphids. The study was conducted in the month of October 2005, which is an ideal month with enough sunlight after the south west monsoon. In addition to *T. platyptera*, eight other dragonflies and two damselflies were also found near the pond (Table 1).

Table 1. The odonata calculated the study site

Damselflies (Zygoptera)
Ceriagrion cerinorubellum (Brauer, 1865)
Libellago lineata indica (Fraser, 1928)
Dragonflies (Anisoptera)
Brachdiplax sobrina (Rambur,1842)
Gynacantha dravida Lieftinck, 1960
Lathrecista asiatica (Fabricius, 1798)
Orthetrum chrysis (Selys, 1892)
Neurothemis fulvia (Drury, 1773)
Rhodothemis rufa (Rambur, 1842)
Rhyothemis v. variegata (Linnaeus, 1763)

Mating Behaviour

The flight season of *Tetrathemis Platyptera* is throughout the year. Mating time was between 10.00 am and 11.30 am. The male *T. platyptera* took position near the pond at around 8.30 am. It perched for a long time in the same place and repositioned its place one or two times. When the female approached from far away moving towards the pond, the male performed a kind of courtship display such as yawning pose (Fig. 1) and made the sidewise motions. The female was receptive so the male captured her by its legs encircling the thorax of the female. Soon they came into wheel position while they were on wings, and finally landed to a perch (Fig. 2). The copulation took only 30 seconds, and after departing from the wheel position, the male flew to a site where the eggs are to be laid. The female simply followed the male. Meanwhile, it was observed that the males of *Orthetrum chrysis* and *Brachdiplax sobrina* tried to attack and push down the female of *T. platyptera* to water two times when it was ovipositing.

Fig 1 Male shows a special type of courtship behaviour in front of female

Fig 2 After making wheel Position, they perch on the flower of water Lilly

Oviposition Behaviour

Immediately after copulation the female started ovipositing. Unlike other dragonflies *Tetrathemis platyptera* oviposits high above the water (Fig. 3). It prefers overhanging objects like roots of epiphytes, under flower petals, leaves and dried substratum. It was observed that it chose dried and loose materials to oviposit from height. The female hovered in front of the ovipositing site for about 50 seconds and pecked its ovipositor into the object where eggs are to be attached. Prolate and spheroid shaped eggs were attached or glued under the objects that were in horizontal position, and on all sides of vertical objects like roots. Female lays up to 250 eggs in a 4 cm stretch (Fig. 4). Each egg mass consists of bunches of 5 to 7 eggs. After the oviposition, the male and female look after the eggs for fear of predators. The hatched out larvae were dropped into water within 16-24 hours. The egg shells of hatched ones were seen as bluish - grey in colour. The same type of oviposition could also be observed in a drinking water well near the pond, which is 2.3 meter in diameter and 10 meters in depth. The female *T. platyptera* also laid eggs on the fern leaves, which were seen on the side walls of the well as well as dried leaves of coconut tree leaning towards the well.

DISCUSSION

In general, most of the females of dragonflies oviposit the eggs just touching the tip of the abdomen on the surface of the water. Very rarely some females attach their eggs on the objects just above the water bodies, so that the nymphs would be dropped into water once they were hatched out from eggs. It has also been observed and described in African dragonflies, *Malgassophlebia aequatoris* Legrand (Legrand, 1979) and *M. bispina* Fraser (Lempert, 1988), both of which belong to the subfamily Tetratheminae. This type of oviposition beghaviour was also noted in other species of the Tetrathemis genus (Clausnitzer, 1995; Clausnitzer & Lempert, 1997; Fraser, 1952; Lempert, 1988; Mccrae & Corbet, 1982). with a variation, through. Similar behaviour combined with gluing eggs to the substratum was also recorded for *Brachythemis lacustris* (Kirby) and *B. leucosticta* (Burmeister) (Corbet, 1962; Miller, 1982).

There are different opinions amongst scientists regarding the epiphytic oviposition above the water surface by female dragonflies : (1) it is an adaptive mechanism to avoid the anoxic conditions of stagnant water bodies (Mccrae & Corbet, 1982), and (2) it is to reduce the predatory pressure from other carnivorous organisms present in pond, such as fishes. (Miller & Miller, 1985; Lempert, 1988). The present study also supports these views in connection with epiphytic oviposition, and it is clearly evident that in the case of the above mentioned cement pond, the second view is more applicable as the hatched out larvae can be easily concealed under the weeds and algae present in the bottom, and complete their further development avoiding much predation. Yet, the first view, which is about avoiding anoxic condition in the stagnant water bodies, is also additionally substantiated by the observation made about the laying of eggs on dried coconut leaves hanging on to the drinking water well.

Fig. 3. The female of *T. platyptera* lays eggs on the dried plant twig

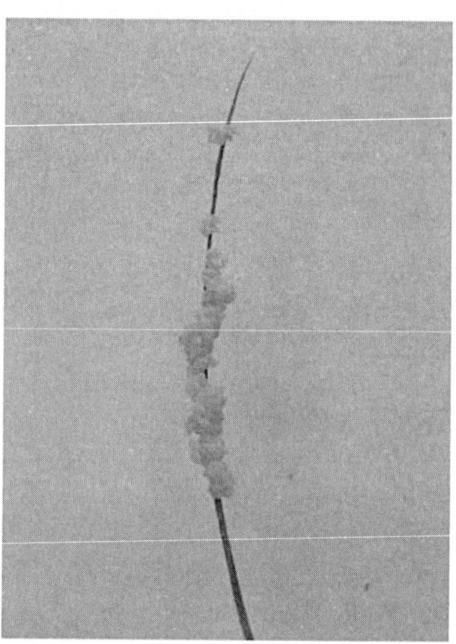

Fig. 4. Magnified view of eggs

ACKNOWLEDGMENTS

The first author is profoundly grateful to Dr. K.G. Dilip for invaluable guidance, persistent encouragement and suggestions given to him throughout this study. The authors wish to express their gratitude to Dr. Allan Brandon, Dr. B.K. Tyagi, and C. Susanth for their assistance and support for the completion of this research study.

REFERENCES

Bick, G.H & J.C. Bick, 1972. Substrate utilization during the reproduction by *Argia plana* (Calvert) *Argia moesta* (Hagen) (Odonta: Coenoagrion-idae) *Odonatologica* 1:3-9.

Clausnitzer, V., 1995. *Verhaltensökologie von Notiothemis robertsi Fraser, 1944 (*Odonata: Libellulidae*)*. DiplArbeit Univ. Marburg.

Clausnitzer, V. & L. Lempert., 1997. Tetratheminae: A comparative approach (Anisoptera: Libellulidae). *J. Afr. Zool.*

Corbet, P.S., 1962. *A biology of dragonflies.* Witherby, London.

Förster, S., 1998. Oviposition high above water in *Micrathyria dictynna* Ris (Anisoptera: Libellulidae). *Odonatologica* 27: 365-369.

Fraser, F.C., 1936. Fauna of British India including Ceylon and Burma: Odonata. Vols. -III. Taylor & Francis, London

Fraser, F.C., 1952. Methods of exophytic oviposition in Odonata. *Ent. mon. Mag.* 88: 261-262.

Heymer, A. 1972. Components social et territorial dae (Odonata: Zygoptera). *Ann. Soc. Ent. Fr.* 8:237-245

Heymer, A. 1973. Etude de compartment reproucter et analyse de mecanines declenhurs innes (MDI) optiques chez less Calopterygidae (Odonata: Zygoptera). *Ann. Soc. Ent. Fr.* 9:219-225

Kumar, A. 1972. The phenology of dragonflies in the Deharadun Valley, India. *Odonatologica* 1(4): 199-207

Kumar, A. 1976. Biology of Indian dragonflies with special reference to seasonal regulation and larval development. *Bull. Ent.* 17:37-47

Kumar, A. 1980. A review of on bioecology of Indian dragonflies. *Proc. First Symposium. odonatol.*: 73-92

Legrand, J. 1979. Morphologie, biologie et écologie de Malgassophlebia aequatoris, n. sp., nouveau Tetratheminae du Gabon (Odonata: Libellulidae). *Rev. fr. Ent.* (N.S.) 1: 3-12.

Lempert, J. 1988. *Untersuchungen zur Fauna, Ökologie und zum Fortpflanzungsverhalten von Libellen (Odonata) an Gewässern des tropischen Regenwaldes in Liberia, Westafrika.* DiplArbeit Univ. Bonn.

Mccrae, A.W.R. & P.S. Corbet 1982. Oviposition behaviour of *Tetrath-emis polleni* (Selys): A possible adaptation to life in turbid pools (Anisoptera: Libellulidae). *Odonatologica* 11: 23-31.

Miller, P.L. 1982. Temporal partitioning and other aspects of reproductive behaviour in two African libellulid dragonflies. *Ent. mon. Mag.* 118: 177-188.

Miller, P.L. & A.K. Miller 1985. Rates of oviposition and some other aspects of reproductive behaviour in *Tholymis tillarga* (Fabricius) in Kenya (Anisoptera: Libellulidae). *Odonatologica* 14: 287-299.

Sangal, S.K., P. Bhandari & A. Saxena 1994. Reproductive behaviour of *Orthetrum sabina sabina* (Drury) and *Neurothemis intermedia intermedia* (Rambur) (Anisoptera: Libellulidae). *Adavances in Oriental Odonatology (Ed. V.K. Srivastava): 63-68*

Tsubaki, Y. & T. Ono 1986. Competition for territorial sites and alternative mating tactics in the dragonfly, *Nannophya pygmaea* Rambur (Odonata: Libellulidae). *Behaviour* 97: 234-251.

Odonata : Biology of Dragonflies
Ed.: B.K. Tyagi
Scientific Publishers (India), 2007, pp. 357-362

22

ABOUT THE ODONATA ETHNIC NAMES IN THE SERBIAN LINGUISTIC AREA

MILOŠ JOVIĆ

Natural History Museum Belgrade, 11000 Belgrade, Serbia
e.mail: milos.jovic@nhmbeo.org.yu

ABSTRACT

Vernacular names of Odonata in Serbian linguistic area are given and discussed. There are only few known expressions for odonates in the central part of Balkans. These names generaly correspond with Odonata names in other European languages but it is interesting that extremely frightful associations were absent. Small number of common names of dagonflies and damselflies in the named area might be result of poorly exploration of the area, small significance of odonates in everyday life of local communities or misplacing during time

INTRODUCTION

Serbian language is one of the southern Slavonic languages which are being used in the central and western Balkans area. The 19th and the 20th century brought significant changes (which are still lasting) in opinion whether languages that are being used in the most of republics of former SFR Yugoslavia, represent variations of one language, or they are all separate languages (Serbo-Croatian or Croato-Serbian, oposite from Serbian, Croatian and Bosnian). Anyhow, these languages can be associated, because of their great relation, with Slovenian and Macedonian (which are being used on territories that had also been parts of former Yugoslavia), and Bulgarian which is being used in the eastern part of the Balkan Peninsula. In this, geographically small, but in cultural and historical sense varied area, dominates no less variety in dialects and local languages, in many viewes. Here named variety is, before all, product of heterogenous influences and mixing with different ethnic groups which are, for this area, native, or they came on the Balkans altogether with the Slavs, or maybe, sometime before or after them. Because of all that, in cultural sense, the Balkan Peninsula is fanta-stic mosaic which still remains unexplored in satisfactory measure.

There are two very important facts in understanding the local nation's apprehension of the world around them – already named appertainance to Slavonic part of Indo-European group of people and remarkable influence of monotheistic religions which had been massively accepted, starting with the 9th and 10th century. Despite the fact that paganism on the Balkans was completely exterminated faster than in other parts of Europe (may be because of the fact that the seats of religious power and the pontiff's dignity in the Mediterranean, were very close), many of aboriginal prepossessions and customs from the foregoing period, remain saved in their original or "washed" state, which was acceptable for new religions. That is why some of the typical christian holydays, like Christmas, during the subsequent centuries, in Serbia were celebrated with ceremonies full of prechristian symbolism which was explained in christianly correct way. Original explanations of ceremonies, just as complete mythology from the period of paganism were, on that way, very efficiantly "erased from collective memory" of the southern Slavs.

In investigation of the origin of ethnic names for plant, animal and fungus species, we should consider cultural specifics of the area where Serbian language is being used, just as the surrounding areas. That is why vernacular names for representatives of Odonata order have similarities with corresponding names in other European languages. Variety of ethnic names for odonates, still is remarkably smaller than in other parts of Europe, where we have to have in mind size of the area which this text refers to, just as the fact that it is almost completely unexplored in this way. Generally, today, odonates are phenomenon which is close only to the people whose profession or hobby is connected with water.

THE OLD NAMES

In Serbian, there are few old names for representatives of Odonata order. They are both original and borrowed from neighbour ethnical languages.

The ethnic name ”vilinski konjic“ (very often found as ”vilin konjic“) is the name which is being used in the biggest part of the central Balkans. Two words which are making this name – ”vila“ (fairy) and ”konj“ (horse) are equally interesting and, most likely, derived from the ancient believes connected to water. Fairies are female demons. There are several kinds of fairies in Serbian tradition, and the most interesting are the water fairies – ”jezerkinja“ (fairy of the lake), ”izvorkinja“ (fairy of the spring)... Water demons are not unusual in mythologies of other European nations, too. Water has been considered for specific soul reservoir, and some of the demons were water protectors. Serbian fairies are example of these demons, drowned young women's souls.

Connection between the names of odonates and horses is much more represented in European languages – for example in German, Spanish, Romanian, French... (Sarot, 1958). Beside the allusion to the appearance and speed and the demonic creatures' need for ”the instrument of transport“, the use of the noun–horse- in building names of odonates on the Balkans, may be explained with special status that horse had in the Balkan people's believes, too. In the Serbian religious tradition, horse is one of the animals which can carry souls of the dead (Čajkanović, 1994). Besides, connection between horse and water is ancient and diverse. Similarly to the story about the Greek god Poseidon who could take over the horse figure. Ilyrians, too, adored water gods and related them with horses (Čajkanović, 1994). The Slavs who had settled in this area, probably inherited some of the similar apprehensions. That is how Duke Momtchilo's (”vojvoda Momčilo“) mythological horse Yabutchilo (”Jabučilo“) was the horse with wings, and had risen as a product of love between a horse elf from one of the lakes on Durmitor mountain (Montenegro) and a mare that had been grazing on the lake front. There is also a few hydrographic objects which are carrying adjective-equine - (”konjski“ – ”Konjsko jezero“ – The Equine lake, ”Konjsko blato“ – The Equine mud) in their names. Usually, in connection with these objects there are legends which are saying that on the bottom of a lake or a spring lives a horse elf, because it is the spot where some horse had drowned.

The name "vilinski konjic" usually refers to all representatives of Odonata order, but in Serbian there is difference between Odonata suborders. In that case, the name "vilinski konjic" refers to representatives of Anisoptera, while the corporate name of Zygoptera is "vodene device" (water virgins). This, second name is very similar to demoiselle (Fr., Eng.), wasserjungfer (German), стрекоза (Russian) and by it's meaning is the nearest term to already named fairies.

In Slavonia (NE Croatia), among the rest of the expressions for odonates, it seems that "zmijska glava" (snake head) is being used, too. I've heard this expression from a lady who escaped from this area in the end of the 20th century, because of the war. This lady was claiming that by this name children had been calling representatives of the genus *Calopteryx*. It is well known that making a connection between odonates' names and snakes is not rare in the other parts of Europe, too, starting from the British Isles all to the western parts of the Balkan Peninsula. Slovenian name "kačji pastir" meaning snake shepherd, is only one of the examples. Analogy is usually derived from the fact that they have peaked abdomen, although the Balkan people, especially the Ilyrians, extremelly respected snakes. Cadmos, the mythical founder of Thebes (Greece) and Budva in Montenegro was conqueror and leader Ilyrians, people of snakes (Srejović, 2002). Cadmos and his wife Harmonia had been converted to snakes. In Serbian, snake with wings is called "zmaj" (dragon), wich is admirable, but most probably accidental similarity with common English name for odonates – dragonfly. In serbian tradition, "zmaj" is often a creature with positive reputation because he is fighting against the firedrake ("ala" – also represented like a snake with wings) that is guiding storm clouds (Čajkanović, 1994).

On the south of the Serbian linguistic area, region around Ulcinj (Montenegro), two cumulative names for odonates are being use. Namely, the population in this area make linguistic difference in Odonata suborders, and the representatives of Anisoptera, they call "patani" and the representatives of Zygoptera "pilivozi" (for the second one, it is certain that is being used among the population of Albania, too).

MODERN NAMES

Modern names for odonates are inspired by technical accomplishments which are resembling to these insects by their characteristics. These are "aviončići" (small airplanes) and "helikopterčići" (small helicopters) as the names for flying devices which are, today, much more present in human consciousness than a not numerous and secret group of insects.

In modern names, some older and rarer names can be conditionally interpolated, too. In Vojvodina, the northern Serbian province, names "šnajder" (tailor) and "makazar" are being used with definite allusion to scissors ("makaze") as an object that reminds of wings of these insects. The name "šnajder" is probably borrowed from German.

Josif Pančić (Čajkanović, 1994), the establisher of modern botany in Serbia, was using name "gospojice" (young ladies) which is translation of some German Wasserjungfer (cf. Pančić, 1998).

CONCLUSION

Not numerous names for odonates in the Serbian linguistic area fit in general picture of the European people's apprehension of these insects as demonical creatures (Cermanović & Srejović, 1996). However, it looks like these names are the expression of a less fatalistic relation of central Balkans' people to odonates, because in the named area, scary stories and negative connotations like satan's helpers, horse slayers or creatures that are sewing together human eyes or mouth, are missing. A minor variety between ethnic names may be a product of poorly exploration of the area, small significance of odonates in everyday life of local communities or misplacing during time.

ACKNOWLEDGEMENTS

I would like to thank to the following persons for their generous help in finding some of the names for dragonflies: Dr. Boštjan Kiauta, MSc Mila Karas, Mr. Svetozar Santovac and Mr. Aleksandar Stojanović.

REFERENCES

Sarot, E. 1958. Folklore of the dragonfly – alinguistuic approach. Letture di pensiero e d'arte. Edizioni di storia e litteratura, Roma.

Cermanović, A. & SREJOVIĆ, D. 1996. Leksikon religija i mitova drevne Evrope. Savremena administracija, Beograd. - [Serb.]

Čajkanović, V. 1994. Stara srpska religija i mitologija. Sabrana dela iz srpske religije i mitologije, knjiga 5. Srpska književna zadruga, BIGZ, Prosveta i Partenon M.A.M., Beograd. - [Serb.]

Pančić, J. 1998. Prvo Pančićevo putovanje sa licejcima po Srbiji. Studije i prepiska. Sabrana dela, knjiga 7. Zavod za udžbenike i nastavna sredstva, Beograd, 9-30. - [Serb.]

Srejović, D. 2002. Drevna Budva u mitu i istoriji. Iliri i Tračani – o starobalkanskim plemenima. Srpska književna zadruga, Beograd, 307-318. - [Serb.].

Odonata : Biology of Dragonflies
Ed.: B.K. Tyagi
Scientific Publishers (India), 2007, pp. 363-366

23

DISCOVERING THE DRAGONFLY WEALTH OF KERALA – THE GOD'S OWN LAND – IN SOUTH INDIA : A TRAVELOGUE

DAN BARTA

Havanska 8, Prague – 7, 170 00, Czech Republic
e.mail: dnlbrt@seznam.cz

(Dedication : I am glad I can share my photographs with you. All the best for your birthday, dear Professor Kiauta. Happiness, health, and let destiny be kind to you).

I am not a scientist, "only" a photographer. To write a treatise about research, its results, assertions or methodology, that is for me a task almost impossible to handle. In fact I relate to the world and especially to Nature only emotionally, as an enthusiastic observer. Somehow, out of all the variety of Nature, I got most captivated by dragonflies, and I decided to photograph them. I have been always thrilled by the endless maneauvres of the adult dragonflies during their roosting, mating, flight and even feeding moments. They present me a world of new information on mechanics and biology which I am ever eager to snap and exhibit for the academic and scientific enrichment of all dragonfly lovers, environmentalists and conservati-onists! My aim is to show them as they are, alive, in their natural environment, and beautiful. Here I am trying to offer, through my photographs, the emotional experience of facing dragonflies in awe (Plates).

Photograpying dragonflies in nature is a great experience to learn about endurance, patience and perseverance. Though it is extremely tiresome, snapping dragonflies in various different moods is an unforgettable experience. It tires a little but pleases boundlessly. The joy of sharing these experiences with others is always a great delight. Beauty, after all, is just one of the faces of truth, I guess.

In December 2004 I spent two weeks in India. I chose Kerala because of the virgin Western Ghat forests, the tropical and humid climate, and because odonatologist Dr. Francy Kakkassery offered me a refuge and help with "little logistics", and, of course, the elegant and enigmatic dragonfly fauna prevalent there (Table 1). And, because I flew from Prague via Colombo, the flight to Trivandrum was short and cheap. I had earlier discussed the plans of visit to Kerala with Dr. Kakkassery, himself an excellent photographer, and we worked out our itinerary to photograph the dragonflies in nature. My decision to do dragonfly photography in Kerala proved to be a right choice since the variety of both the fauna and habitats were overwhelmingly abundant (Fraser, 1933, 1934, 1936) - to a photographer's pleasure, of course! The south Indian dragonfly fauna are close to that of Sri Lanka, an island country in the neighbourhood where dragonflies have been profusely studied (*cf.* de Fonseka, 2000). I had never been in India before, so none of the local dragonfly species seemed commonplace or boring. They were all new, different and unknown to me.

In Kollam, the lighthouse keeper let me into his garden and for the first time I could see through my lens *Diplacodes trivialis*, perching on dry fallen leaves and palm fronds. In Aleppey and Kottayam I had to face lots of water, but it was impossible to take pictures from a boat. In Aleppey I could only watch the channels and endless rice paddies. In the Thekkady National Park I gave up on watching tigers and elephants, and in the nearest moist thicket photographed *Vestalis gracilis* or young *Acisoma panorpoides* on the lake shore, as well as *Ceriagrion coromandelianum*, and above muddy pools among trees I discovered magnificent red males of *Neurothemis fulvia*. And in Munnar...

Munnar, December 21, 2004

In the morning I was woken by my alarm clock. Though still in bed, with my eyes closed and determination in mind, I planned for the day ahead. I decided to shoot the dragonflies with my camera (Nikon F

Plate - 1

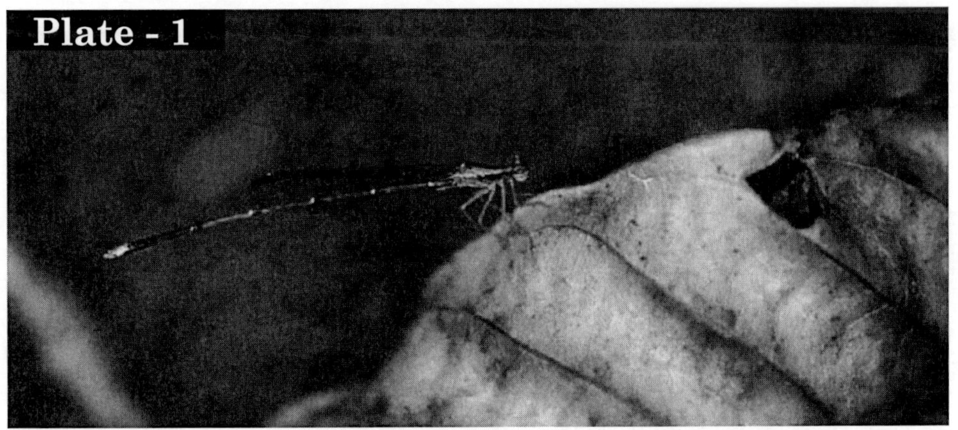

Copera marginipes

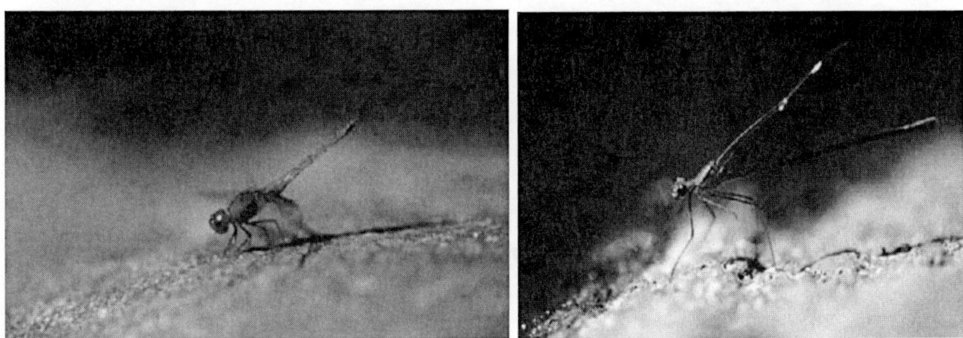

Trithemis aurora *Neurobasis chinensis*

Diplacodes trivialis

Plate - 2

Neurothemis tullia

Neurothemis fulvia

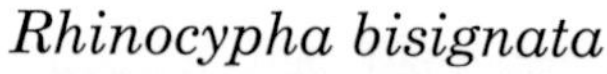

Rhinocypha bisignata

100, Micronikkor 105 mm, 2,8 f, colour reversal film Fuji Velvia 50 ASA; UV filter) sooner than eight in the morning. I picked up my camera bag, tripod, fisherman′ trousers, and went to the bus stop to pick up the first leg of my peregrination. The bus was forty-five minutes late, but at least I could drink a couple of pints of hot sweet tea with milk and smoke a few cigarettes, all the while however thinking of my encounters with the new species of dragonflies and the modes of photographing them. I asked the driver to stop "on that big bridge", about thirty kilometres from Munnar, from where I watched the tea and cardammon plantations. I got off (the place was called Poopara) and walked upstream from the bridge, about a kilometre and a half, before finding a place where the creek, about half a metre deep, bent and zigzagged, full of pebbles, tree trunks and sticks, and where I could expect light until late hours – the forest opened just above the water. I washed my trousers and put on the fisherman′s ones. There I started taking pictures, focusing on two species, *Neurobasis chinensis* – about five females and two males – and small and colourful *Rhinocypha bisignata*.

Table 1. Dragonfly species photographed during my most enriching sojourn in Kerala

S.N.	Species	Sex	Place	State (Country)	Date of collection
1	*Acisoma panorpoides*	Male	NP Thekkady	Kerala, INDIA	XII. 04
2	*Neurobasis chinensis*	Male	Poopara, Idduki	Kerala, INDIA	XII. 04
3	*Neurothemis fulvia*	Male	NP Thekkady	Kerala, INDIA	XII. 04
4	*Rhinocypha bisignata*	Male	Poopara Idduki	Kerala, INDIA	XII. 04
5	*Rhinocypha bisignata*	Female	Poopara, Idduki	Kerala, INDIA	XII. 04
6	*Diplacodes trivialis*	Female	Kollam	Kerala, INDIA	XII. 04

There was about a hundred of them, mating, the females laying eggs, the males fighting for the females and territories, hunting! The females of *N. chinensis* sat on stones close to the water and discouraged the males from mating by raising their abdomens and spreading wings, resembling flowers. The males were flying very fast above the creek, flashing the metallic mirrors on front wings. They would sit only occasionally. The *R. bisignata* males flew around the females and showed them their snow-white tibias and purple tips of whirling wings, or chased each other away in countless aerial duels... I shot seven films, 7x36 shots. Water got into my rubber trousers, I drowned my cigarettes and the correction filter 81A, but nothing deterred me in my resolution to snap these beautiful creatures in nature. This meant a lot of patience and deciding the right moments for the photography.

Therefore, I spent a lot of time just watching their behaviour, enjoying the atmosphere of this space and time. Later in the late afternoon the sun dropped behind the tree tops as I decided to go back.

As a last leg of my travel plans, I then went to Kotamangalam and to Thrissur to meet Francy and his family and friends. Thanks to them I spent another unforgettable week in Kerala, with people and dragonflies. But that is another story, to be told on some other occasion.

ACKNOWLEDGEMENT

My thanks are due to my friend Dana Habova who helped me with the translation, and to Dr. Francy Kakkassery for his warm hospitality during my sojourn in Kerala – a dragonfly paradise, and for the help in photographing these most fascinating creatures in different moods. My thanks are also due to Dr. B.K. Tyagi for his much persuasion and determination to stimulate my photohraphic skills merge with scientific writing – to contribute to this Festschrift in the profound honour of the 70th glorious birthday of Prof. Dr. B. Kiauta.

REFERENCES

Fraser, F.C., 1933-136. *The Fauna of British India, including Ceylon and Burma. Odonata.* Pt. I-III. Taylor & Francis, London.

de Fonseka, Terrence, 2000, *The Dragonflies of Sri Lank*a. WHT Publication (private) Limited, 95 Cotta Road, Colombo 8, Sri Lanka.